Agricultural Extension: Farmer Education and Rural Development

Agricultural Extension: Farmer Education and Rural Development

Edited by **Salvador Flores**

SYRAWOOD
PUBLISHING HOUSE
New York

Published by Syrawood Publishing House,
750 Third Avenue, 9th Floor,
New York, NY 10017, USA
www.syrawoodpublishinghouse.com

Agricultural Extension: Farmer Education and Rural Development
Edited by Salvador Flores

International Standard Book Number: 978-1-68286-330-5 (Hardback)

Printed in China.

Contents

Preface

This book has been an outcome of determined endeavour from a group of educationists in the field. The primary objective was to involve a broad spectrum of professionals from diverse cultural background involved in the field for developing new researches. The book not only targets students but also scholars pursuing higher research for further enhancement of the theoretical and practical applications of the subject.

The objective of agricultural extension is to provide education to farmers regarding production and distribution of crops. Extension programs are dynamic and flexible. They not only include methods to increase efficiency in agricultural production but also conservation of natural resources, developing leadership, implementing various plans for rural development, etc. This book elucidates the concepts and innovative models around prospective developments such as water use efficiency, effect of irrigation on consumptive use, etc. Researchers and students in this field will be assisted by this book.

It was an honour to edit such a profound book and also a challenging task to compile and examine all the relevant data for accuracy and originality. I wish to acknowledge the efforts of the contributors for submitting such brilliant and diverse chapters in the field and for endlessly working for the completion of the book. Last, but not the least; I thank my family for being a constant source of support in all my research endeavours.

Editor

Dimension of food security in a selected state- Uttar Pradesh

Rukhsana

Department of Geography, Jamia Aliah University, Kolkata 700091, India. E-mail: rakshandaansari@yahoo.co.in.

Development agencies regard the concept of household food security-often defined as "access for all people at all times to enough food for an active, healthy life"-as a guiding principle for designing interventions in rural areas because poverty is the major determinant of food insecurity and achievement of food security is the goal of development. It is a known fact that the poor do not have adequate access to food in quantities needed for a healthy life, hence food security is an absolute necessity for development. In this paper, database related to food security for the Western Uttar Pradesh region of India were analysed for food security indices such as food availability, food stability, and food accessibility to classify the region. The result classify the southern portion of the western Uttar Pradesh as highest food security region; and the northern portion as low food security region due to lack of fair price shops, low level of purchasing power, lack of storage capacity and lack of food grains production.

Key words: Food availability, food stability, food accessibility, food security.

INTRODUCTION

The present study examines the dynamism of availability of food to the people. Need of food security arises primarily due to the fluctuation in food production and insufficient availability of food from domestic sources. Food is essential for life. It is one of the most basic needs. According to an ancient Hindu script 'all actions stem from food - food depends on food'. Unless there is enough food today for all – men, women, young and old, the prospects for food security tomorrow are greatly diminished (Singh, 2003). One of the most commonly accepted definitions of food security is adequate access to food at all times, throughout the year and from year to year. Access is ensured when all households and all individuals within those households have sufficient resources to obtain appropriate foods for a nutritious diet.

A minimum level of income below which a person is officially considered to lack adequate subsistence and to be living in poverty. The World Bank definition of poverty line for under-developed countries like India is US$ 1 per person per day. As per the Government of India, poverty for urban areas is Rs. 296 per month and for rural areas Rs. 276 per month that is, people who earn less than Rs. 10 per day. People are said to be poor if their incomes fall below a certain level called a threshold, also known as poverty line. In this definition, they do not have enough income to purchase and also do not have easy access to basic goods and services, such as food, clothing, housing, transportation and education (Encarta, 2005).

Today, availability of food is associated with purchasing power and food insecurity is caused by poverty. If people do not have purchasing power, they have substitute of food reserves. Food security and poverty are directly related to each other. So the needs of the poor should be protected by improving their purchasing power, through employment and income generation programmes (Srinivasan, 2002). A large proportion of the world's underfed population starves not because of general food shortage but because of insufficient access to food supplies or insufficient consuming power of people. During 2001-2002, India faced drought conditions but with the huge buffer stocks of foodgrains, the situation was under control. There are number of steps which have been taken to promote food security in India. India's population in 1951 was 361 million with a growth rate of 1.3% per annum, but in 1961 the population drastically rose to 439 million with a growth rate of 2.16% per annum. In 1971, it jumped to 548 million with a growth rate of 2.48%. In 1981, it rose to 683 million with a growth rate of 2.13%. The production of food grains in the country in 1999-2000 was 209 million tones and it worked out to about 60 g per person per day. Theoretically $^1/_2$ kg food, if made available to each individual, is inadequate.

Moreover, the availability of food to an individual depends on the purchasing power of the person concerned, otherwise it will lead to under-nutrition, malnutrition and hunger (Shafi, 2006). Swaminathan (1995) argued that for too long, food has been considered primarily as a means to abolish hunger and minimum needs for most people, overlooking its greater role in the process of economic development. He examined the agricultural production roles in a number of developing countries and focused on the present number of malnourished people. Swaminathan (1996) has studied the role of science and technology for sustainable food security.

The agricultural productivity associated with the 'Green Revolution' in India has become possible with the introduction of high yield variety of seeds (HYVs), irrigation facilities, fertilizers and manures etc. He found that major goals of developing countries should be productivity improvement, natural resource conservations and poverty eradication. Qamar (1996) has emphasized the public distribution system in India covering about 560 million people through 135 million ration card holders. Despite the financial constraints faced by the government, provision has been made to safeguard the public distribution system (PDS) food security against inflations and maintain a buffer stock to ensure food security for the lower sections of the population. Radhakrishna (1996) examined food grain production and food consumption trends in India. India has achieved moderate success in combating transitory food insecurity caused by crop failures due to droughts or floods. Ayres and McCall (1996) examined the progress in rural, agricultural development and food security over the past 25 years. They identified those groups that are going hungry still and illustrated policies to improve food security, supplies, utilization and access.

Suryanarayana (1997) has argued that the assessments by conventional macro measures show considerable improvement in food security in India, the level of cereal consumption and calorie intake of even the general populations are still below the normal thres-hold limits. Empirical evidence is used to raise a number of issues and recommendations are presented for an emphasis on investment in human capital by way of improvement in food and calorie intake for efficiency and economic growth. Data covered a variety of sources and aspects of food security during the period 1954 - 1992. Chaturvedi (1997) measured food security with the help of three components; availability, stability and access. Availability of enough food for all can be attained through efficient domestic production. Stability can be ensured by taking appropriate pre-emptier steps. The vast majority of malnourished either cannot reduce or afford natural resources, jobs, or social support. Thus, sustainability, efficiency and equitable distribution of food requires a comprehensive approach at the country level within the available infrastructure and traditional practices so that food is produced on a sustainable basis and is made available to the poor.

THE STUDY REGION

Keeping in view the importance of food security, the author has selected 'Western U.P.' as an area for detailed study because the area under investigation is rapidly changing from agriculture to industrial activities and commercialization of crops and as a result, food security is declining due to commercialization which has reduced food grain production. Western U.P. is the most prosperous, developed and rich in agriculture region due to the great impact of the Green Revolution, though there is regional variation in food security.

The technology used by farmers is labour intensive and production is mainly dependent upon the amount of labour that the cultivator is able to put. Farmer's lack of knowledge about water management, soil fertility and land practices is prevalent in those areas of Western U.P. which is facing the problems of food insecurity. Western U.P. lies in the upper Ganga plain, which is endowed with fertile soil, level topography and suitable climatic conditions. Western U.P. lies between 26° 20' to 30° 20' north latitudes and 77° 45' to 80° 22' east longitudes. It covers an area of about 80.076 sqkm and has a population of about 61.06 million (Census of India, 2001). The region includes the districts of Saharanpur, Muzaffarnagar, Meerut, Baghpat, Bulandshahr, Gautam Bddha Nagar, Ghaziabad, Aligarh, Hatras, Mathura, Agra, Firozabad, Mainpuri, Etah, Bareilly, Budaun, Shahjahanpur, Pilibhit, Bijnor, Moradabad, Jyotiba Phule Nagar, Rampur, Farrukhabad, Kannauj, Auraiya and Etawah (Figure 1).

OBJECTIVES OF THE STUDY

In view of the complexity of the problem, the aims and objectives of the present work are as follows:

1. To determine the spatial variation of food availability, stability and accessibility in order to ascertain food security.
2. To assess the spatial variation in food security
3. To examine the interrelationship between the variables of food security.

MATERIALS AND METHODS

The raw data relating to food security (Table 1) were collected from various published and unpublished records of government and non-governmental organizations.

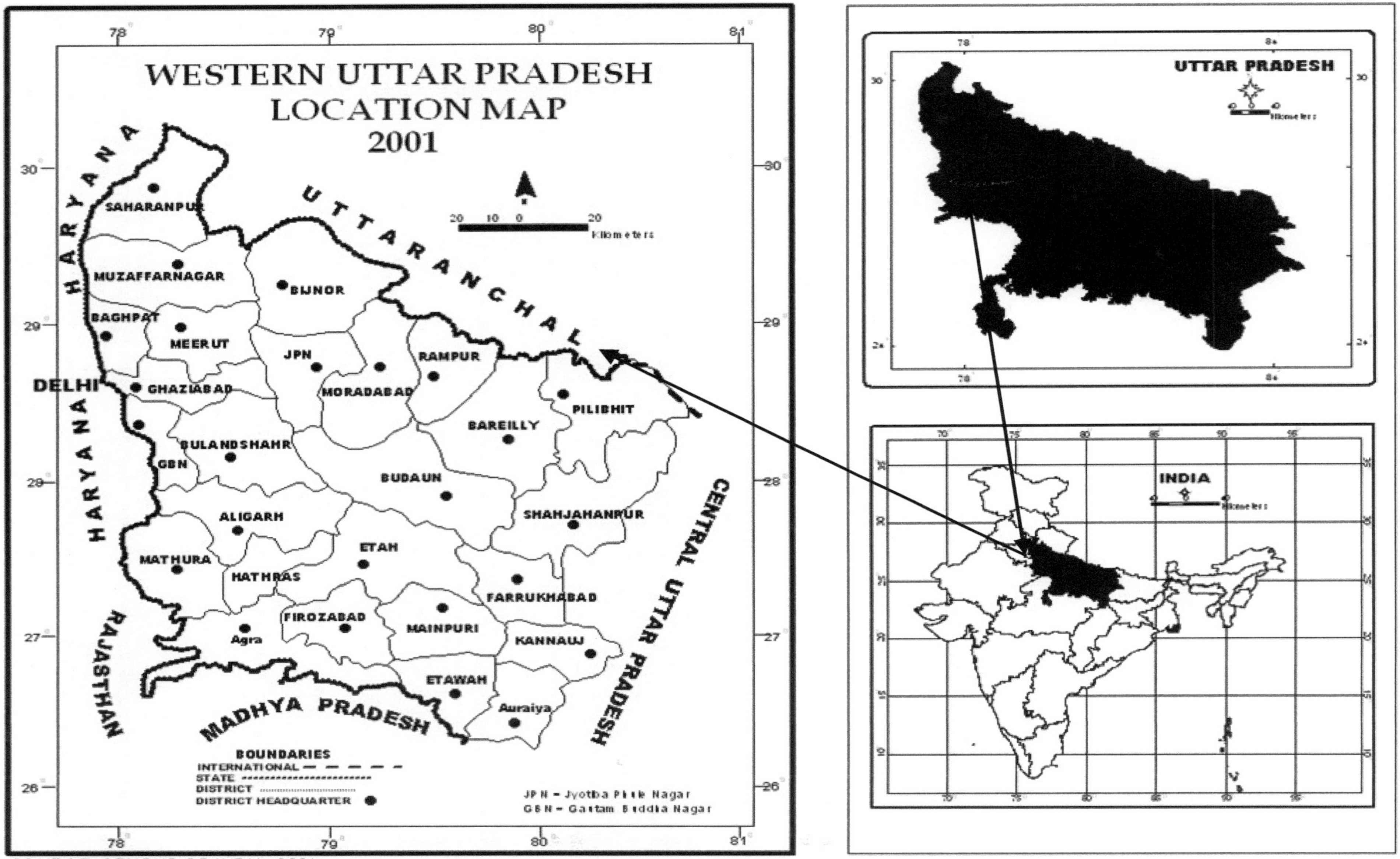

SOURCE: CENSUS OF INDIA, 2001

Figure 1. Map of the study area.

Table 1. Data relating to food security.

Selected Indicators for the study	Sources
Livestock population	Economics and Statistics Institute
Food grains availability	Ministry of Agriculture Department
Caloric availability of food grains	Ministry of Agriculture Department
Yield of food grain	Ministry of Agriculture Department
Number of fair price shop	Economics and Statistics Institute
Percentage of storage capacity	Food Supply Department
Percentage of main worker	Census of India
Percentage of literacy rate	Census of India
Percentage of families below poverty line	Food Supply Department

Agriculture data (unpublished) have been collected from the Ministry of Agriculture, Department of Statistics and Economics, Lucknow and; the Government of Uttar Pradesh, Lucknow, U. P., India. Poverty and storage capacity data have been taken from the Directorate of Food Supply Department, Lucknow, U. P., India while population data has been collected from Census of India (2001). Socio-economic data have been taken from the Economics and Statistics, State Planning Institute, Lucknow, U. P., India.

Food grains availability has been calculated in grams per head per day to examine the regional aspects of surplus and deficit areas of foodgrains availability at district level, taking 482 g/head/day as standard requirement (food intake) of the Indian Council of Medical Research (ICMR, Hyderabad). Caloric availability of foodgrains has been calculated in calorie per head per day. The Indian Council of Medical Research (ICMR) has worked out that on an average, the standard caloric requirement (calorie intake) is 2400 calories per person per day. The average daily caloric requirement in the world is 2800 calories per person. More than 70% of caloric intake in India is derived from food grains. Considering the distribution of persons in different age and sex groups and in different activities in India, the average per capita requirement of calories would be about 2400 calorie (Kravdal, 2001).

Yield is the main factor to increase foodgrains production. Yield level in kg/ha has been selected as a unit to measure the productivity and stability of production. Yield is a measure of productivity. In other words, it is a measure of output from a given amount of input. Thus, the productivity of foodgrains can be examined with the help of the yield method, which is expressed as follows:

$$Y = \frac{P}{A} \times 1000$$

Where Y= Yield, P= Production of food grains, A= Area under food grains

The production of food grains has been divided by area under food grains district wise and then obtained value multiplied by 1000.

For the purpose of standardization, the raw data for each indicator were computed into standard scores commonly known as z-score or standard score. In the first step, district-wise z-score for each indicator were calculated (Smith, 1973). The values so obtained were added district-wise and standardized scores taken out of this composite z-score (Cs) for each district and each set of indicator like food security region. Again these results were transformed back into z-score, so that 'zero' indicates average performance and unity (+ or -) represent one standard deviation, + or - indicates high or low values respectively. Thus, the levels of food security can be examined with the help of composite z-score technique, which is expressed as follows:

$$Z = \frac{X - \overline{X}}{SD}$$

Z= Standard score, X=Original values of the score, $\overline{X}$ = Mean of variables, SD= Standard deviation of variables

The obtained Z-score of each indicator is added district-wise to be known as composite Z-score(s) for each spatial unit of the study area.

$$Cs = \sum Z_{ij}$$

Cs denotes composite Z-scores
Zij indicates the sum of Z-scores of indicators j in district i.

This paper is divided into four sections. The first section deals with food availability which includes distributional pattern, indicators of food availability and food availability regions. The second section focuses on food stability containing distributional pattern of indicators of food stability and food stability regions. The third section highlights food accessibility and food accessibility regions. The fourth section briefly analyses food security regions in the study area.

The districts with reference to composite z-scores have been arranged into five categories of very high (more than 1.5), high (0.5 to 1.5), medium (-0.5 to 0.5), low (-1.5 to -0.5) and very low (less than -1.5). A careful selection of class intervals to decide the categories drawn on the maps are based on the mean and standard deviation. Corropleths maps have been drawn with the help of GIS technique.

SPATIAL DISTRIBUTIONAL PATTERN OF FOOD AVAILABILITY

Food security has three dimensions; available food, having enough food for the entire population, at all times and to sustain human life (Bush and Lacy, 1984). In this section, therefore, an attempt has been made to find out the trends and distributional pattern of food availability which encompasses food grains availability, caloric availability of food grains and livestock availability. Foodgrains availability (in calories/head/day), caloric

Table 2. District-wise distribution of availability, stability and accessibility with indicators in Western U. P. (2002).

Districts	Food availability				Food stability		Food accessibility		
	Livestock per 1000 population	Food grains availability in gram/head/day	Caloric availability of food grains in calories per head/day	Yield of food grains in kg/ha	Total number of fair price shop / 1000 population	Storage capacity in MT / 1000 population	Percentage of main workers to the total population	Percentage of literacy rate to the total population	Percentage of families below poverty line
Saharanpur	345	488.61	181.84	9571	0.34	38.01	23.9	62.61	6.12
Muzaffarnagar	325	462.78	153.15	1998	0.28	58.71	25.39	61.68	3.85
Meerut	300	281.43	99.95	3266	0.22	19.92	24.59	65.95	1.04
Baghpat	266	522.09	175.71	3325	0.27	2.15	24.9	65.65	0.86
Bulandshahr	283	953.73	323.5	3933	0.39	85.82	27.01	60.19	2.78
Ghaziabad	381	264.91	88.95	3071	0.18	37.49	24.31	70.89	1.42
Goutam Buddha Nagar	266	869.08	282.81	2920	0.29	0	25.7	69.78	1.08
Aligarh	417	971.01	364.03	2753	0.38	63.02	23.31	59.7	2.76
Hathras	201	809.11	357.71	1917	0.38	29.44	22.11	63.38	1.89
Mathura	365	1050.8	393.14	2912	0.32	67.74	25.94	62.21	4.64
Agra	435	442.71	164.18	2726	0.33	25.7	22.49	64.97	1.06
Firozabad	335	667	252.53	2503	0.38	44	22.16	66.33	3.6
Mainpuri	340	1190.08	237.71	2592	0.44	44.09	22.87	66.51	5.14
Etah	362	977.23	404.07	2452	0.45	38.24	23.6	56.15	6.18
Bareilly	275	676.51	338.81	1957	0.45	75.27	23.84	47.99	11.17
Badaun	377	1092.37	232.97	2406	0.48	32.86	25.05	38.83	5.84
Shahjahanpur	365	149.65	414.31	2756	0.47	125.56	24.16	48.79	4.39
Pilibhit	356	1437.14	489.32	2940	0.39	63.15	21.9	50.87	3.96
Bijnor	420	453.49	502.05	2816	0.33	11.43	22.62	59.37	2.3
Moradabad	267	723.77	153.14	2627	0.39	64.5	26.07	45.74	5.18
Jyotiba Phule Nagar	331	700.85	232.11	2729	0.49	4.34	25.53	50.21	2.32
Rampur	444	1161.66	396.95	2800	0.4	100.56	23.76	38.95	3.88
Farrukhabad	472	685.88	398.28	2873	0.42	31.31	24.25	62.27	4.29
Kannauj	456	804.87	240.52	2607	0.34	6.5	24.99	62.57	3.3
Etawah	492	943.22	307.86	2497	0.37	103.85	22.3	70.75	5.16
Auraiya	423	1170.42	365.21	3542	0.47	12.84	24.17	71.5	5.77
Total	325	768.37	2710.1	2647	0.37	48.36	28.38	58.44	100.00

Source: Calculation is based on unpublished data obtained from Directorate Ministry of Agriculture, Department of Statistics and Economics, Economics and Statistics, State Planning Institute and Directorate of Food Supply Department, Lucknow, U.P., Directorate of census operation 2001, District census handbook of U.P. Lucknow.

availability of foodgrains (in g/head/day), livestock availability (per thousand population) are variables of food availability from which food availability regions have been developed (Tables 2 and 3). The technique of z-score has been used, wherein all the variables are added in order to get the

Table 3. District-wise distribution of food availability, stability and accessibility (in Z-Scores) in Western U. P. (2002).

Districts	Food availability				Food stability		Food accessibility		
	Livestock per 1000 population	Food grains availability in gram/head/day	Caloric availability of food grains in calories per head/day	Yield of Food grains in kg/ha	No. of fair price shop per 1000 population	Percentage of storage capacity per 1000 population	Percentage of main workers to the total population	Percentage of literacy rate to the total population	Percentage of families below poverty line
Saharanpur	-0.18	-0.88	-0.93	-1.19	-0.43	-0.23	-0.19	0.43	1.37
Muzaffarnagar	-0.45	-0.96	-1.19	0.70	-1.29	0.40	1.15	0.31	0.01
Meerut	-0.80	-1.53	-1.65	0.50	-2.14	-0.78	0.43	0.88	-1.68
Baghpat	-1.27	-0.77	-0.99	0.53	-1.43	-1.32	0.71	0.84	-1.78
Bulandshahr	-1.04	0.59	0.31	0.08	0.29	1.22	2.61	0.11	-0.63
Ghaziabad	0.32	-1.58	-1.75	0.03	-2.71	-0.25	0.18	1.54	-1.45
Goutam Buddha Nagar	-1.27	0.32	-0.05	0.04	-1.14	-1.38	1.43	1.39	-1.65
Aligarh	0.82	0.64	0.66	0.10	0.14	0.53	-0.72	0.04	-0.65
Hathras	-2.17	0.13	0.61	-0.60	0.14	-0.49	-1.80	0.53	-1.17
Mathura	0.10	0.89	0.92	0.08	-0.71	0.67	1.65	0.38	0.48
Agra	1.07	-1.02	-1.09	2.05	-0.57	-0.60	-1.46	0.75	-1.66
Firozabad	-0.31	-0.32	-0.31	0.28	0.14	-0.05	-1.76	0.93	-0.14
Mainpuri	-0.24	1.33	0.43	-0.61	1.00	-0.05	-1.12	0.95	0.78
Etah	0.06	0.66	1.01	-0.58	1.14	-0.22	-0.46	-0.43	1.40
Bareilly	-1.15	-0.29	-0.44	-1.49	1.14	0.90	-0.24	-1.52	4.39
Badaun	0.27	1.02	1.27	-0.17	1.57	-0.39	0.85	-2.74	1.20
Shahjahanpur	0.10	-1.94	1.10	-3.10	1.43	2.43	0.05	-1.41	0.33
Pilibhit	-0.02	2.11	1.76	-0.01	0.29	0.53	-1.99	-1.13	0.07
Bijnor	0.87	-0.99	-1.20	0.26	-0.57	-1.04	-1.34	0.00	-0.92
Moradabad	-1.26	-0.14	-1.19	-0.42	0.29	0.57	1.77	-1.82	0.80
Jyotiba Phule Nagar	-0.37	-0.21	-0.49	-0.18	1.71	-1.25	1.28	-1.22	-0.91
Rampur	1.20	1.24	0.95	-0.23	0.43	1.67	-0.32	-2.72	0.02
Farrukhabad	1.59	-0.26	0.96	1.86	0.71	-0.43	0.13	0.39	0.27
Kannauj	1.36	0.12	0.46	1.54	-0.43	-1.19	0.79	0.43	-0.32
Etawah	1.86	0.55	0.17	0.06	0.00	1.77	-1.63	1.52	0.79
Auraiya	0.91	1.27	0.67	0.47	1.43	-1.00	0.05	1.62	1.16

Calculation is based on Table 2.

weight of food availability and to delineate homogenous regions (Table 4). These scores are grouped and mapped into five grades of very high, high medium, low and very low (Figure 2). Very high concentration in food availability is found in two districts, namely, Rampur and Pilibhit which lies in the eastern portion of the study area.

High concentration of food availability region comprises of seven districts, namely, Aligarh, Farrukhabad, Auraiya, Etawah, Etah, Bijnor and Mathura which are found to be scattered throughout the study area. These districts are observed to

Table 4. Districtwise Distribution of Food Availability, Food Stability and Food Accessibility (in composite standardized Z-Scores) in Western U.P.

Districts	Food availability	Food stability	Food accessibility	Food security
Saharanpur	-0.92	-1.33	0.98	-0.79
Muzaffarnagar	-1.20	-0.15	0.87	-0.49
Meerut	-1.83	-1.75	-0.14	-1.98
Baghpat	-1.40	-1.60	-0.06	-1.60
Bulandshahr	-0.07	1.12	1.27	0.94
Ghaziabad	-1.38	-2.11	0.22	-1.69
GBN	-0.46	-1.79	0.76	-0.73
Aligarh	0.98	0.54	-0.76	0.54
Hathras	-0.66	-0.69	-1.40	-1.29
Mathura	0.88	0.01	1.47	1.18
Agra	-0.48	0.61	-1.34	-0.61
Firozabad	-0.44	0.25	-0.58	-0.42
Mainpuri	0.70	0.23	0.32	0.41
Etah	0.80	0.23	0.24	0.71
Bareilly	-0.86	0.38	1.37	0.42
Badaun	1.18	0.71	-0.45	0.34
Shahjahanpur	-0.34	0.53	-0.62	-0.26
Pilibhit	1.77	0.57	-1.81	0.63
Bijnor	-0.61	-0.98	-1.30	-0.42
Moradabad	-1.19	0.30	0.42	-0.48
JPN	-0.49	0.18	-0.45	-0.43
Rampur	1.56	1.32	-1.78	0.83
Farrukhabad	1.06	1.51	0.45	1.49
Kannauj	0.89	-0.07	0.55	0.51
Etawah	1.19	1.29	0.35	1.45
Auraiya	1.31	0.63	1.62	1.77

Calculation is based on Table 3.

be high in food grains availability due to high productivity, high infrastructural facilities and low population density and high food grains availability. Ten districts come under the medium grade which spreads in the south-west and -east portion of the study area. Six districts including Saharanpur, Muzaffarnagar, Baghpat, Ghaziabad, Moradabad and Hathras form the low food availability regions. Only one district that is, Meerut falls in the very low concentration of food availability and lies in the western portion of the study area. Food availability has been recorded in these districts due to high population density, high rate of urbanization and high concentration of cash crop production. Pilibhit district has been reported to have the highest food availability because of its low population density and low rate of urbanization.

SPATIAL DISTRIBUTIONAL PATTERN OF FOOD STABILITY

Food stability calls for undertaking appropriate pre-emptive steps through which harmful, suspended and inter-annual instability of supplies of food can be reduced. Built-in stability needs to be installed in the production, pricing, marketing and distribution systems. Natural and man- made disasters can often be anticipated and even prevented (Ghosh, 2000). In this section therefore, an attempt has been made to find out the distributional pattern of food stability. The analysis of food stability in spatial dimension has been worked out with three indicators such as food grains productivity, storage capacity and public distribution system (Table 2).

Food grains productivity (yield kg/ha), public distribution system (total number of fair price shop per thousand population) and storage capacity (storage capacity in metric tonnes per thousand population) are the chief determinants to food stability (Table 3). The regions of food stability have been marked out on the basis of the value of standardized z-score (Table 4). Figure 3 shows that the region of very high food stability is recorded only in one district that is, Farrukhabad which lies in the eastern portion of the study area. High grade of food stability districts constitutes Pilibhit, Shahjahanpur, Budaun, Aligarh, Agra, Rampur, Etawah, Bulandshahr and Auraiya with major concentration in the eastern and

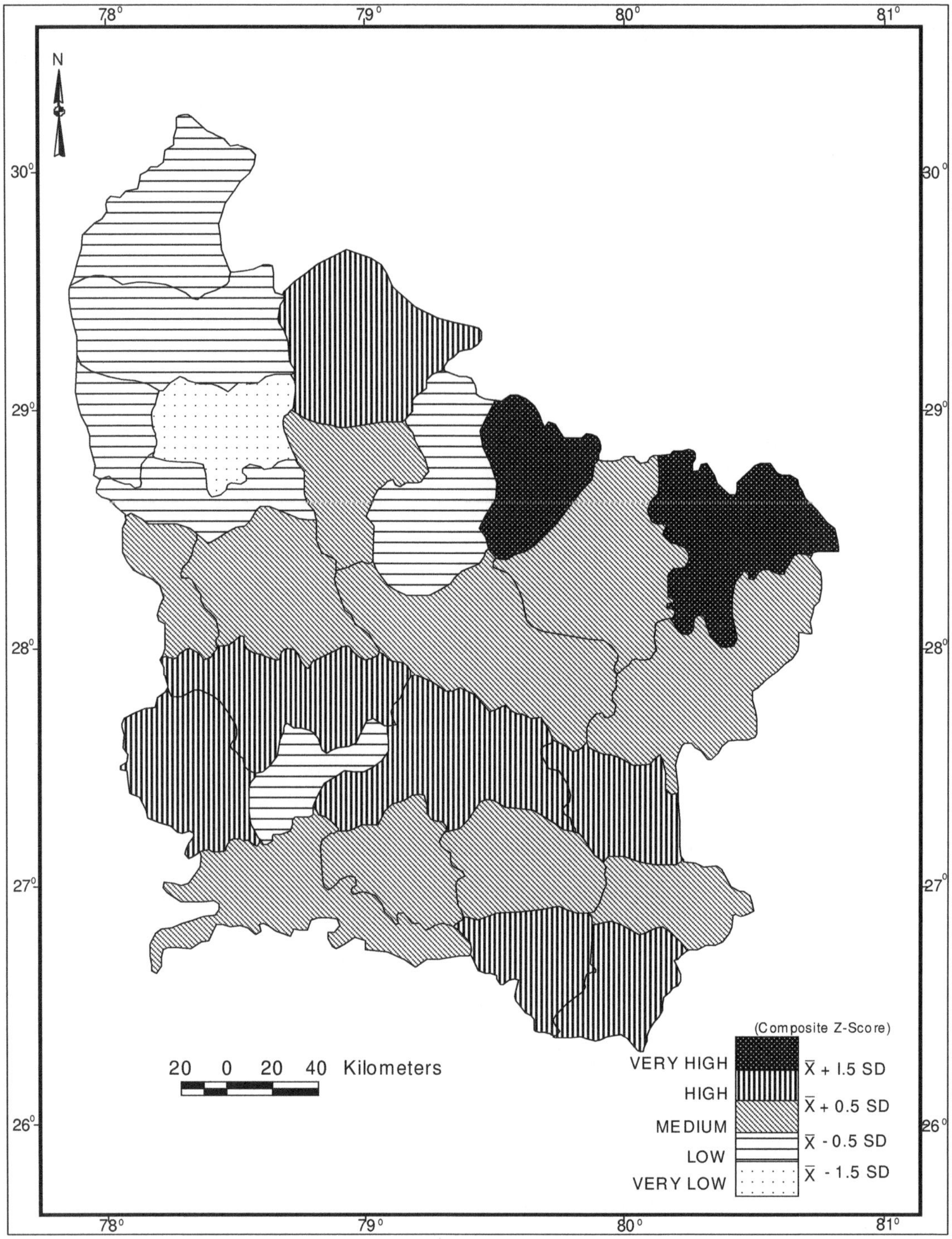

Figure 2. Food availability regions in Western Uttar Pradesh.

and southern portion of the study area.

All indicators of stability are high in these areas except productivity. Public distribution system is the most important indicator to identifying a food stable region. Medium food stability region makes a fairly extensive area on the north eastern portion and southern portion of the study area, both regions contain nine districts. Three districts, namely, Saharanpur, Bijnor and Hathras fall under the low concentration of food stability region. Very low concentration of food stability region forms four districts of Meerut, Baghpat, Ghaziabad and Goutam Buddha Nagar which lie in the western pocket of the study area. These districts are low due to low productivity and high density of population.

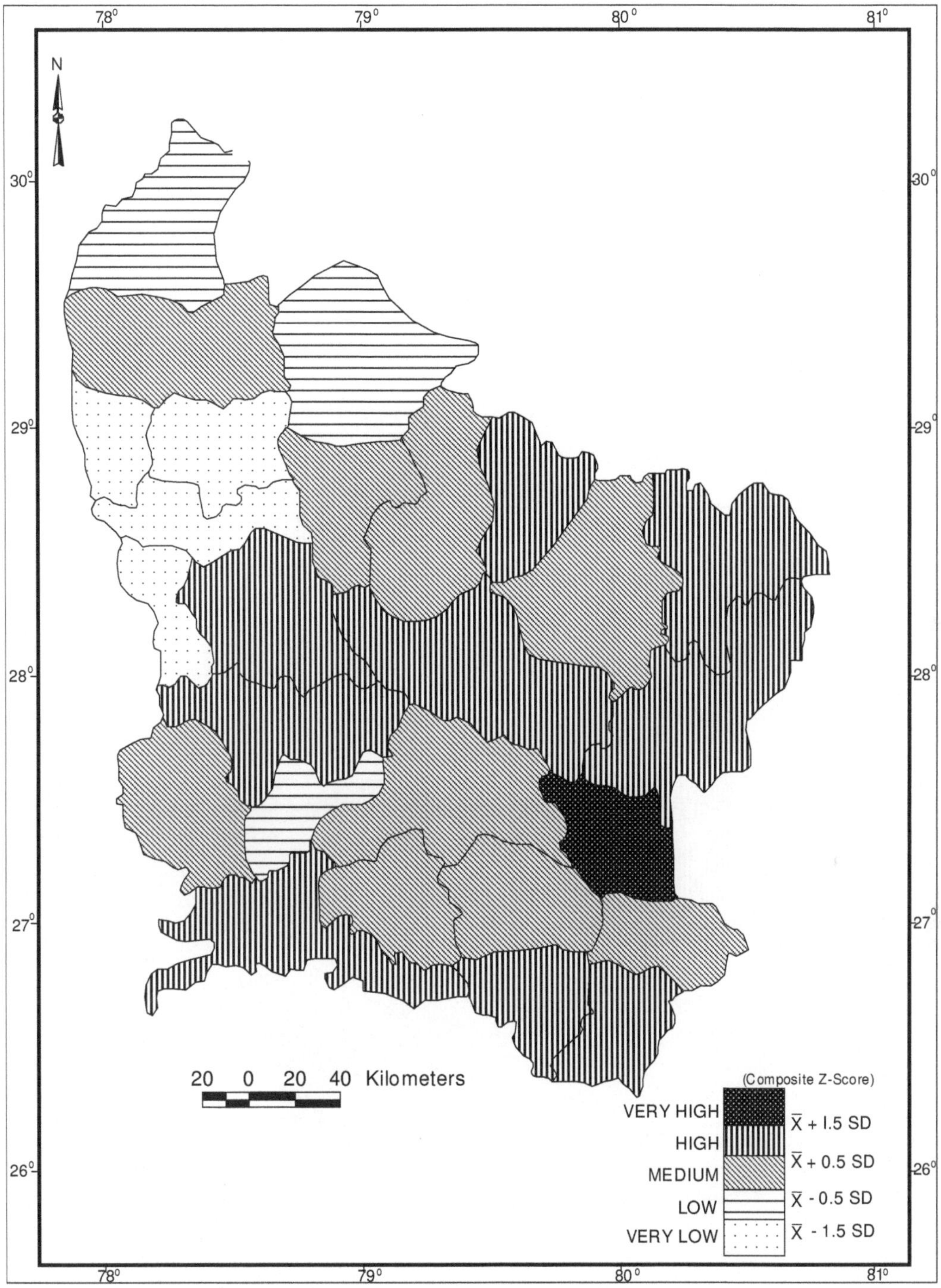

Figure 3. Food stability regions in Western Uttar Pradesh.

ASPATIAL DISTRIBUTIONAL PATTERN OF FOOD ACCESSIBILITY

'Access to adequate food for all people at all times is defined as 'food security' (World Bank, 1986). Food security is not only dependent upon food availability and stability but also upon accessibility. The majority of malnourished people cannot produce or afford to buy enough food. They have inadequate access to natural resources, jobs, income or social scores (Chatarvedi, 1997). Food accessibility is the composite result of indicators like purchasing power, employment and literacy rate (Tables 2 and 3). Purchasing power, employment and literacy rate are the variables of analysis of food

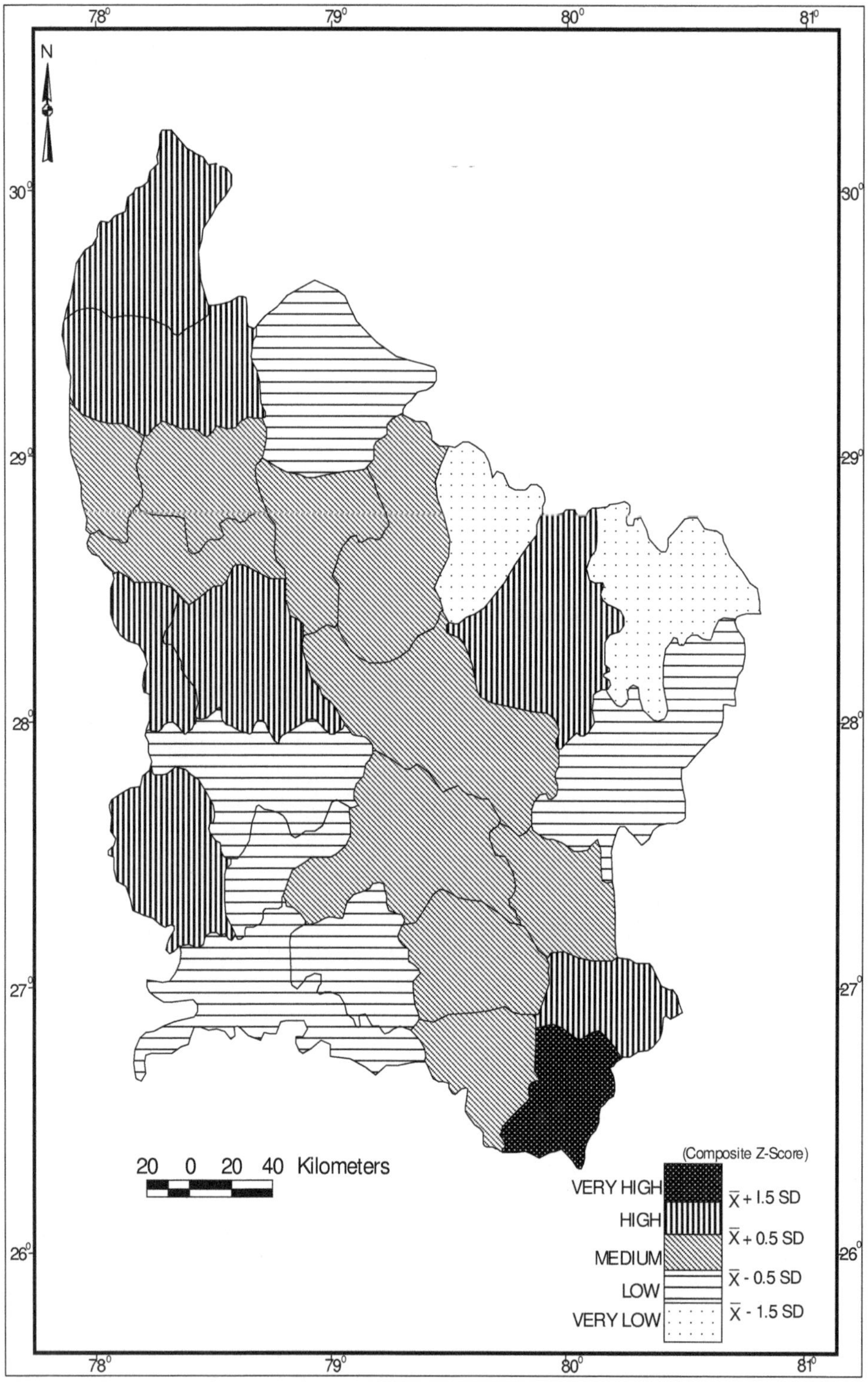

Figure 4. Food accessibility regions in Western Uttar Pradesh.

accessibility. The standardized z-scores of these variables are added in order to get the weight of food accessibility regions (Table 4).

Figure 4 reveals that Auraiya district is the only district with a very high grade of food accessibility. High grade of food accessibility region includes seven districts namely; Saharanpur, Muzaffarnagar, Bareilly, Goutam Buddha Nagar, Bulandshahr, Mathura and Etawah. These

Table 5. Matrix correlation between the indicators of Food Security

	Food security	Food availability	Food stability	Food accessibility
Food security	1			
Food availability	0.86*	1	1	
Food stability	0.81*	0.70*	-0.07	1
Food accessibility	0.27	-0.11		

*Correlation is significant at the 1 per cent. ** Correlation is significant at the 5%.

districts witnessed high accessibility of food due to high purchasing power and employment which leads to increase in food security. But the region of high accessibility of food is dotted over some areas with lack of crop production and high population density. Food accessibility is most positively correlated to purchasing power. Literacy rate is the most important factor in eliminating poverty and to increase the knowledge of farmers to use new technology in a proper way. Medium grade of food accessibility regions are demarcated in ten districts of the study area. Six districts, namely, Shahjahanpur, Aligarh, Hathras, Bijnor, Agra and Firozabad lie in the low grade of food accessibility region. Very low grade food accessibility region form two districts, namely; Pilibhit and Rampur which lie in eastern portion of the study area. These areas are low in food accessibility due to unemployment and high poverty.

A REGIONAL ANALYSIS OF FOOD SECURITY

The foregoing analysis clearly reveals that there is a lot of variation in the level of food availability, food stability and food accessibility across the districts in Western U.P. In order to assess overall food security, composite z-scores of nine variables, including three each of food availability, three stability and three of accessibility have been added (Table 4). To divide the western Uttar Pradesh into food security regions, the composite standardized z-scores values are categorized into five grades of very high, high, medium, low and very low.

Figure 5 reveals that the district of Mathura, Etawah and Auraiya in the southern part of the study area are regions of very high food security. High food security regions are in the concentrated part of the study area and are characterized by high concentration of food stability consequent upon high storage capacity, public distribution system, high food grains production, low population density and high large size land holdings which results in higher rural development. So, purchasing power and employment are the main factors of high level food security in these regions. Three districts, namely, Bijnor, Jyotiba Phule Nagar and Bareilly come under the medium grade of food security regions. Goutam Buddha Nagar, Muzaffarnagar, Saharanpur, Moradabad, Firozabad and Agra come under the low concentration food security region scattered throughout the study area. Very low grade of food security region are concentrated in four districts including Ghaziabad, Hathras, Meerut and Baghpat which lie in the western part of the study area. These districts are low in food security due to lack of fair price shops, low purchasing power, lack of storage capacity and lack of food grains production. Low level of food security decreases rural development.

DIMENSIONS OF CORRELATION AND LEVELS OF SIGNIFICANCE BETWEEN THE VARIABLES OF FOOD SECURITY

Finally to find out the relationship between the variables of food security, an effort has been made to work out on the basis of correlation matrix. These have been symbolized and presented in Table 5 Karl Pearson's Correlation Coefficient technique has been used to find out the relationship among them and the level of significance has been tested by employing students't' test technique.

Thus it have been concluded through Table 4 which reveals that the food security is positively correlated with food availability (r=0.86) and food stability (r=0.81) at 1 per cent level of significance. However, it is positively correlated to food accessibility (r=0.27) but none is significant up to the accepted level. It may be asserted from the analysis that food security of the region depends on food availability and food stability. But food availability does not significantly affect 'the food stability. It is mainly due to the fact that, with the industrialization, urbanizetion, commercialization and technological innovation in agriculture, farmers tend to cultivate cash crops for quick earning. The large scale cultivation of cash crops, hampers the production of foodgrain crops, subsequently the foodgrain availability decreases. Food availability is strongly correlated to food stability (r=0.70) at 1 per cent level of significance. It is negatively correlated to food accessibility. It is clear from the analysis that food availability and stability are interdependent and interrelated. It may be said that, food stability may increase as much as the food availability increases.

Conclusion

Food security is positively correlated to food availability stability and accessibility. Food availability and stability

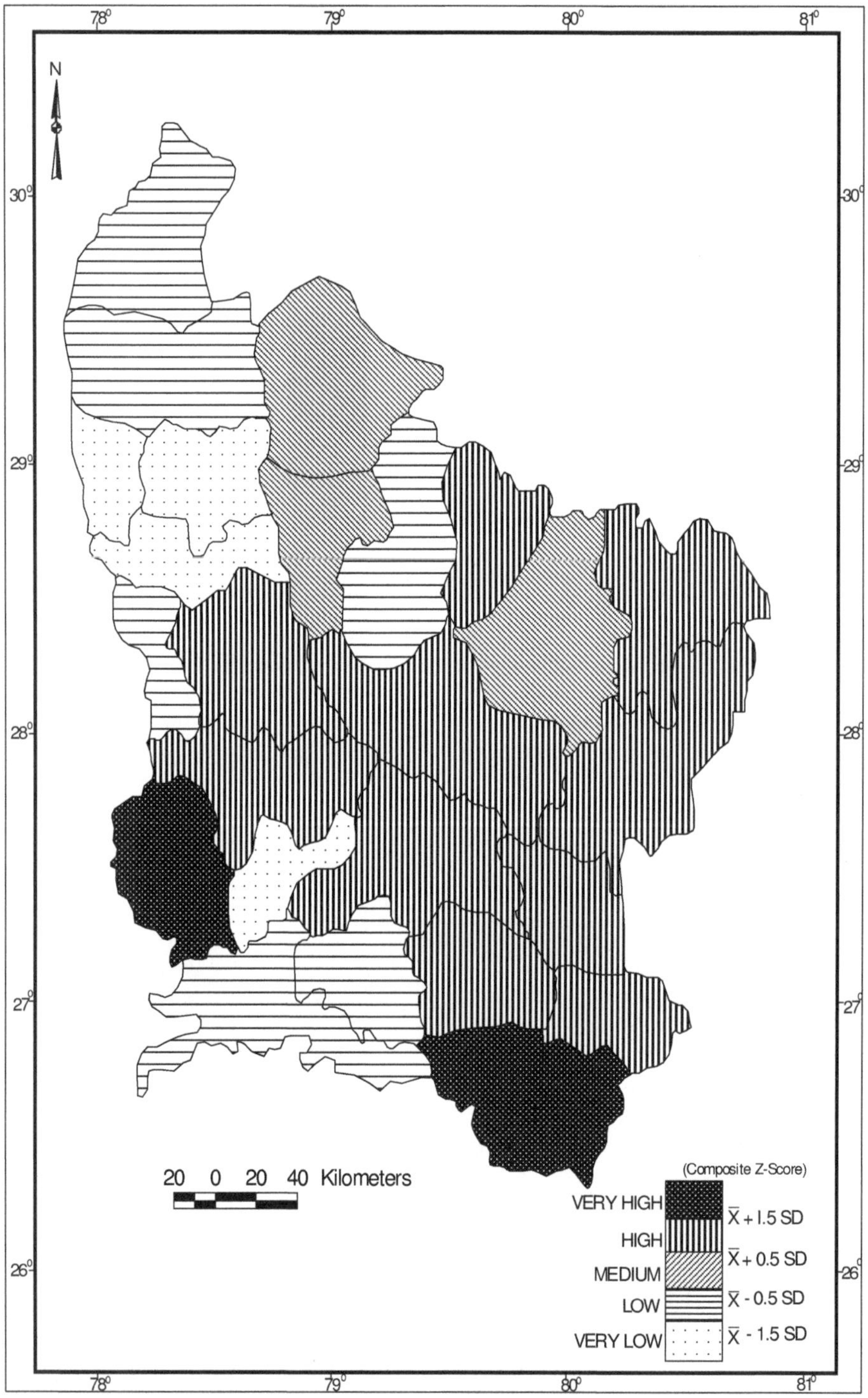

Figure 5. Food security regions in Western Uttar Pradesh.

are interdependent. But food availability is reducing mainly due to the transformation of cereal crop cultivation into commercial cropping. Food stability may be reduced due to changing land use pattern associated with Industrialization, commercialization and globalization. Food accessibility merely depends on the purchasing power of inhabitants and is different from food availability and stability. It is mainly due to the fact that, with industrialization, urbanization, commercialization and technological innovation in agriculture, farmers tend to cultivate cash crops for quick earning. The large scale cultivation of cash crops severely hampers the production of food grains and subsequently, food grains availability is gradually reduced.

RFERENCES

Ayres WS, MeCalla AF (1996). Rural Development, Agricultural and Food Security. Finance and Development. Washington D.C. 33(4): 8-11.

Bush L, Lacy WB (1984). Introduction, What does Food Security mean In Bush L. et al. (eds) Food Security in the United States. Boulder. Co. Westview. Press Inc. Census of India (2001). Primary Census Abstract. Total Population. Series-1. Part 11. B: (1).

Chaturvedi R (1997): Food Security and Panchayat Raj, Pub, Concept, New Delhi: 5-7

Encarta (2005). Encarta Reference library, by Philip Oldenberg, 2003 (It is a software of Encarta).

Ghosh GN (2000): Food Insecurity, the Greatest Challenge of the Millennium, Indian Farm., 50(7): 7-9.

Kravdal O (2001). Has Population Growth Restricted Improvement in Food Availability per head. 1970 95. Population Stud., 55(2): 105-117.

Shafi M (2006). Agricultural Geography, Pub: Dorling Kindersley, India, Pvt. Ltd. Licenses of Pearson Education, South Asia: pp. 1-8

Singh RL (2003). Food Security Issues in South Asia, in Chaturvedi, P., (ed) Food Security in South Asia, Pub: Indian Association for the Advancement of Science, by Concept Publishing Company, New Delhi, 110059: 1-7.

Smith DM (1973): The Geography of Social Well Being in the United States, New York, Amol Heinman: p.78

Srinivasan PV (2002). Poverty an Agricultural Research, in (ed) by Parikh, K.S. and Radhakrishna, R., India Development Report, Oxford University Press: pp. 234-238

Suryanarayana MH (1997). Food Security in India, Measures, Norms and Issues, Dev. Change, 28(4):771-789

Swaminathan MS (1995). Agriculture, Food Security and Employment, Changing Times, Uncommon Opportunities, Nature Resources, 31(1): 2-15.

Swaminathan MS (1996). Science and Technology for Sustainable Food Security, Indian J. Agric. Econ., 51(112): 60-75.

Qamar A (1996). PDS, Food Security on Politics of Food, Bihar, J. Agric. Marketing, 4(4): 393-395.

Radhakrishna R (1996). Food Trends, Public Distribution System and Food Security concerns, Indian J. Agric. Econ., 51(1/2):168-183.

World Bank (1986). Poverty and Hunger; Issues and Options for Food Security in Developing Countries, World Bank, Washington D.C.:7-8

2

Differences in the perception on leadership of the village development and security committee

Jeffrey Lawrence D'Silva, Bahaman Abu Samah and Hayrol Azril Mohamad Shaffril

Laboratory of Rural Advancement and Agricultural Extension, Institute for Social Science Studies, Universiti Putra Malaysia, Malaysia.

Previous studies revealed that the socio-demographic factors are vital and would sway the perception of rural community towards the leadership of their community leaders. Nevertheless, the outcome of these studies had shown inconsistent results. Hence, it is pertinent to investigate further on the socio-demographic factors that influence village development and security committee (VDSC) perception towards the VDSC leadership in Vision Village Movement. Data was gathered through a survey questionnaire and a total of 235 respondents from three states in Malaysia participated in this study. Interestingly, overall the members of VDSC had a high level of positive perception towards the VDSC leadership and the factors that had significant difference with perception towards VDSC leadership were age, zone and monthly income. The findings of this study would help to develop a better understanding on some of the pertinent socio-demographic factors that rural development leaders need to be aware of to become effective leaders in the future.

Key words: Village development and security committee (VDSC), leadership, rural community.

INTRODUCTION

Malaysia is often cited as one of the front-runners in implementing rural development programs among the developing nations. Evolution and transformation in rural areas have started since 1957 with Pre-new Economic Policy to reduce the level of poverty among the poor and the marginalized. In the era of Second Transformation of Rural Development (1994 - 2000), the focus was on further developing the rural areas so as to balance the economic growth in both rural and urban areas, in line with the establishment of the Malaysian Vision Policy, that would be able to drive the nation towards achieving an equitable society (Fatimah and Mad, 1997). One of the major ways that the government had undertaken to enhance rural development is by establishing the Village Development and Security Committee (VDSC) and launching the Vision Village Movement (VVM) program.

The VDSC was established with its main objectives to create government representatives in order to initiate more successful rural community programs and to strengthen the relationship between the government and the rural community. Based on the recent statistic, there are 68,000 VDSC members all over Malaysia administering 14,503 villages throughout the country. The Vision Village Movement (VVM) concept was introduced for the purpose of establishing more competitive, self-reliance, educated and discplined rural community. Presently, there are 7,808 VVM villages in Malaysia. This program was initiated to give specific focus on village development with the aim of creating developed and attractive rural areas. This program hope to strengthen the ability of rural community whereby it can be a catalyst for the change of attitude among rural community towards development. VVM emphasizes villagers involvement in their planning to create more developed, attractive and profitable villages. A research done by Mohd (2005) claimed that VVM is one of the mechanismes to make a paradigm shift in attitude, practice and inspiration among rural community in accordance with the objectives of the Second Transformation of Rural Development and Vision 2020.

*Corresponding author. E -mail: hayrol82@gmail.com.

The development of rural areas and communities are essential for the well being of rural people in developing countries such as Malaysia since many rural areas are still characterized by low population density, poor infrastructure and limited access to public services (Fletcher and Fatimah, 2007). According to Brennan (2006), the need for rural development and local leadership to address traditional obstacles and facilitate development in the face of new challenges is therefore vital. Due to the tremendous pace that development is taking place, leadership aspect is important to ensure that rural residents are not left behind and able to compete with the urban residents especially when it comes to development issues. The conditions facing rural areas also show the need for development efforts to be focused on the community and territorial elements of rural areas as opposed to focusing solely on various sectors of the economy in the urban areas. Effective stewardship therefore is essential to guarantee the availability of creative and innovative response as it will help to produce stable organization thus smoothening the running of organizations such as VDSC.

Before we go any further, let's first define leadership. According to Astin and Astin (2007), leadership is a process that is ultimately concerned with fostering change. Leadership implies a process where there is movement to some future place or condition that is different. Leadership also implies intentionality, in the sense that the implied change is directed toward some future end or condition which is desired or valued. Accordingly, leadership is a purposive process which is inherently value-based. Consistent with the notion that leadership is concerned with change; the leader is basically a change agent. Leaders, then, are not necessarily those who merely hold formal leadership positions; on the contrary, all people are potential leaders meaning here, VDSC members without considering their position are the leaders of the community. Furthermore, since the concepts of "leadership" and "leader" imply that there are other people involved; leadership is, by definition, a collective or group process and therefore will not work alone. Good VDSC leaders are needed as they hold the responsibility to bring development and safety to villages. There are numerous theories that explain about effective leadership and according to Bass (1990) one of the basic ways to explain effective leadership is through analysing the personality traits of individual leaders. Mitchell (2008) suggests that several factors account for differences in the attitudes and behaviour of leaders. The level of education and age were identified as important determinants. Mitchell (2008) mentioned that the level of education influences people's values, wants and needs and makes them think and behave differently. Age, on the other hand tends to give greater or lesser degree of individualism among the leaders with the younger generations feeling more comfortable exhibiting individualistic behaviours. Older leaders were identified to maintain a calmer and more understated demeanour besides showing a greater degree of empathy and concern for others in society.

Several studies too have been conducted that addressed the relationships between selected demographic variables such as region or zone and period of staying and leadership. One such study was Wicks (2007) who found a weak relationship between zone and leadership. Conversely, in a study on rural leadership, Stevens (2006) found that almost 60% of the rural leaders in their study believed that a region where they stayed had an impact in their leadership styles. Besides, a number of researchers had singled out the phenomenon of period of staying as one of the factors that serve to influence leadership (Horwitz, 2006). However, not all studies drew this conclusion. Rotter (2006) found that the relationships between period of staying and leadership were not especially strong.

One of the pertinent factors contributing to the leadership is the type of occupation that a leader is undertaking (Gray, 2008). This is due to the fact that cultures are inherited from where a particular place where the leader is working and as the culture is embedded with the individual, the culture shapes the styles of leadership (Schein, 2004).

On top of it, monthly income of an individual does affect leadership styles of individuals (Robbins et al., 2008). In today's dynamic world, leaders must be able to cope with increasing volatility and turbulence of the environment and they need to be financially sound to overcome numerous obstacles. Even though there were many references to a relationship between leadership and monetary values in many researches, little systematic research has been conducted to investigate the link between the nature of relationship of these concepts among the agricultural leaders (Parry, 2008). Since these two concepts are important in the development of an organization, further research is required to provide significant information to leaders.

Based on the above literature, this study aims to explore the level of perception among the VDSC members towards the VDSC leadership and to seek the differences between selected demographic factors and perception towards VDSC leadership. Consequently, it is hoped that this study would add to the body of knowledge pertaining to rural leadership.

METHODOLOGY

This research employed a quantitative survey method to obtain data. The data were gathered from 235 respondents through a face-to-face interview based on stuctured questionnaire. The data collection process started in April and ended in May 2008. The instrument was designed by the researchers based on previous literature and was pre-tested in order to ensure the items realibilty and validity and also to ensure it can easily be comprehended by the respondents. The pre-test was conducted at Sungai Ramal Dalam Village located in city of Kajang in February 2008. The Cronbach alpha value obtained for all the constructs were between 0.80 - 0.85, exceeding Nunnally's (1978) recommended threshold

Table 1: Respondents profile (n = 235).

Respondents profile	Frequency	Percentage
Gender		
Male	202	86.0
Female	35	14.0
States		
Central	90	38.3
Southern	75	31.9
East Coast	70	29.8
Village grade		
Grade A	57	24.3
Grade B	69	29.3
Grade C	109	46.4
Occupation		
Work on their own	105	43.8
Government sector	57	24.3
Private sector	33	14.0
Not working	32	17.9
Education achievement		
Primary school	55	23.4
PMR/SRP/LCE	43	18.3
SPM/SPMV/MCE/certificates	105	44.7
Pre University/ University	32	13.6
Position in VDSC		
Chairman/Vice Chairman	19	8.1
Secretary	17	7.2
Treasury	15	6.4
Committee	184	78.3

of 0.7, indicating high reliability. Sampling frame was obtained from Department of Community Development (DCD) based on the states selected. The sample list was developed by selecting villages that were listed in VVM. VDSC members selected were from three states. For each state, a district was selected and from each district, six villages were chosen as a research sample based on stratified random sampling. Before the data collection process, the reseachers have contacted leaders of VDSC to determine the date of data collection session to ease the process.

The study sample was categorized in three zones. A total of 90 VDSC members represented central zone (Selangor), 70 VDSC members represented southern zone (Negeri Sembilan) and 75 VDSC members represented east coast zone (Pahang). In the earlier stages, the number of respondents determined for each states was same but because of some obstacles and problems, the number of respondents determined earlier could not be fulfilled. Among the problems faced were some of VDSC leaders cannot give the actual number of VDSC members, there were absentees among VDSC members during the session of data collection. To overcome this problem, the researchers left a number of questionnaire forms to be fulfilled by the absentees. Even though this action was taken, the VDSC members on certain location failed to return the questionnaire as required. SPSS software was used to analyse the data collected, statistics such as frequency, percentage, mean, standard deviation and ANOVA were used.

FINDINGS AND DISCUSSION

There are many studies that showed the effect of socio-demographic factors on leadership. This section presents the results of the study based on the stated objectives. The first part gives a description of demography profile of respondents, followed by the level of VDSC members perception towards VDSC leadership and factors that affect it.

Respondents demography profile

A total of 235 respondents were involved in this study. Out of this total, 38.3% respondents were from Selangor, 31.9% from Negeri Sembilan and 29.8% from Pahang. A total of 46.4% VDSC members that responded to the questionnaires were from grade C village, 29.3% from grade B and the remaining 24.3% from grade A village.

The majority (44.7%) of respondents had attended their education to SPM/MCE/SPMV/Certificates, while primary school and PMR/SRP/LCE was at 23.4% and 18.3% respectively. There were also respondents who had received pre-university and university (13.6%) education. In terms of employment, a total of 24.3% of repondents are working in the government sector, 14.0% in the private sector and the majority (43.8%) are self-employed. In this study, it revealed that a large majority (78.3%) were VDSC members, while the remaining percentage consists of chairman/vice chairman (8.1%), secretary (7.2%) and treasury (6.4%) (Table 1).

As in Table 2, it signals that the majority (37.9%) of respondents were in the 56 years and above age group followed by 33.6% 46 - 55 years age group and the remaining one 28.5% fall in 45 years and below age group. The mean age of respondents was 52.61 years. For monthly income, this study reveals that the majority of respondents earned between RM1,001-RM2,000 per month (40.4%). A total of 39.9% earned below RM1,000 per month while 19.7% earned more than RM2,001 per month. Their mean monthly income was RM1,663.08.

The majority of respondents had stayed at the village for a long period (M = 37.1 years). Most of the respondents had stayed the same village between 26 - 45 years (38.7%), followed by those who had stayed for 46 years and more (33.0%) and lastly those who had stayed for 25 years and less (24.3%). Most of the respondents had some experience in VDSC based on the mean period of holding a position in VDSC (7.33 years). Slightly more than half (50.7%) of VDSC members had 5 years or less experience, one third of respondents (29.7%) had 6 to 10 years experience while the rest (19.6%) had 11 years and more experience in VDSC.

Table 2. Respondents profile.

Respondents profile frequency	Percentage	Mean	S.D
Age (n = 235)		52.61	10.38
≤ 45 years	67	28.5	
46 - 55 years	79	33.6	
≥ 56 years	89	37.9	
Income per month (n = 213)		1663.08	1239.85
≤ RM1,000	85	39.9	
RM1,001 - RM2,000	86	40.4	
≥ RM2,001	42	19.7	
Staying period (n = 221)		37.10	15.10
≤ 25 years	57	24.3	
26 - 45 years	91	38.7	
≥ 46 years	73	33.0	
Period of holding a position in VDSC (n = 209)		7.33	6.09
≤ 5 years	106	50.7	
6 - 10 years	62	29.7	
≥ 11 years	41	19.6	

Table 3. Distribution on perception towards CVD leadership (n = 235).

Level	Frequency	Percentage
Low (0 - 1.33)	-	-
Moderate (1.34 - 2.67)	3	1.3
High (2.68 - 4.0)	232	98.7

As depicted in Table 3, the majority of the respondents have very good perception towards the VDSC leadership. This is a positive finding whereby it indicated VDSC leadership received positive perception among the VDSC members thus enhance their opportunity in creating effective rural administration. Only three respondents or 1.3% found to have fair perception towards their VDSC leadership. It was interesting to know that none of the respondents have poor perception on their VDS leaderhip.

On top of it, Table 4 provides further input on VDSC members perception towards the leadership in VDSC organization. To reveal the perception on VDSC leadership, 16 items were measured. Based on the overall mean score of 3.90 from the maximum 5.0 mean score, there was a high perception among VDSC members towards the leadership aspects. Results presented in Table 4 obviously signal that there were three items that recorded the highest perception towards leadership aspects in VVM village which were "I am satisfied with VDSC members relationship" (M = 4.15), followed by "VDSC leaders take care of its members charity" (M = 4.11) while "team spirit in this village is high" recorded the third highest mean score which was 4.09 while the item of "there was unhealthy relationship in VDSC organization recorded a mean score of 2.76.

Leadership aspect is highly needed for sustainable development as stressed by Uphoff (1992) and Shorthall and Shucksmith (1998) in their study that revealed that leadership aspect is important in order to create participative institution thus creating positive relationship between leaders and local institution members. Results gained signalled that there were healthy leadership and relationship among VDSC members in VVM village. A positive VDSC leadership proved to produce positive relationship among the VDSC members. This is based on the high perception recorded on items measuring VDSC members relationship, VDSC members team spirit, acceptance of VDSC way of administration and satisfaction on VDSC leadership while items measuring unhealthy relationship in VDSC members recorded lowest perception. Besides this, effective and influential leadership will also create good acceptance and involvement from the organization members and the community (Salawu and Abu Bakar, 2008) and it fits with the finding of this study where items measuring their satisfaction

Table 4. Perception towards leadership aspects (n = 235).

Statement	Absolutely not agree	Not agree	Not sure	Agree	Absolutely agree	Mean
Overall mean score for perception towards VDSC leadership						3.90
I am satisfied with the VDSC members relationship	-	0.4	4.7	74.5	20.4	4.15
VDSC leaders take care of members charity	-	0.9	7.7	70.6	20.9	4.11
Team spirit in this village VDSC is high	-	0.4	11.1	67.7	20.9	4.09
The way of this village VDSC administration can be accepted by any party.	-	1.7	10.6	65.5	22.1	4.08
Generally i'm satisfied with the this village VDSC leadership	-	2.1	8.5	68.9	20.0	4.07
There are coorperation between VDSC and government/ private / NGO agencies	-	1.3	9.8	69.4	19.6	4.07
VDSC is alert on changing of government policy.	-	1.3	10.2	71.1	17.4	4.05
Members opinion are accepted by all VDSC members in any discussion or meeting	-	1.3	9.4	75.7	13.6	4.02
VDSC has initiative to overcome isues regarding social problem, family problems, drugs addict problems and community unity problem	0.9	3.4	8.5	72.3	14.9	3.97
All VDSC members have equal chances of attending training and courses.	2.1	3.4	10.6	70.2	13.6	3.90
There are scheduled program/ activity conducted by VDSC for the purpose of changing the mindset and attitude of the villagers.	0.9	1.7	14.0	73.2	10.2	3.90
There are efective information dessiminations to the villagers.	0.9	3.8	14.9	69.4	11.1	3.86
This village leadership provides complete and useable village profile and action plan	0.4	1.7	23.8	60.0	13.6	3.85
There are management of emergency issue such as unexpected programs that must be conducted instantly.	0.9	1.7	24.7	63.8	8.9	3.78
There are complete and useable village activity schedule.	1.3	4.3	20.0	65.5	8.9	3.77
There are unhealthy relations in VDSC organizations.	9.4	36.2	28.1	22.1	4.3	2.76

towards their leadership, coorperation between VDSC and other agencies, acceptance of members opinion in a

Table 5. Differences on perception towards VDSC leadership (n = 235).

Variables	N	Mean	S.D	F	p
Zone				5.094	0.007
Central	90	3.92	0.377		
East coast	75	3.98	0.478		
Southern	70	3.78	0.346		
Age				7.066	0.001
< 45 years	67	3.75	0.432		
46 - 55 years	79	3.93	0.324		
> 56 years	89	3.98	0.407		
Income per month				3.707	0.026
< RM1,000	85	3.98	0.335		
RM1,001 - RM2,000	86	3.86	0.386		
> RM2,001	42	3.79	0.379		
(Value in Ringgit Malaysia)					
Period of staying in VVM village				2.733	0.067
< 25 years	57	3.89	0.431		
26 - 45 years	91	3.84	0.402		
> 46 years	73	3.98	0.390		
Occupation				1.380	0.250
Work on their own	103	3.92	0.396		
Goverment staff	57	3.82	0.438		
Not working	42	3.97	0.365		
Private company staff	33	3.90	0.399		
Education achievement				1.690	0.170
Primary school	55	3.99	0.407		
PMR/SRP/LCE	43	3.91	0.321		
SPM/MCE/ SPMV/CERTIFICATES	105	3.88	0.371		
Pre-University and University	32	3.80	0.536		

VDSC meeting recorded mean score exceeded 4.0.

Differences in the perception towards VDSC leadership

An ANOVA test was employed to determine whether zone, age, income per month, period of staying, occupation and education achievement influence the different perception of VDSC members towards the VDSC leadership and the results were presented in Table 5.

The ANOVA test results showed that there was significant difference on the perception towards VDSC leadership and for three zones studied [F (3. 235) = 5.094, P < 0.05].

A post hoc multiple comparison test employed showed that there was a significant difference on the perception towards VDSC leadership between the east coast and southern zone. However, results revealed that there was no significant difference between central zone and east coast and southern zone. The highest mean score recorded for those from grade east coast zone (M = 3.98) followed by central zone (M = 3.92). The lowest mean score was recorded by southern zone (M = 3.78).

Based on the data presented in Table 5, obviously it concludes that there is difference in the perception of VDSC members towards VDSC leadership by age group. The ANOVA results [F (3. 235) = 7.066, p < .05] revealed that there was significant difference on perception towards VDSC leadership among the three age groups studied. There is a possibility that the older group will have more positive perception towards their leadership (Rickets et al., 2002) and among the reasons related to this are experience and loyalty to their leader.

The post hoc analysis done showed that there was a significant difference between VDSC members whose ages were < 45 years old, those whose ages were between 46 - 55 years and > 56 years. However, there was no difference between VDSC members whose ages were 46 - 55 years with those whose ages were > 56 years.

In terms of monthly income category, the highest mean score (M = 3.98) was recorded for those who received < RM1,000, followed by those who gained between RM1,001 - RM2,000 (M = 3.86) and those who gained > RM2,001 (M = 3.79). Results gained showed VDSC members perception towards the VDSC leadership differ according to the monthly income category [F (3. 235) = 3.707, $p < 0.05$]. Post Hoc test suggested that there was difference on those who received < RM1, 000 and those who gained > RM2, 001.

On the other hand with regards to staying period in VVM village, the results presented in Table 5 depicted that there was no significant difference between the three groups studied on their perception towards VDSC leadership [F (3. 235) = 2.733, $p > 0.05$]. The highest mean score recorded by those who had stayed in the village for a period of more than > 46 years (M = 3.98) followed by those who had stayed for < 25 years (M = 3.89) and those who had stayed for 26 - 45 years (M = 3.89).

Moreover, the results also showed that types of respondents' occupation had the possibility of not influencing perception towards VDSC leadership. This was based on ANOVA reading that showed F (4, 235) = 1.380, $p > 0.05$, indicating that there was no significant difference between four types of occupation group studied. Based on the result, there is equality in perception towards leadership between the junior and senior village residents. This supports what have been done by Farinde et al. (2004) where he found that occupation will not influence people's perception towards leadership.

The findings of this study also pictured that members of VDSC who do not work recorded the highest mean score which was 3.97. Followed by those who work on their own (3.92) and private company staff (M = 3.90). The lowest mean score was recorded by government staff (M = 3.82).

The highest mean score recorded for those who achieved primary education with the mean score of 3.99 while the lowest mean score recorded for those who achieved pre-university and university level of education (M = 3.89).

Based on the results, the level of education may not influence the level of competency of VDSC members. ANOVA analysis done proved that there was no significant difference between four groups of education based on the value of F (4. 235) = 1.690, $p > 0.05$. This is pertinent with the study done by Sirsikar (1964) where he found that education attainment is not a factor that influences leadership and he stressed that there are people without higher education who were able to become a leader if he has the commitment, capability and interest to become a leader.

Conclusion

The empirical findings of this study showed that there is a high perception among the CVD members towards the VDSC leadership. It is a good indicator that VDSC members have positive and healthy relationship with their leaders. It is suggested that the responsible authorities should maintain the existing VDSC leaders and their organization based on the high acceptation and positive relationship that occured among them.

Besides, the factors of zone, age and income per month were found to have significant difference with the perception towards VDSC leadership. Age was found as the most contributing factor to competency level. The study indicates that senior VDSC members have the possibility to have more leadership competency compared to the younger VDSC members. Thus, the study implies that agriculture agencies have the responsibility of creating various programs whereby the senior VDSC members can share and impart useful knowledge and experience to the younger members so that it would be able to create a pool of future leaders with great calibre.

Previous research had proved the importance of region or zone as a positive indicator towards leadership in creating effective organization. The finding of this study showed the importance of region or zone and it has clear implication for policy makers to understand the curent trend of leadership in rural areas. It is vivid that zone plays an important role perhaps due to the fact that these leaders of the particular area or zone might have a deeper knowledge of the range of elements that might be effective in their organization to improve performance. Besides, they might know best the changes that is needed in their own area so that they would be able to build a dynamic and successful organization in their particular zone. Hence, their leadership is greatlly needed to ensure the smooth performance of village development.

The findings also indicated the influence of monthly income towards rural leadership. Since leadership is about coping with change, the people in the rural areas believe that money has a major role in ensuring projects will be carried out without any hitches. Thus, they realize that they need leaders who are wealthy and influential in delivering strong leadership that would enable to get the support of the public policy implementers to execute more projects in their respective villages.

In conclusion, the influence of age, zone and monthly income is enormous in generating rural leaders who are able to match the needs of rural people. However, it is believed that more effort is needed from the policy makers to develop university graduates to take up the challenges of being rural entrepreneurs so that more rural development programs could be implemented in a creative and innovative manner.

It is suggested that future research could be carried out on other factors, such as culture, effective communciation skills and leadership styles, in determining its influence on rural leadership. Moreover, it is recommended that future research could use different methods employing a combination of quantitative and qualitative approaches in further understanding the construct of leadership among the rural leaders.

REFERENCES

Astin AW, Astin HS (2007). What is leadership. Paper presented at F21 National Assembly: Leaders Developing Leaders Developing Leader. November. Virginia, United States of America. pp. 2-4.

Bass B (1990). From transactional to transformational leadership: learning to share the vision. J. Organ. Dynam., 18: 19-31.

Brennan MA (2006). IFAS Community Development: The Continued Importance of Rural Development. http://edis.ifas.ufl.edu/pdffiles/FY/FY82500.pdf.

Fatimah MA, Mad NS (1997). Rural Development Model in Malaysia. Avaialable on; http://www.econ.upm.edu.my/~fatimah/rural.pdf.

Farinde AJ, Okunade EO, Loagun EA (2004). Community Perception of Women Occupying Leadership Position in Rural Development Projects of Osun State, Nigeria. J. Anthropol., 6: 273-278.

Fletcher LB, Fatimah A (2007). Malaysia's Agriculture and Rural Development: A Research Perspective. In Fatimah Arshad et al. (Eds.). 50 Years of Malaysian Agriculture: Transformational Issues, Challenges and Direction. Serdang, Selangor: UPM Publisher. pp. 729-741.

Gray SR (2008). Occupation and leadership in the workplace. Admin. Q. 45: 3.

Mitchell SP (2008). A Competing Values Approach to Effective Leadership. Farmers Productivity Rev. 5: 122-140.

Mohd KMB (2005). Rural development in Malaysia: Vision Village Movement (VVM) as mechanism for rural development. J. Technol., 42: 31-48.

Nunnally J (1978). Psychometric Theory. New York: McGraw-Hill.

Parry TJ (2008). Applying the monetary values approach to leadership. Leadersh. Psychol., 45(4): 385-415.

Rickets JC, Osbourne EW, Rudd RD (2002). Female Leaderhip among rural adolescent. Paper presented at the International LeadersipAssociation conference. Seattle, Washington, United States of America. http//www.ila-net.org/publications/proceedings/2002/jricketts. Pdf.

Robbins SP, Judge TA, Millett B, Waters-Marsh T (2008). Organisational Behavior 5th. Ed. Frenchs Forest, NSW: Pearson Education. http://issuu.com/mylibrary.

Salawu JA, Abubakar TB (2008). Introduction to Agricultural Extension and Rural Sociology. National Open University of Nigeria. Available on: http://www.nou.edu.ng/ noun/NOUN-OCL/pdf.

Schein EH (2004). Organizational Culture and Leadership. San Francisco, C.A.: Jossey-Bass. http://www.amazon.co.uk/gp/reader/0787975974/ref=sib_dp_pt=reader

Shorthall S, Shucksmith M (1998). Integrated rural development: Issues arising from the Scottish experience. J. Eur. Plan. Stud., 6: 73-88.

Sirsikar VM (1964). Leadership Patterns in Rural Maharashrata. J. Asian Surv. 4: 929-939.

Standing Conference for Community Development. Available on: http://www.idealist.org/if/i/en/av/Org/91363-185.

Stevens K (2006). The attributes of a high performance leader. Leadersh. Manage., 6(3): 10-18.

Uphoff N (1992). Local institution and participation for sustainable development. Gatekeeper Series Number SA31. International Institute for Environment and Development. www.ctu.edu.vn/institutes/mdi/ntth/VN/PRA/PRA-IIED002.pdf

Wicks P (2007). Personality determinants of effective leader. J. Manage., 15(2): 28-39.

Zainalabidin M, Mohammad RH, Mohd MI, Norsida M, Asnarulkhadi AS, Azimi H, Hanina H, Zamre Y (2005). Micro Credit Finance and Poverty Eradication amongst the Rural Poor. Paper presented at International Borneo Business Conference 2006 organized by UNIMAS and UMS, Hilto Hotel, Kuching Sarawak. pp. 1-14.

Socioeconomic impact of SARO agro allied organic cocoa programme on beneficiary cocoa farmers in Nigeria

Faturoti B. O.[1]*, Madukwe M. C.[2], Ogunedojutimi O.[1] and Anyanwu L.[1]

[1]SARO Agro Allied Nigeria Limited, Plot C66A, Oba Kayode Akinyemi way, Amuwo P. M. B. 1837, Apapa, Lagos, Nigeria.
[2]Department of Agricultural Extension, University of Nigeria, Nsukka, Enugu State, Nigeria.

Organic cocoa farming involves cultivation of cocoa with rational use of allowable inputs (fungicides and bio-insecticides). SARO Agro Allied Nigeria and her foreign partner ADM/Schokinag started the organic cocoa project (OCP) in 2006/2007 cocoa season in four of their major operating states in Nigeria (Ondo, Osun, Ekiti and Cross River). This study was designed to evaluate the income profile, biological and socioeconomic impact of OCP on the pilot farmers who were purposively selected for the study. Data were analyzed using frequency, percentages, mean and T-test statistics. Significant difference was recorded in 5 socioeconomic variables between 2007 and 2010, while a positive relationship was established between organic cocoa production, yield and farmers income profile. Three major sources of income (yield increase, savings from pesticide use reduction and premium) were identified from the 471 respondents. The cocoa yield increased from 0.5 mt/ha in 2007 to 0.73 mt/ha in 2010, there was 40% reduction in fungicide use which translated to N13.2 m savings for the 471 farmers, revenue from yield increase was N112.7 m, while revenue from premium for organic cocoa production was N24.7 m. Total revenue as a result of organic cocoa production per farmer was thus N319,745.22. Moreover, land under organic production by the pilot farmers was 1687 ha which the same remained over the study period. The study concluded that the OCP impacted positively in increased yield, input cost reduction, environmental and health improvement and increased revenue.

Key words: Organic cocoa, yield, premium, fungicides, bio-insecticides, International Federation of Organic Agriculture Movements (IFOAM).

INTRODUCTION

Production of cocoa in Nigeria has been an unstable agricultural and economical venture; it was at a time the main stay of Nigeria economy before the discovery of crude oil, when its production was neglected and farms became moribund. Renewed efforts on the part of government and private individuals have resuscitated the cocoa industry in Nigeria to its now fledging status where it has regained its major economic base for Nigeria. Revenue from cocoa exports is the country's second biggest foreign exchange earner, after crude oil (World Bank, 2010). Over 620,000 ha of land is under cocoa cultivation, with national production put at a little above 245,000 MT of cocoa beans in 2010/2011 cocoa season. Low farm productivity and poor cocoa bean quality as well as poor quality of farmers and farm hands working condition has adversely affected increased output and quality of Nigeria's cocoa. Small scale farmer that produces the bulk of cocoa lacks the knowledge and skills that can lead to improve productivity and quality of output. The current market conditions do not allow farmers to increase their income even if they offer better products or improve their production process. In the same vein, increased output through increased land area

*Corresponding author. E-mail: tundefaturoti@yahoo.com. or tunde.faturoti@saroafrica.com.ng.

remains largely difficult and not sustainable. Therefore, attempts to increase output/income must focus on increased productivity per land area, quality of produce and quality of operator's life (Faturoti, 2010).

Organic agriculture is a health and environment conscious farming system which avoids the use of synthetic compounds and seeks to minimize pollution of air, water and soil. It also preserves the health and productivity of operators and consumers. Organic cocoa farming involves cultivation of cocoa with rational use of allowable inputs (fungicides and bio-insecticides). It is a regulated way of farming that mimics nature [International Federation of Organic Agriculture Movements (IFOAM, 2002)].

Organic agricultural methods are internationally regulated and legally enforced by many nations, based in large part on the standards set by IFOAM (2002), an international umbrella organization for organic farming established in 1972 (Paull, 2010). IFOAM defines the overarching goal of organic farming as: "Organic agriculture is a production system that sustains the health of soils, ecosystems and people. It relies on ecological processes, biodiversity and cycles adapted to local conditions, rather than use of inputs with adverse effects. Organic agriculture combines tradition, innovation and science to benefit shared environment and promote fair relationships and a good quality of life for all involved (Organic Monitor, 2009)

Standards regulates production methods and in some cases final output for organic agriculture. Standards may be voluntary or legislated. As early as the 1970s, private associations certified organic producers. In the 1980s, governments began to produce organic production guidelines. In the 1990s, a trend toward legislated standards began, most notably with the 1991 EU-Eco-regulation developed for European Union (EU), (Control Union World Group, 1991), which set standards for 12 countries, and a 1993 UK program. The EU's program was followed by a Japanese program in 2001, and in 2002 the U.S. created the National Organic Program (USDA, 2008). IFOAM (2005) created the Principles of Organic Agriculture, an international guideline for certification criteria. Typically, the agencies accredit certification groups rather than individual farms.

SARO Agro Allied, a major cocoa exporting company started operations in Nigeria in 1996, with her operations more intensified in four states of the country, namely Ekiti, Ondo, Osun and Cross River states. The company's cocoa procurement strategy is origination and innovativeness, as such it has deep rooted interactions with buying agents located in villages, but with fringe interactions with farmers; prior to introduction of organic cocoa project (OCP) her cocoa export was majorly to Germany through the then Schokinag Schokolade.

In 2006, because of the increased demand for organic cocoa and increasing attention and demand for food with known origin and known inputs (organically grown cocoa) from Schokinag's customers in Europe and America, SARO Agro Allied was approached for a venture in organic cocoa which was in tandem with the vision of SARO to innovatively add value beyond sourcing of cocoa and exporting cocoa. The OCP was therefore started in the 2006/2007 cocoa season in four states of Nigeria namely, Ondo, Osun, Ekiti and Cross River. Farmers were organized into groups for ease of project delivery, internal control system was appointed and trained to oversee conformities to the organic codes as it was a novel project. At inception, there were apprehensions on farmers understanding and compliance to the organic codes as well as the outcome of the project on the yield, income and sustainability of the project.

This study, therefore, assessed the impact of the organic cocoa production in Nigeria. Specifically, the study:

1) Describe the personal characteristics of the small holder farmers, who were involved in the organic cocoa program,
2) Ascertain the impact of the organic cocoa production program on respondents farmers yield,
3) Ascertain the income profile of small holder farmers, who are involved in the organic cocoa program,
4) Examine sustainability of the OCP

Hypotheses:

1) There is no significant increase in respondents' status and income profile in 2007 and 2010
2) There is no significant impact of the organic cocoa program on the respondents
3) There is no significant difference in agro-economic yield in 2007 and 2010.

METHODOLOGY

Study area

The study area was the four SARO Agro Allied OCP pilot states in Nigeria (Ekiti, Ondo, Osun and Cross River). The four states accounts for 70% of Nigeria total cocoa production and 68% of available land mass for cocoa cultivation (Omolaja et al., 2011).

Population and sample

The population of the study was taken from the pilot organic cocoa farmers in the four pilot states (Ekiti, Ondo, Osun and Cross River). The project started with 501 farmers in 2006/2007 cocoa season which were reduced to 471 with the removal of non-compliant farmers after the first round of organic inspection certification in 2007. Purposive sampling of 471 compliant farmers was done, thus, a total of 471 respondents were included in the study (Table 1). Data were collected from the respondents through the use of questionnaire and interview schedule.

Data analysis

Data collected were subjected to statistical analysis using the

Table 1. Composition of population and sample.

Location	No. of farmers in 2006/2007	No. of violators	No. of farmers sampled
Ikom	74	10	64
Erinjo	44	3	41
Idanre	110	6	104
Ekiti	93	3	90
Ondo	70	4	66
Ilesha	110	4	106
Total	501	30	471

Source: Field survey (2010).

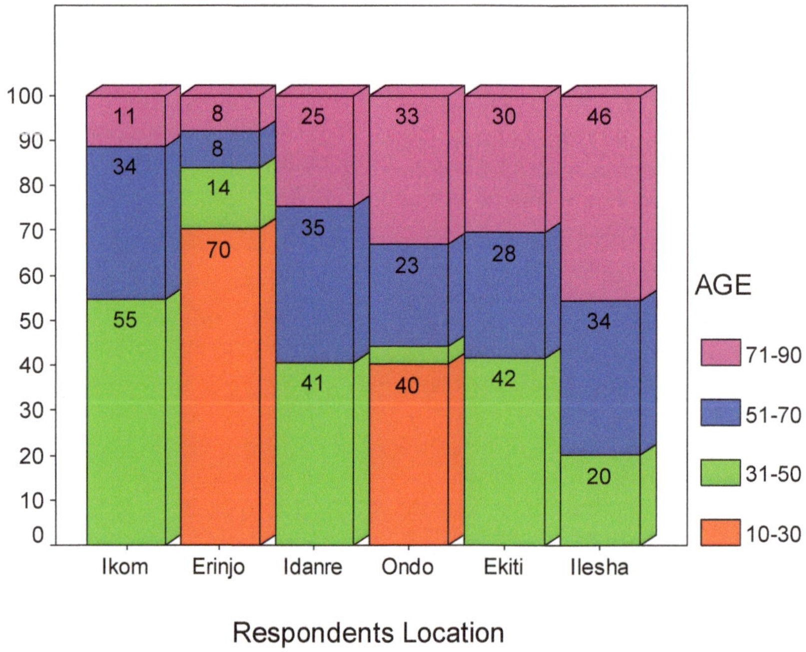

Figure 1. Organic farmer's age distribution.

Statistical Package for Social Science Research (SPSS) as a tool of analysis. Descriptive statistics (percentages and mean statistics), graphs and tables were used to describe farmers personal and socioeconomic characteristics, T-test statistics was used to compare the baseline data collected in 2006/2007 season (yield, pesticide usage, input cost and income generated) with results obtained in 2010/2011 season (yield, pesticide usage, input and income generated).

RESULTS AND DISCUSSION

Age distribution

Data shows that 29.3% of the farmers are in the age class of 31 to 50 years (highest class), 27% are 51 to 70 years. Also, 25.7% are within the ages of 71 to 90%, while only 18% are below 30 years (Figure 1). In terms of productivity and sustainability, 47.3% are within the ages of 10 to 50 years. The mean age of the farmers was 55 years. This shows a relatively growing and productive group with capacity for farm expansion and dynamism for adoption of innovation. Erinjo and Ondo locations had greater percentages of their farmers 70 and 40%, respectively below 30 years; this may be a pointer for greater sustainability and adoption. Ilesha had the oldest farmers in the study and this reflects in their low production. The younger the farmer, the more likely he is to adopt innovation in its early cycle. Older farmers rely less on external information, and therefore, do not get in touch with innovations in the market as early as their younger colleagues (Diederen et al., 2002; Schinitkey et al., 1992)

Respondents' sex

Data in this study shows that there are 88.1% male

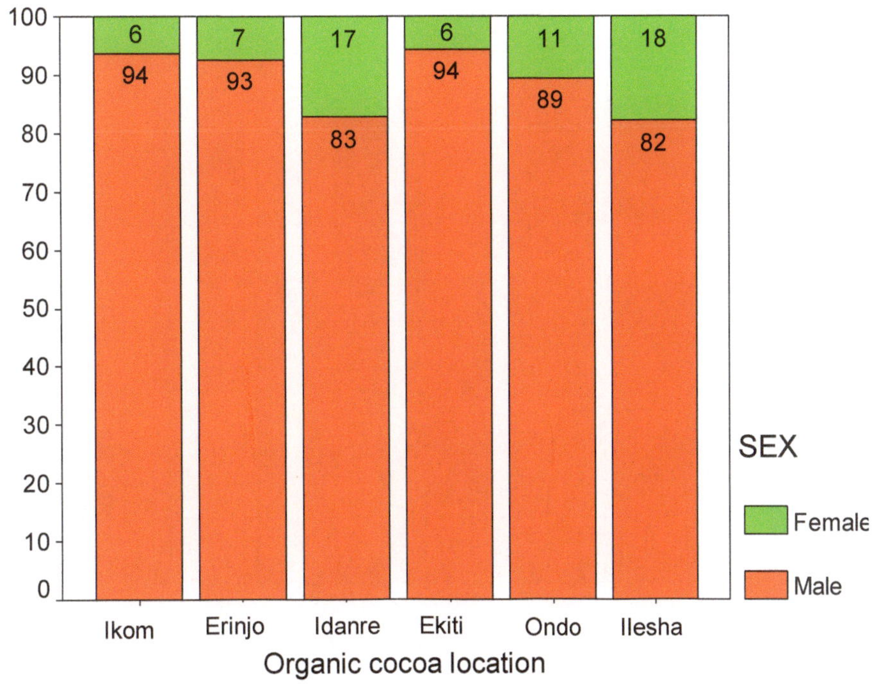

Figure 2. Respondents percentage sex distribution.

respondents and 11.9% females (Figure 2). This is not un-expected as land tenure and inheritance rights mostly reside in custody of males in Nigeria as such decision over land is vested largely in males who had control over the family by native law and custom. However, with cocoa becoming a cash crop and a plantation that can be contracted out for maintenance, more and more women are becoming cocoa owners. The initial large number of females especially in Ilesha and Idanre may also be as a result of willingness of more women to join the program at inception as most men in the area adopted a wait and see attitude causing many of them to join after seeing positive results from pioneers (Faturoti et al., 2008). Adoption is a copying process which diffuses to other users when the superiority of the innovation becomes visible to others (Rogers, 2003).

Educational status

Data collected on the respondents educational status shows that one third of the respondents (27.6%) had no educational background with majority coming from Ilesha and Erinjo. 29.5% had primary education, 25.3% secondary education, while 17.6% had university education. Primary and secondary education accounted for 54.8% of the respondents (Figure 3). The result shows relatively high literate respondents with over 70% literacy. This has implication for adoption as educational level has been found to have a positive correlation with adoption (Lemchi et al., 2005a).

Land ownership

Result shows that 49.67% respondents bought their land and owned it personally, 21.83% derived their title ownership from family and has unlimited control over the land, 20.5% rented their lands with renewable tenors that can be revoked, while 8% lease their farms from government this can also be revoked (Figure 4). The majority 71.5% of the respondents which has unlimited long-term tenor on their land has serious implication and positive effects on adoption of organic and other certified farming innovations that requires undisturbed access and control over a long period of time to actualize sustainable certification, as land is the certified object on which the crop derived its certification (IFOAM, 2004).

Land size

Data collected on land distribution show that the respondents were predominantly small holder farmers with 47.2% having land holding between 1 to 10 ha. Only 14% had land larger than 20 ha (Figure 5). This suggested that aggressive aggregation has to be done to increase land under organic/certified produce as more farmers will continuously be needed for production increase. Land size was found to have positive correlation with innovation adoption where farmers can try an innovation on a smaller scale before full scale adoption when satisfied with the outcome (Dielderen et al., 2002). The predominantly small landholders' farmers

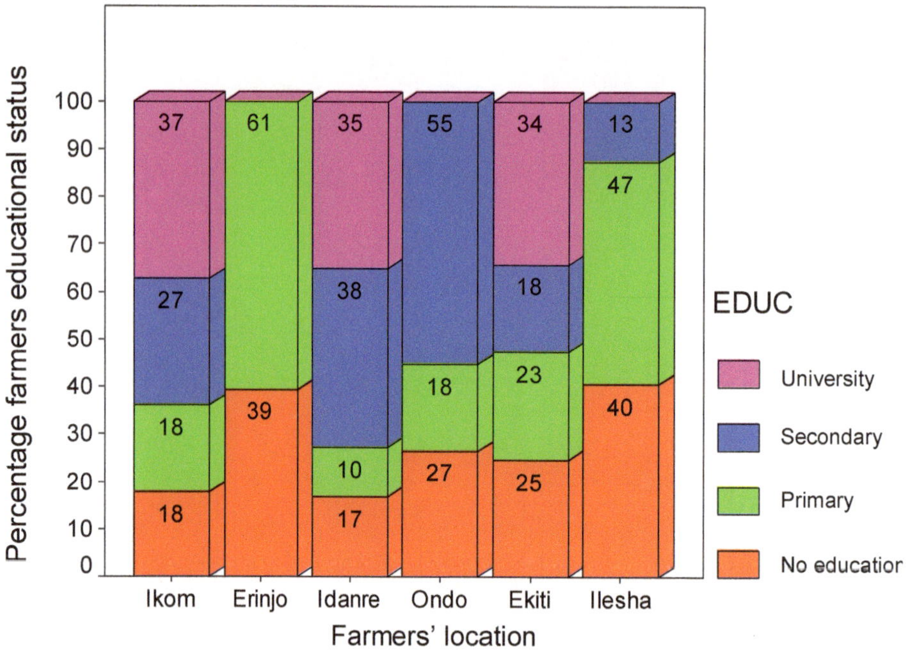

Figure 3. Organic farmers' educational status.

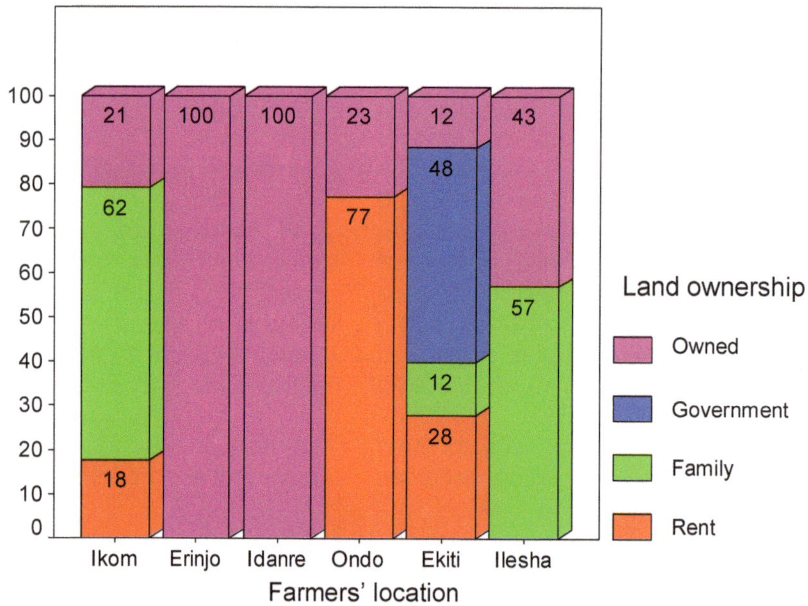

Figure 4. Organic farmers' land ownership status.

in this study are a reflection of the generally small holder cocoa production system in Nigeria and fragmented land inheritance pattern.

Average yield per hectare under organic production

The result of baseline yield per hectare of the compliant organic cocoa farms monitored over a 4-year period show 100% yield increase in Ekiti (0.3 mt in 2007 to 0.6 mt in 2010). Erinjo and Ondo also recorded appreciable yield increase of 50 and 40%, respectively. A below average yield increase of 25 and 14% was recorded in Ilesha and Idanre, respectively from 2007 to 2010 (Figure 6). However, a marginal yield increase of 9% was recorded at Ikom (1.1 mt/ha in 2007 to 1.2 mt/ha) over

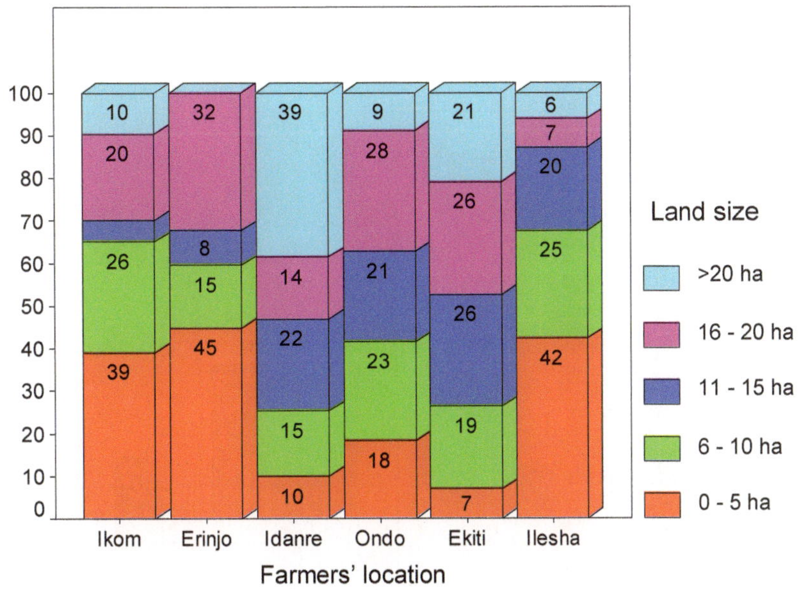

Figure 5. Organic farmers' land size distribution.

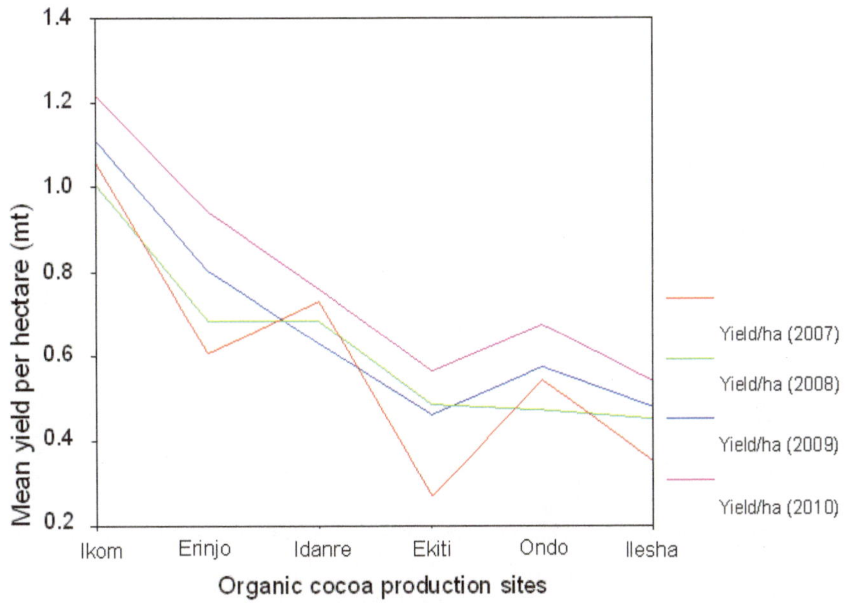

Figure 6. Yield per hectare of organic production.

the study period. It was also observed that the current average yield per hectare in the study area is well above the national average of 0.4 mt/ha, whereas Ekiti, Ondo and Ilesha were around the national average in 2007 when the project started but has surpassed the current national production average (Gain Report, 2011). This may not be unconnected with the impact of the good agricultural practice that resulted in lesser black pod disease and better soil management which were basic concern of the project (Roundtable for a sustainable cocoa economy, 2009).

Fungicide consumption in the study

Data in Figure 7 shows that there has been general decrease in allowed fungicide usage in the study area. Results show that there was more than 40% reduction in fungicide usage across study location. The highest reduction in fungicide usage was recorded in Idanre with 40.6%, while the lowest reduction in fungicide usage was recorded in Ilesha 34.6%. This is a substantial economical savings and environmental friendliness for the project area. Idanre was noted for her high intensification

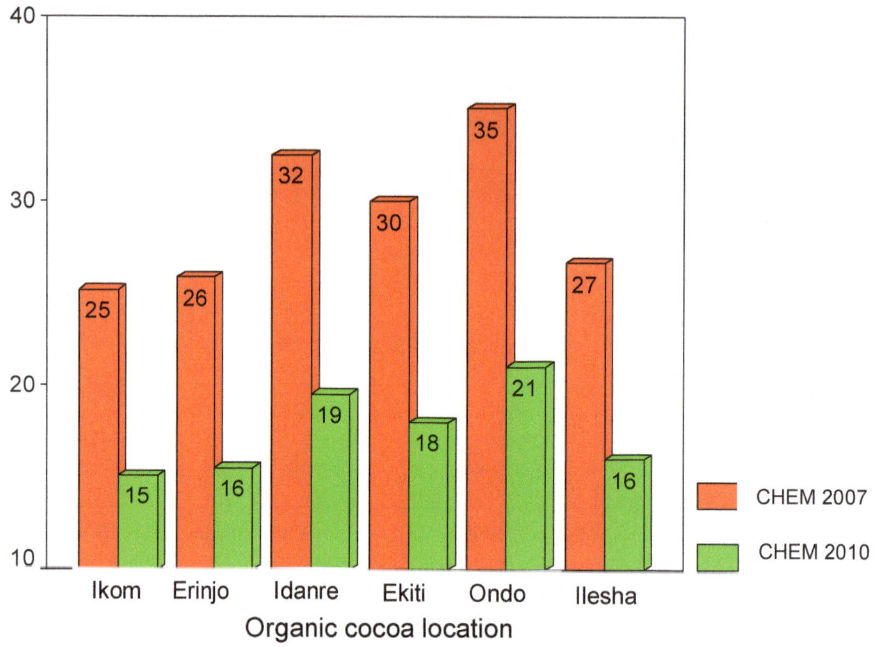

Figure 7. Fungicide usage hectare 2007 and 2010.

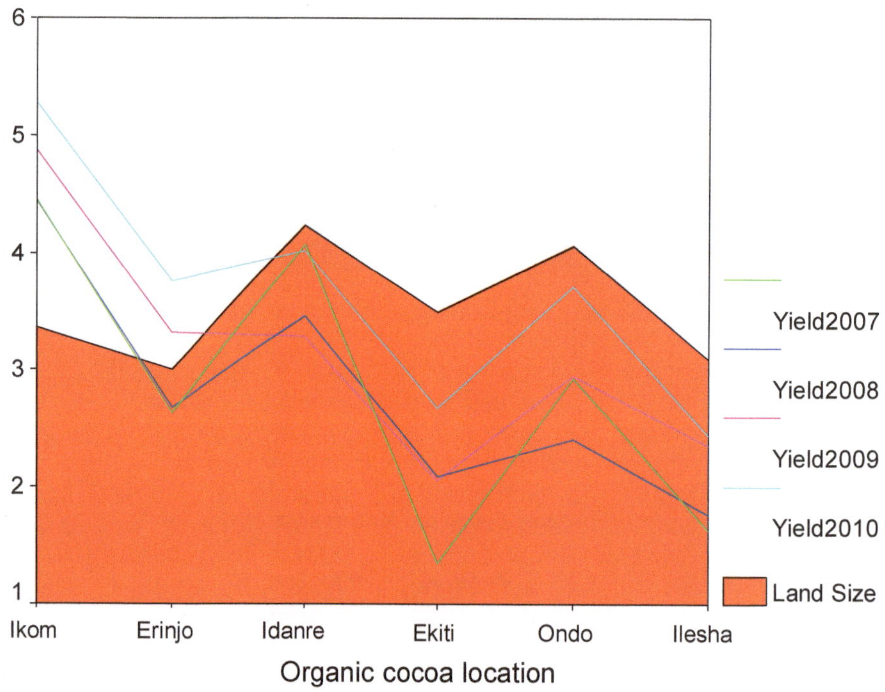

Figure 8. Respondents mean yield per organic land area.

of cocoa production with emphasis on pesticide usage as such adoption of organic cocoa production resulted in lower pesticide use which translates to higher monetary savings for farmers in the study with more financial benefits accruing to Idanre farmers.

Average yield per land area available to organic farmer

Results obtained (Figure 8) when respondents yield was compared to average land area available in the project in

Table 2. Differences in means of farmers economic returns in 2007 and 2010 as a result of organic cocoa intervention.

Variable	Mean 2007	Mean 2010	T-value
Mean yield mt/farmer	2.77	3.51	9.20*
Mean yield mt/ha	0.56	0.73	11.01*
Mean fungicide kg/farmer	29.48	17.68	70.12*
Mean fungicide kg/ha	8.20	4.90	2.00*
Mean fungicide cost (N)	15,886.76	44,129.89	2.00*

Source: Field survey (2010). *Significant (P<0.05).

Ikom project area, 40.7% yield increase at Erinjo 27%, increase in ondo, 60% increase at Ilesha, while Ekiti recorded 115% yield increase over the four season period. This result has accelerated willingness to be part of the program with a resultant effect of adjoining villages persuading operators of the program to come over and work in their area too. On the contrary however, there was no significant yield increase at Idanre during the period of study, which accounted for the less cooperative attitude of Idanre farmers to the program. This reactive process is explained by the established theory of adoption, that adoption is favored and enhanced where the gains are visible and more profitable than existing innovation (Rogers 2003; Lemchi et al., 2005b).

Impact assessment variables

Result revealed significant statistical difference in the means of five impact related variables considered in the study. The variables are yield per farmer, yield per hectare, fungicide used per farmer, fungicide rate per hectare and fungicide cost; the means were significantly higher in 2010 than 2007 when the project commenced. Significant yield increase was recorded in the study, yield per hectare increased from 0.56 to 0.73 mt/ha; this was well above the national average despite the generally unrestrained use of pesticide in the conventional cocoa production in Nigeria. However, the significant difference recorded for fungicide rate and cost was not strange as organic production regulates allowable use of copper fungicides and places emphasizes on good agricultural practices (pruning, farm sanitation, removal of infected pods), which in turn reduces black pod incidence and consequent reduction of fungicide usage (IFOAM, 2011). There was however, general increase in fungicide cost which impacted less on the respondents' farmers as a result of their reduced fungicide use resulting from good agricultural practice in the study (Table 2).

Revenue generation profile in organic cocoa production 2007 versus 2010

Table 3 shows the income generation profile of the organic cocoa program in Nigeria, contrary to the widely held believe that organic cocoa initiative will reduce farmers yield and consequent income stream; the study accepts the alternative hypotheses as remarkable revenue increase from 2007 to 2010 was recorded. There was 27.63% increase in yield during the study period, which may have come from good agricultural practices (timely protection with allowable fungicides and general farm sanitation) on which the project was formed and operated. Fungicide application rate also reduced by 40% over the study period; this fungicide reduction and yield increase has monetary implication as recorded in this study.

Despite the 362% increase in the national cost of copper fungicide between 2007 to 2010, a savings of N13,201,971.60 m was recorded as a result of reduction in rate of fungicide application between 2007 and 2010; this is no doubt a substantial amount of money considering the hectares involved. Price of cocoa increased globally by 50% over the study period, therefore, the 27.63% yield increase recorded translates to N112,652,400.00 in 2010 compared with 2007. Another source of income in the study is premium paid for organic certified cocoa farmers, most of which were paid in kind through, training, subsidy on fungicide, protective gears and warehouse separation and packaging. However, the total cash premium paid to the farmers was to the tune of N24,780,000 m. All these streams of income put together amounted to N158,107,950 m accrued to the organic farmers in this study. This translated to N127, 609.32/mt, and N93,676.95 extra income generated per hectare of organic cocoa production. This is no doubt a morale booster that can ignite further participation and investment in organic cocoa production (Rundgren, 2006).

Respondent's benefits from organic cocoa

Results obtained as recorded in Table 4 show that increased yield (34.4%) and increased income (24.6%) were the two most visible benefits accrued to farmers in this study, though benefits related to health were equally ranked high by beneficiary farmers (reduced sickness

Table 3. Comparison between 2007 and 2010 revenue generation profile in organic cocoa production.

Variable	Mean 2007	Mean 2010	Difference	% Change
Land area	1,687.80	1,687.80	0.00	0.00
Total yield	970.78	1,239.00	268.22	27.63
Price (naira/mt)	280,000.00	420,000.00	140,000.00	50.00
Fungicide application rate (kg/ha)	8.20	4.90	-3.30	-40.24
Mean fungicide cost/farmer	15,886.76	44,129.89	28,243.13	177.78
National cost of cupper fungicide (naira/kg)	540.00	2,500.00	1,960.00	362.96
Fungicide cost/ha	4,442.00	12,341.00	7,899.00	177.83
Total fungicide cost	7,473,578.40	20,675,550.00	13,201,971.60	176.65
Total fungicide cost at 2007 rate of application	7,473,578.40	34,599,900.00	27,126,321.60	362.96
Actual revenue gain from fungicide reduction	7,473,578.4	20,675,550	13,201,971.60	176.65
Revenue from sales of cocoa	271,818,400	520,380,000	248,561,600	91.44
Cash premium	0.00	24,780,000	24,780,000	100
Total revenue	279,291,478	541,055,550	261,763,571	93.72
Revenue from yield increase at 2010 price	407,727,600	520,380,000	112,652,400	27.63
Total revenue gain/loss	407,727,600	565,835,550	158,107,950	38.78

Source: Field survey (2010), $1 = N130 in 2007 and N150 in 2010.

Table 4. Respondents benefits from organic cocoa project.

Code	Location	Benefits					Total
		Increased income	Increased yield	Less sickly	Health awareness	All combined	
Farmers' code	Ikom	22	14	11	17		64
	Erinjo	15	5	6	4	11	41
	Idanre	37	35	15	16	1	104
	Ekiti	13	44	29	3	1	90
	Ondo	8	14	13	27	4	66
	Ilesha	21	50	14	14	7	106
Total		116	50	14	14	7	471
Percent farmers benefits		24.6	34.4	18.7	17.2	5.1	

Source: Field survey (2010).

frequency 18.7%, health awareness 17.2%), In most of the operating zones, increased income and yield pooled more than half of accrued benefits except in Ondo where health benefits (health awareness 41% and less sickly 20%) ranked highest; this may not be unconnected with the fact that Ondo had the highest pesticide consumption at inception of the program and has been found in this study to also have highest rate of reduction of pesticide usage over the study period (Figure 7), this may have proved the negative effects of pesticide usage to the respondents. However, increased productivity as represented in terms of income and yield increase remain the main motivating factor to organic cocoa production as recorded in this study.

Dissemination of organic cocoa information by beneficiary

Result obtained as shown in Table 5 explained that the highest number of farmers informed by respondents was 1 to 5 farmers (53.9%), 6 to 10 farmers (18.1%) and only 7.6% informed more than 10 farmers each. Sadly, a very high proportion of the farmer 20.4% has not informed any other farmer about the program. This result is not strange as one third of the respondents had no formal educational background which obviously explained the poor dissemination of the program by these set of farmers (Figure 3).

CONCLUSION AND RECOMMENDATIONS

Massive economic and environmental rewards associated with organic cocoa production program were recorded in the study area. The significant increase recorded in the bioeconomic indices evaluated, where five variables (Yield per farmer, yield per hectare, fungicide per farmer, fungicide per hectare and fungicide

Table 5. Dissemination of organic cocoa information by beneficiary farmers.

Code	Location	No. informed				Total
		None	1-5 farmers	6-10 farmers	>10 farmers	
Farmers' code	Ikom	11	28	12	13	64
	Erinjo	4	27	1	9	41
	Idanre	14	75	12	3	104
	Ekiti	28	34	28		90
	Ondo	5	46	4	11	66
	Ilesha	34	44	28		106
Total		96	254	85	36	471
Percent informed farmers		20.4	53.9	18.1	7.6	

Source: Field survey (2010).

cost) showed significant difference in 2010 over 2007 pointed to positive impact of the program. Environmental safety, health consciousness and yield increase were tripartite gain of the project; this is achieved through rational and reduced use of pesticide which reduces risk of contamination and cost incurred on inputs which accounted for high revenue savings of N13.2 m, the initial fear of the respondents of yield loss as a result of non allowed use of fertilizer or other synthetic inputs/additives were allayed with the consistent yearly increase in yield and commensurate decrease in black pod disease as a result of good agricultural practice which is a major mandatory deliveries of organic cocoa production and certification codes. Despite the fixed land area used by the pilot farmer over the study period yield increase of 28% was recorded with total revenue of N112.6 m in 2010 compared to 2007. Revenue from premium to organic (N24.8 m) was also a major revenue boost that contributed to the project impact.

Therefore, the OCP impacted positively in increased yield, input cost reduction, environmental and health improvement and increased revenue. Perhaps, the most sustainable gain of the project is the reduced use of pesticide and health awareness of the respondents which keeps the environment safe for all to use and safe food from cocoa and associated food crops on the organic land. It was also ascertained in the study that sustainability has to do with fairness interaction of productive factors (man, crop and environment) on each other which ensures that today actions preserves tomorrow needs. Therefore, the OCP as sponsored by SARO Agro Allied and her foreign partner ADM/ Schokinag, has ensured sustainability bringing yield per hectare above the national average of 0.45 mt/ha; intensification and scaling up of these program will no doubt bring huge foreign exchange to the economy of Nigeria and raise the farmers income level while also providing assured produce for SARO and her foreign partners. Finally, integrating organic agriculture into overall agricultural policies and poverty reduction strategies, and building organic agriculture supply capacities through education, research, extension services, local and regional market development and export facilitation, are key to realizing the benefits that organic agriculture offers.

REFERENCES

Control Union World Group (1991). EEC. Regulation No. 2092/91.

Dielderen PJM, Van JCM, Wolters AM (2002). Innovation and farm performance: The case of Dutch agriculture in: Innovation and farm performance: Econometric explorations of survey data, Klein-Knecht A., Mohnen P., London, Palgrave Macmillan.

Faturoti BO, Agwu AE, Igbokwe EM, Tenkouano A (2008). International Institute of Tropical Agriculture plantain and banana programme: An Insight into the contribution of farmer-to-farmer extension paradigm. Afr. J. Biotechnol. 7(13):2137-2146.

Faturoti BO (2010). Private sector led cocoa extension delivery in Nigeria: A case study of SARO Agro Allied Organic Cocoa Program. Paper presented at GIZ organized conference on midwiving the Farmer Business School In Nigeria.

International Federation of Organic Movement (2002). IFOAM Training Manual For organic Agric. In the Tropics, Theory, Transparencies, Didactic Approach Compiled by FIBL.

International Federation Of Organic Movement (2004). IFOAM Small holder group certification training curriculum for producer organizations + Guidance manual.

International Federation Of Organic Movement (2005) The IFOAM Norms.

International Federation Of Organic Movement (IFOAM) (2011). Organic Agriculture Worldwide: Key results from the survey on organic agriculture worldwide.

Lemchi JI, Tshiunza M, onyeka U, Tenkouano A (2005). Factors driving the adoption of cooking banana processing and utilization methods in Nigeria. Afr. J. Biotechnol. 4(11):1335-1347.

Lemchi J, Ezedinma CI, Tshiunza M, Tenkouano A, Faturoti BO (2005b). Agroeconomic evaluation of black sigatoka resistant hybrid plantains under small holder management systems. Afr. J. Biotechnol. 4(10):1045-1053.

Organic Monitor (2009). Organic Monitor Gives 2009 Predictions, press release, www.organic monitor.com.

Omolaja S, Iremiren G, Aikpokpodion P (2011). Implication of the "CRIN 8" for the Nigeria Cocoa Economy. Varietal release support paper, presented at Centre for Cocoa Development Initiatives, Akure, Nigeria.

Rundgren Gunner (2006). Best practices for organic policy: What developing country governments can do to promote the organic sector, UNEP-UNCTAD CBTF.

Rogers EM (2003). Diffusion of Innovations (5th edition) New York. N.Y: Free Press. p 152

Roundtable for a sustainable cocoa economy (2009). Guidelines for best known practices in the cocoa value chain. www.roundtable.cocoa.org. RSCE 2/3.
Schnitkey G, Batte M, Jones E, Botomogno J (1992). Information preferences of Ohio commercial farmers: Implementation and extension, Am. J. Agric. Econ. 4: 486-496.
USDA (2008). NOP standards.USDA Foreign Agricultural Service Global Agric. Information Network. GAIN REPORT NUMBER: N111018 (2011).
World bank (2010). "Doing Business in Nigeria". http://. Doing business .org/explore economics.

4

Survey on women access to agricultural extension services at selected districts of Mid Rift Valley of Ethiopia

Gurmesa Umeta*, Felekech Lemecha and Taha Mume

Adami Tulu Agricultural Research Center, P. O. Box 35, Zeway, Ethiopia.

The study was conducted at administrative Zone of Oromia Region in Central Rift Valley of Ethiopia namely; Dugda, Boset, Ada'a and Fentale districts during 2009 with two objectives: Assessing women farmers access to agricultural extension services and Identifying problems constraining women farmers' participation in agricultural extension package program. The district was selected purposively based on farming system of the zone. Sampled kebele was selected purposively based on house hold size and suitability of the area for transportation. Respondents were selected based on probability proportionate to sample size (PPS). A total of 201 respondents were selected for interview. Women farmers headed by male (MHHLD) and households headed by women (FHHLD) were included in the sample for comparison purpose. The data was collected through structured interview schedule. The data was analyzed by using descriptive statistics namely; frequencies, mean, and standard deviation by using SPSS.13 version. Chi-square and t- test were also employed for data analysis. The current study result indicates that participation of FHHLD in agricultural extension package program is still very low (44.4%) when compared with MHHLDs (75%). The major constraints to access and utilization of extension packages identified by this study includes; low supply related problems, cost of the technology (expensive), delay of inputs (input is not available on time), low awareness about technology recommendations and biasdness of extension agents towards progressive farmers can be mentioned. Participation of women farmers in extension events like training, field days and demonstration is also very low (<21%). On the other hand, FHHLDs' access to productive resources is low when compared with MHHLDs. FHHLD farmers owned a mean of 1.43 ha where as MHHLDs owned a mean of 2.03ha of farm size and their difference is significant at 1% significant level (t = 3.28, p= 0.001). MHHLDs are better access to oxen than FHHLDs and their difference is significant at 1% probability level(x^2= 6. 88, p = 0.009). MHHLDs are also better access to other productive resources like TLU and credit services than FHHLDs. Generally, the study recommends that women farmers /FHHLDs/ needs to be encouraged so as to participate in agricultural extension package program. Specifically, strengthening of women participation in extension events like training, field days and visits needs to be re-considered. Also, strengthening of linkages between development agent and women farmers is one of the other critical factors that requires giving due attention. Furthermore, the study recommends that there is a need to diversify women's livelihood options.

Key words: access, male headed households, female headed households.

Back ground and justification

In 2002, three out of four poor people in developing countries lived in rural areas, with the majority of them relying, either directly or indirectly on agriculture for their livelihoods (World Bank, 2007). Agriculture plays an important role in both poverty reduction and economic growth. Agriculture remains the main source of income for around 2.5 billion people in the developing world (FAO, 2003). Moreover, the poorest half of the population

*Corresponding author. E-mail: gurme2010@yahoo.com.

benefits significantly more from agricultural growth than growth in other sectors of the economy (UN, 2008; World Bank, 2007). Nevertheless, despite evidence that investment in agriculture has beneficial impacts on agricultural growth and poverty reduction (Fan and Rao, 2003), since 1980 there has been a decline or stagnation in public expenditure on agriculture in most developing countries (Akroyd and Smith, 2007). Likewise, the proportion of official development assistance (ODA) going to agriculture has also declined from around 18% in 1979 to 3.5% in 2004 (World Bank, 2007).

Agricultural extension and advisory services play an important role in agricultural development and can contribute to improving the welfare of farmers and other people living in rural areas. Anderson (2007) defines the terms agricultural extension and advisory services as "the entire set of organizations that support and facilitate people engaged in agricultural production to solve problems and to obtain information, skills and technologies to improve their livelihoods". Extension services can be organized and delivered in a variety of forms, but their ultimate aim is to increase farmers' productivity and income. According to Anderson and Feder (2003) productivity improvements are only possible when there is a gap between actual and potential productivity. They suggest two types of 'gaps' contributing to the productivity differential – the technology gap and the management gap.

Extension can contribute to the reduction of the productivity differential by increasing the speed of technology transfer and by increasing farmers' knowledge and assisting them in improving farm management practices (Birkhaeuser et al., 1991; Feder et al., 2004b). Additionally, extension services also play an important role in improving the information flow from farmers to scientists (Anderson, 2007; Birkhaeuser et al., 1991). The Ethiopian economy depends almost on agriculture. The agricultural sector is dominated by mixed farming where crops and livestock play major roles in serving dual purpose for domestic consumption and foreign currency earnings.

Agriculture contributes about 50% of gross domestic product (GDP) and 90% of the national export earnings. In general, more than 85% labor force is based on "agriculture" (Alene et al., 2000). In the economy of Ethiopia, as many of African countries, women are the back bone of food production system (EARO, 2000). Gender related constraints reflect gender inequalities in access to resources and development opportunities. Although class, poverty, ethnicity and physical location may influence these inequalities, the gender factor tends to make them more severe (Kabeer, 2003).

Access to productive assets is a major issue in the gender empowerment discourse. Despite the significant roles women play in agriculture and food security in many developing countries, they continue to have a poorer command over a range of productive resources, including education, land, information and financial resources (World Bank, 2001; Odame et al., 2002; Welch et al., 2000). Women's productivity in agriculture is highly dependent on their opportunity to having access to productive resource such as land, credit fertilizer and to other agricultural technologies. However, many rural women lack access to land or to have insecure land tenure due to customary laws, culture and tradition. According to land utilization survey (CSA, 2000) the total number of private peasant landholders in Oromia was 4.3 million. Out of which the proportion of female landholders was 18%. Concerning farm resources this survey clearly indicated the disparity of access to resource by female and male-headed household that is average farm size, number of oxen and other animals owned by female heads were 0.97, 0.84 and 3.20 ha respectively while the corresponding figure for male heads was 1.34, 0.96 and 3.40 ha, respectively.

Objectives

i) To assess women access to agricultural extension services.
ii) To identify and recommend problems constraining women farmer participation in agricultural extension package program and related support services.

METHODOLOGY

The study area and sampling techniques

The study was conducted at administrative zone of Oromia region in Central Rift Valley of Ethiopia namely; Dugda, Boset, Ada'a and Fentale districts during 2009. The districts were selected based on the farming system of the zone/main agro - ecology of the woreda (Figure 1). Kebele was selected purposively based on respondents' size. Respondents were selected based on probability proportionate to sample size (PPS).

Characteristics of sampled households

A total of 201 respondents were drawn from the four districts and two kebele were selected from respective districts. Two categories of respondent's namely male headed household and female headed household were selected for an interview. In this context FHHLDs are those farmers in which a husband is not present due to divorce and death and MHHLDs are those household in which husband is present. For both cases, women farmers were interviewed.

Data collection techniques

Data were collected from both primary and secondary sources. Primary data were collected from respondents through individual interviews, key informant interviews and group discussion where as secondary data were collected from BoARD and literatures. Before collecting of actual data, the questionnaire were pre-tested which is used to re-modify the questionnaire which are either irrelevant or to add missed part.

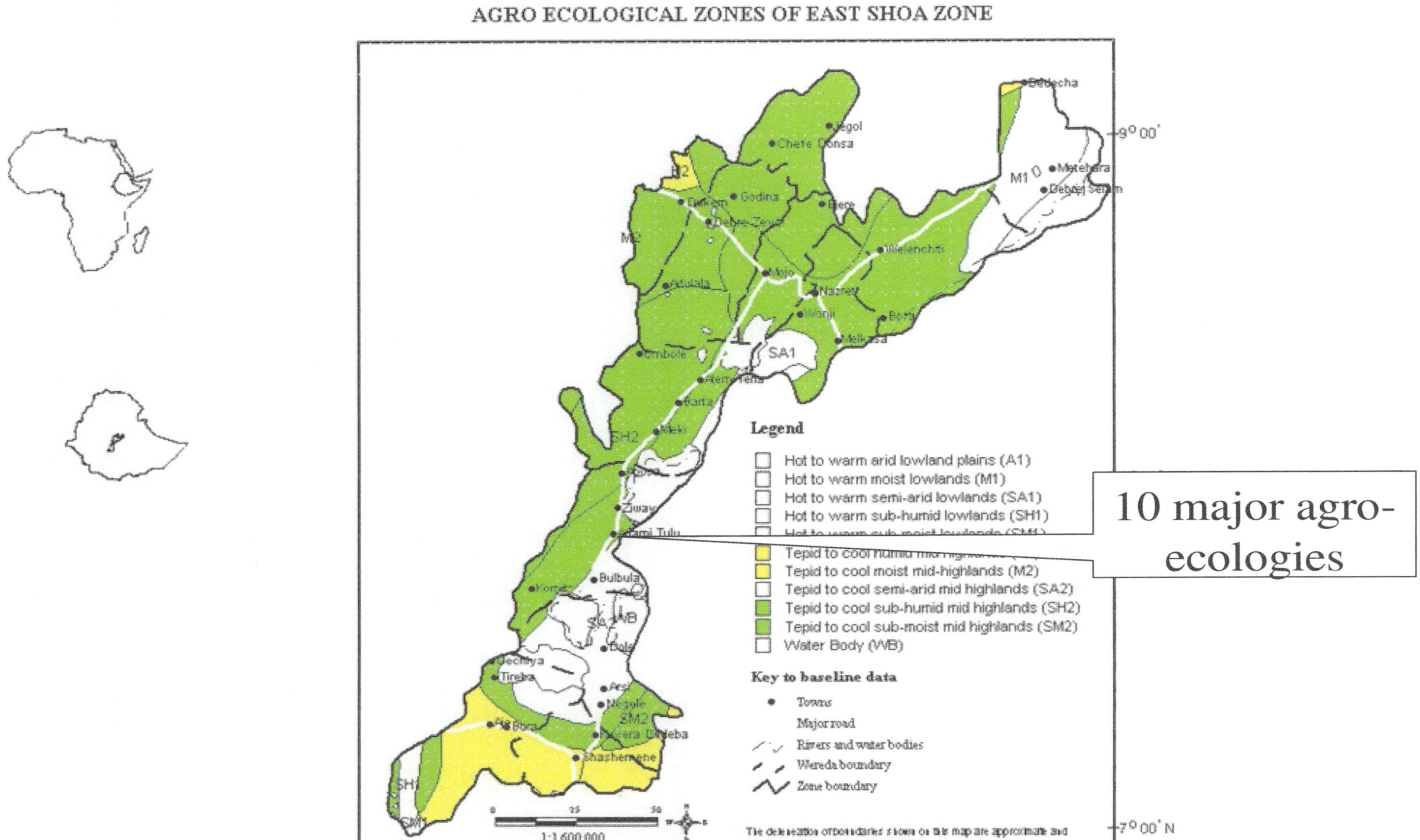

'ces: AEZ, MoA, 2000
Administrative data CSA & WFP, 1998

Figure 1. Major agro- ecology of the zone.

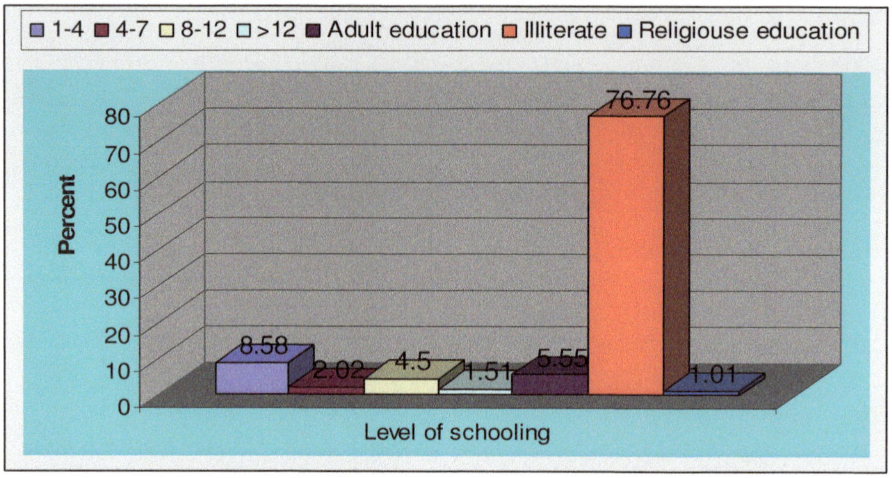

Figure 2. Respondents level of schooling

Analytical procedure

After data collected, it was entered and cleared for analysis. The data was analyzed by SPSS 13 version. The data was analyzed by using descriptive statistics: frequencies, mean, standard deviation, independent sample t- tests and paired sample t – test.

RESULTS AND DISCUSSION

Households' demographic characteristics

Age of respondent

From the sampled households, the majorities (45.9%) have an age category between 21 to 41 years, 40.8% have an age category between 42 to 62 years, and 10.3% have an age greater than 60 and 3.09% of respondents have an age category between 1 to 20 years.

Household size

From the sampled households, 21% of respondents have six household members where as the least percentage of respondents have 12 household members. This might be due to polygamous marriages. The household member reported between the districts is statistically significant at 1, 5 and 10% significance level (df = 3, F = 4.220).

Households' level of schooling

As it is explained by Figure 2, around 77% of sampled households are illiterates, 9% have level of schooling ranging from 1 to 4, 2% have level of schooling ranging from 4 to 7, 5% have level of schooling ranging from 8 to 12 and 2% have level of schooling more than 12 level of schooling (Figure 2).

Religion of household

From the sampled households, 71% are protestant, 23% are Muslim and 7% are "wakefeta".

Socio-economic characteristics

Land size owned by household type

The proportion of land size owned by house hold type was also assessed by the current study. To see this variation, land size owned by house hold type was categorized under different sizes (Table 1). Across the different land size indicated by this table, the majority of both households have less than 1ha of land. Even though the majority of both households owned less than 1ha, the proportion of households who owned less than 1 ha is different across house hold type. Accordingly, the proportion of FHHLDs who owned less than 1ha is higher (36.7%) than that of MHHLDs (29.1%). This finding also indicates that, the proportion of FHHLDs who owned 3-4 ha is lower than that of MHHLDs. From this result one can also understand that, the number of landless farmers is high at FHHLDs. This somehow indicates that FHHLDs are less access to productive resource when compared with MHHLDs. Generally, respondents' access to different land size is totally different as explained by table 1.

When we come back to the overall mean, MHHLD owned a mean of 2.03 ha with a standard deviation of 1.45 whereas FHHLD owned a mean of 1.43 ha with a standard deviation of 1.1 and their mean difference is

Table 1. Proportion of land size owned by house hold type during the 2008/9.

Land size owned in ha	Proportion of MHHLD (%) who owned respective land size	Proportion of FHHLD (%)who owned respective land size
Land less	2.8	8.5
<1	29.2	36.7
1-2	14	20.3
2-3	27.8	21.8
3- 4	19.5	14
>4	7	-
Total	100	100

Source: Own survey data, 2009.

Table 2. Significance of land size owned by house hold type.

House hold type	Mean of land size in ha	Standard deviation	t-value
FHHLD	1.43	1.1	
MHHLD	2.03	1.45	3.28*

The mean difference is significant at 1, 5 and 10% respectively
Source: Own computational survey data, 2009.

Table 3. Proportions of households who have oxen during the survey period.

Response of house holds	House hold type		X^2
	MHHLD (%)	FHHLD (%)	
Yes	75	50	6. 88**
No	25	50	
Totals	100	100	

**= The difference is significant at 1%, 5% and 10%probability level.

significant at 1, 5 and 10% respectively (Table 2).

Oxen ownership

Oxen ownership is one of the critical factors for crop production. So the current study was also tried to look into house holds' access to it. Frequency analysis was employed to see the proportion of households accessed to oxen. Chi-square test was also employed to see whether the significance difference exists between the two household types. As indicated by frequency analysis, 50% of female headed household were not owned oxen where as 75% of male headed households are owned either single or a pair of oxen during 2008/9 and the difference in access to oxen is significant at probability level of 1%, 5% and 10% respectively as indicated by chi-square tests(Table 3). This indicates that FHHLDs are less accessed to oxen when compared with MHHLDs. This disparity can be one of the factors that hinder FHHLD from use of agricultural extension packages.

Options available to plow land by house hold type

Respondents have been using different available options in order to plow their land for crop production purpose (Table 4). These options include, using own oxen, borrowing of oxen from relatives, using of local farm tools like hand held hoe, renting in lands and giving of lands to "hitra" which means a person who rent in the land will share large proportion of the yields achieved (the most common one is two third of yields achieved).

Livestock size owned by household type

In the study area live stock production is one of the components of the farming system/livelihood activities for farmers. So this study assessed the number of livestock

Table 4. Options available to plow their land in percentages by house hold type.

No	Option available for plowing their land	Proportion of MHHLD who owned respective land size (%)	Proportion of FHHLDs who owned respective land size (%)
1	Borrowed oxen from relatives	18.3	24.4
2	Using hand held hoe	1.4	6.7
3	Renting of oxen	-	1.7
4	Giving of their lands to "hirtaa"	4.2	14.3
5	Own ox	74.6	50.3
6	Rent out	-	4.4
7	Others	1.4	-
	Totals	100.00	100.00

Source: Own survey data, 2009

Table 5. Live stock size owned by house hold type in TLU during the survey period

No	Type of live stock owned	House hold type	N	Mean	t-values
1	Number of cows owned currently in TLU	MHHLD	72	1.38	2.294*
		FHHLD	129	0.93	
2	Number of oxen owned currently in TLU	MHHLD	72	0.19	0.416NS
		FHHLD	129	0.15	
3	Number of goats owned currently in TLU	MHHLD	72	0.18	0.416NS
		FHHLD	129	0.14	
4	Number of sheep owned currently in TLU	MHHLD	72	0.31	0.819*
		FHHLD	129	0.15	
5	Number of poultry owned currently in TLU	MHHLD	72	0.04	0.013NS
		FHHLD	128	0.03	
6	Number of donkeys owned currently in TLU	MHHLD	72	0.70	2.27*
		FHHLD	129	0.45	
7	Number of horse owned currently in TLU	MHHLD	72	0.10	1.600NS
		FHHLD	129	0.02	
8	Number of camel owned currently in TLU	MHHLD	72	0.09	0.984NS
		FHHLD	128	0.02	

Source: Own computational data, 2009, *= the mean difference is significant at 5 & 1%, Ns: Not significant

owned by sampled respondents in TLU by using Storck et al. (1991) conversion factor. The mean of TLU calculated for both household type indicated that MHHLD owned more live stock size than FHHLD indicating that FHHLDs are less owned livestock than MHHLDs. From these livestock components, significant difference was found only for cows, sheep and donkeys At 1% significant level (Table 5)

Sources of income for house hold type

The agricultural extension system aims to transfer technologies and capacities that should finally impact on farmer's productivity and hence increasing of farmer's income. So this study tried to assess respondents' major sources of income and level of income as well. Respondents were asked to list up production objectives.

Table 6. Proportion s of respondents participated in off- farm and non- farm activity during 2008/9.

Response	Male headed house hold (%)		Female headed house hold (%)	
	Non-farm	Off-farm	Non-farm	Off-farm
Yes	19.7	27.8	17.3	22.0
No	80.3	72.2	82.7	78.0
Total	100.0	100.0	100.0	100.0

Source: Own survey data.

Table 7. Mean annual income of respondents during the survey period.

Major sources of income	House hold type	N	Mean(EB)	Std. deviation	t-value
Sell of crops	MHHLD	71	3050.95	4758.89	2.334*
	FHHLD	128	1580.39	3955.17	
Sell of livestock	MHHLD	71	926.47	1929.12	1.218^{Ns}
	FHHLD	127	623.68	1520.68	
Off-farm activities	MHHLD	71	669.80	1818.56	0.065^{Ns}
	FHHLD	127	651.02	2186.56	
Non - farm activities	MHHLD	72	292.36	1177.92	0.166^{Ns}
	FHHLD	128	330.75	1749.78	

*= The mean difference is Significant at 5 and 1%, Ns: Not significant.

The major farmers' production objectives mentioned by respondents include both for household consumption and income generation purposes. The major sources of income assessed here include; crop production, livestock production, off farm and non - farm activities. Finally, the amount of crops sold per annual was estimated by respondents in terms of quintals. Then, the mean annual sales of crops by respondents during the survey period were calculated. For livestock; type, number of livestock sold and annual income generated from sale of live stock were assessed.

Also farmers involvement on other income generating activities like off – farm and non- farm activities were assessed. Off-farm activities considered here includes vegetable trading, cattle trading, grain trading, hiring of donkey cart etc where as non-farm activities considered here includes different activities like daily laborer, hand craft and remittance. The result of the study indicates that, participation of both households on non- farm and off farm activities is very low (Table 6). Mean of income generated by both household were also analyzed by this study.

The mean income that has been generated from available sources of income was also assessed by this study. Across the districts, the mean of income generated from sell of crops and live stock by the sampled respondents was calculated. The maximum of income generated from sell of crops and live stock were $ 29,840.00 and $ 14,100.00 EB respectively. The study also assessed the mean of income size achieved by both house hold type. Accordingly, the mean size of income generated from on farm activity like crop production is stronger at MHHLDs than that of FHHLDs and it is slightly similar for the left activities considered for comparison purpose. Paired sample t-test was also employed to see their significant difference. The analysis result showed that, the income generated from sell of crops is significantly different for household type at 1% significant level implying that MHHLD were generated more income than FHHLD from crop production. No significant difference was reported between the other three activities (Table7).

Problems hindering women participation in non - farm and off - farm activities

Different livelihood strategies can help farming households to diversify their income which finally help them to join agricultural extension package program. Diversification of income can also help farming households as a risk mitigation strategy. Respondents have been phasing different problems that inhibit them from participating in both non- farm and off farm activities. The

Table 8. Problems limiting respondents 'from participating in non-farm and off -farm activities.

No	Major constraints	Off farm activities (%)	Non – farm activities (%)
1	Shortage of money	27.6	6.1
2	Lack of knowledge and skills	26.7	50.4
3	Lack of interest	1.7	1.7
4	Lack of time	13.8	11.3
5	No problem	5.2	2.6
6	Shortage of labor	9.5	12.3
7	Age related problem	12.9	13
8	Market related problem	2.6	2.6
Totals		100.00	100.00

Source: Own survey data, 2009.

Table 9. Proportion of house hold who ever received a sort of credit.

House hold type	Ever received (%)	Never received (%)	Totals
MHHLD	52.9	47.1 100	100
FHHLD	39.8	60.2	100

Source: Own survey data, 2009.

Table 10. Percentage of households participated in extension package program during the two years from the survey period.

Response of HHLDs	House hold type	
	MHHLD (%)	FHHLD (%)
Yes	75	44.4
No	25	55.6
Totals	**100**	**100**

Source: Own survey data, 2009

current study tried to look into these problems. According to the current finding shows, lack of knowledge and skills (27.6%), shortage of money (26.7%) and lack of time (13.8%) are the major problems identified by both households (MHHLDs and FHHLDs) that have been limiting households from participating in off-farm activities where as lack of knowledge and skills (50.4%), shortage of labor (12.3%) and lack of time (11%) were the major problems identified by both households respectively as first, second and third for non – farm activities. The major problems limiting women farmers' participation in non-farm and off farm activities are explained by Table 8.

Respondents' access to credit services

Access to favorable credit service can encourage farmers to utilize agricultural extension packages. So, the study tried to look into proportion of respondents who ever received any sort of credit from credit providers. According to this survey result, 52.9 % of MHHLD have received a sort of credit at least for one or more than one times whereas only 47.1% of FHHLD received it (Table 9).

Proportion of respondents' participation in extension package programme

By the current study, proportion of respondents who received a sort of extension packages available in the woreda was assessed based on the two years from the survey time. As indicated by the survey result, 75% of MHHLDs were received either one or more packages of technologies where as only 44.4% of FHHLDs received either one or more packages of technologies available in the area (Table 10).

This study result shows that, participation of FHHLD in extension package program is lower than MHHLDs (Table 10). During the study, respondent were asked the

Table 11. Rank of constraints to access and utilization of improved seeds.

No	Types of problems	Their rank in order of importance	
		MHHLD	FHHLD
1	Low supply problems	2nd	2nd
2	Not timely available	4th	4th
3	Poor quality	5th	6th
4	Cost of seeds is expensive	1st	1st
5	Low level of awareness	3rd	3rd
6	Biasness towards progressive farmers	6th	5th

Source: Own survey data, 2009.

Table 12. Constraining to access and utilization of fertilizers.

No	Types of problems	n	Yes (%)	No (%)	Total
1	Low supply related problem	175	23.9	76.1	100
2	Fertilizer is not available on time	175	14.3	85.7	100
3	Fertilizer cost is expensive	176	66.5	33.5	100
4	Low level of awareness about its recommendation	176	14.3	85.7	100
5	Biasness towards progressive farmers	176	4.6	95.4	100

Source: Own computation survey data, 2009.

major types of extension packages available in the sampled area.

Accordingly, the major extension packages identified includes: improved crop production technologies (pulses, cereals), horticultural crops (onion bulb production, mango, onion seeds production, tomatoes, potato, carrots, sweet potato production), dairy packages (breed improvement/cross breed dairy cows, management improvement, AI services, veterinary services), fattening extension packages (fattening through concentrate supplementation/industrial byproducts, veterinary services, housing management), improved animal feed technologies (lablab, cowpea, cajanus cajan, rhodes grass, elephant grass) and improvement of market linkages, poultry production packages (improved breeds, feeding, health improvements and housing management), bee keeping packages (improved box hive with its accessories, transitional bee hives), organic (compost) and in organic fertilizers are the major extension services available in the area. The major agents of GOs currently using to disseminate these technologies to farmers is through development agents/extension workers.

According to data taken from East Showa Zonal Rural Development and Agricultural office indicates, 292 FTC were constructed at nearly all kebeles of the woreda of the zone (East Showa Zone, 2010, Research Extension Advisory council second quarter meeting report).At many kebeles, three development agents combined from animal science, plant science and natural resource were assigned to give modular and non modular training, facilitate agricultural input delivery and giving advisory services for farmers.

Constraints to access and utilization of some agricultural inputs

Improved seed and related problems

In the study area different improved crop varieties have been distributed by different actors working in the area. The main sources of improved crop varieties for farmers are oARD, unions and local markets. So, this study assessed problems associated with access to and utilization of available improved crop production inputs by household type. Accordingly, the sampled households mentioned that, problems like low supply related problems, low level of awareness about technology recommendations and costs of improved seed variety (expensive) are the main problems limiting their access and utilization of crop production inputs (Table 11).

Access to and utilization of fertilizer and related problems

Respondents' access to fertilizers was also examined by the current study. The study further tried to look into problems associated with use of fertilizer by household type. Accordingly, the majority of respondents (66.5%) expressed that the higher price for fertilizer is discouraging them to effectively utilize it. About 24% expressed that the present supply system is inadequate while 14.3% have indicated untimely availability of fertilizer and lack of appropriate awareness about the technology recommendation (Table 12). Due to these

Table 13. Rank of constraints to access and utilization of fertilizers by house hold type.

No.	Type of problems	Their rank in order of importance	
		MHHLD	FHHLD
1	Low supply related problem	2nd	2nd
2	Not available on time	3rd	4th
3	Fertilizer is not available on time	1st	1st
4	Low level of awareness about the technology recommendation	4th	3rd
5	Biasness towards the rich	5th	5th

Source: Own survey data, 2009.

Table 14. Major constraints to access and utilization of livestock packages.

No	Problems	Sample	Yes (%)	No (%)	Total
1	Low supply problems	176	34.7	65.3	100.00
2	Cost of technology is not affordable	168	36.9	63.1	100.00
3	Low level of awareness	178	22.0	78.0	100.00
4	Biasness towards the rich	169	5.3	94.7	100.00

Source: Own computation survey data, 2009.

Table 15. Proportion of women farmer access to different sources of information.

No	Types of mass media	Count	Yes (%)	No (%)	Total
1	TV ownership	200	7.0	93.0	100
2	Radio ownership	199	45.2	54.8	100
3	Access to extension materials	200	3.0	97.0	100

Source: Own survey data, 2009.

problems, only 64% of respondents used both UREA and DAP whereas 46% of sampled farmers were not used fertilizers (both UREA and DAP) during the cropping season of 2008/09. Both house hold types were also asked to rank problems associated with use of fertilizers (Table 13).

Constraints to access and utilization of livestock inputs

Different live stock production technologies are available in the zone. According to the result, the survey indicates different livestock technologies like improved fattening practices, improved poultry breeds, improved dairy breeds, improved animal feeds are available and used by the sampled households. So the study assessed major problems constraining access to and utilization of live stock production packaged in general. Accordingly, the majority of respondents expressed that, cost of the technology is not affordable; secondly, supply related problems (low) and thirdly, low level of understanding about the technology recommendations (Table 14).

Sources of agricultural information and women access

One of the means of increasing the effectiveness and efficiency of agricultural extension program is through the application of improved and innovative extension communication methods. The current extension approach (PADETES) employs varies forms of extension and communication methods to achieve its objective. So this study evaluated major sources of agricultural information Agricultural information has also been disseminated to women farmers through different extension methods like mass extension methods (Radio, TV, and printed media), individual extension methods (Office contact, farm visit) and different extension events (training, field days, demonstration).

Hence, this study examined women access to these sources of agricultural information available in their woreda. The current finding indicates that about 45.2% of respondents have radio and 7% have TV. From the sampled respondents, very few of them used extension materials (Table15).

Table 16. Frequency of use of radio, TV by women farmers for agricultural activities.

No	Sources	Never (%)	Rarely (%)	Occasionally (%)	Often (%)	Very often (%)
1	Radio	63.5	12.2	18.8	5.1	0.5
2	TV	91.14	1.5	5.6	1.5	-
3	Extension materials	95.4	2.5	2.0	-	-

Source: Own survey data, 2009

Table 17. Percentage of women participated in extension events during five years from the survey period.

No	Type of events	MHHLD		FHHLD	
		YES (%)	NO (%)	YES (%)	NO (%)
1	Field day	11.4	88.6	4	96
2	Training	20.8	79.2	14.1	85.9
3	Demonstration	16.9	83.1	7	93

Source: Own computation survey data, 2009.

Frequency of use of radio and TV by women farmers

This study also evaluated the frequency of use of media by women farmers for agricultural related activities. According to frequency analysis result indicated the majority of women farmers were never used agricultural information that has been disseminated through different sources of agricultural information like radio, TV and extension materials. The study figures showed that 63.5%, 91.14% and 95.4% of respondents were never used agricultural information that has been designated through radio, TV and extension materials respectively (Table 16).

Women farmer access to training, field days and demonstration

Farmers ability to use their land more effectively and efficiently is influenced by a variety of factors including personal views, family views, technology, profitability, complex public opinion, research, change agents and marketing (Kotile and Martin, (1998). Some of these factors can be alleviated through provision of adequate training. So this study assessed women access to extension events on the bases of five years from the survey period. From this study it was realized that, participation of women on training, field day and demonstration is very low (Table 17).

Extension contact

Development agent is one of the major sources of agricultural information for farmers and an effective linkage between development agents and farmers is critical in disseminating of agricultural information. Currently there are about 292 FTCs in the "zone" and many development agents combined from different disciplines are employed there to offer agricultural extension services. The FTCs are constructed with the participation of farmers in the kebele. The FTCs are expected to serve as; centers of extension, places where modular training to farmers for up to six months are given, demonstration of entrepreneurship and sources of advice. So, this study assessed proportion of women farmers receiving advisory services from extension agents during production season of 2008/9. As the result of the survey finding indicates, 57.58% of women farmers had been received advisory services from development agents whereas 42.42% of respondents had been never received any advisory services from DAs during the this period. This indicates that the existing linkage between development agents and women farmer is still very low. Nearly similar findings were reported by Rajak (1990).

Frequency of extension agents' visit

Frequency of visits made by DAs for women farmers was also assessed by this study. The study result indicated that, the majority of women (42.2%) had not been visited by Development workers during the cropping year of 2008/9, 15.2% of women expressed that, "extension agents" contact them on weekly bases, 11.5% expressed that development agents contact them on monthly bases (Table 18).

CONCLUSION AND RECOMMENDATIONS

In the study area different agricultural extension package

Table 18. Frequency of visits made by extension agents for women during 2008/9.

No	Frequency of contact	n	Response (%)
1	Once in a week	29	15.5
2	Fortnightly	18	9.6
3	Monthly	22	11.8
4	Only during plantation	18	9.6
5	During input provision	17	9.1
6	During credit collection	2	1.1
7	Not contacted by DAs at all	79	42.2
8	Any time when technical advice required	2	1.1
	Total	187	100.00

Source: Own computation survey data, 2009.

are available. The study identified that participation of FHHDs in agricultural extension package program is very low when compared with MHHLDs. The major constraints to access and utilization of agricultural extension packages are; cost of input related factors (cost of input is expensive), low supply related factors, delay of inputs, lack of giving attention for those farmers who can't afford cost of technology which farmers explained it as 'biasness towards the rich farmers. Women access to productive resources like land, credit, TLU, off-farm and non- farm activities is also very low. In line with these, Participation of women on extension events like training, field day, visits /demonstration is also very low. On the other hand, the existing linkage between development agent and women farmers is very weak. These factors can be the other major constraints to access and utilization of agricultural extension packages.

The study therefore recommends that, female headed households'/women needs to be encouraged to participate in agricultural extension package program by stakeholders working in the area. One of the methods used to put this into action can be through encouraging them and strengthening of their participation in extension events like training, field days, visits and arranging of favorable credit services. In addition to this, strengthening of linkages between development agent and women farmers is one of the other critical factors that should be considered. Furthermore, the study recommends that there is a need to diversify women's livelihood options like off-farm and non-farm activities which finally helps them to overcome their financial problems and join agricultural extension package programme.

Abbreviations: MHHLD, Male headed households; **EARO,** Ethiopian agricultural research organization; **FHHLD,** female headed households; **HA,** hectare; **OARD,** office of agriculture and rural development; **SPSS,** statistical packages for social sciences; **CSA,** central statistical authority; **ZoARD,** zonal office of agriculture and rural development; **FRG,** farmers research group.

REFERENCES

Akroyd S, Smith L (2007). The decline in public spending to agriculture – does it matter?', Briefing Note, No. 2, Oxford Policy Management Institute, Oxford.

Alene A, Poonyth D, Hassan RM (2000). Determinants of adoption and intensity of improved maize varieties in central high lands of Ethiopia.

Anderson JR (2007). Agricultural Advisory Services', Background Paper for the World Development Report 2008.

http://siteresources.worldbank.org/INTWDR2008/Resources/WDR_00_ book.pdf 1191427986785/Anderson_AdvisoryServices.pdf

Anderson JR, Feder G (2003). Rural Extension Services', World Bank Policy Research Working Paper 2976, World Bank, Washington D.C. http://econ.worldbank.org/files/24374_wps2976.pdf

Birkhaeuser D, Evenson RE, Feder G (1991). The Economic Impact of Agricultural Extension: A Review', Econ. Dev. Cultural Change, 39: 607-650.

CSA (Central Statistical Agency) (2000). land utilization survey, Ethiopia

EARO (2000). Institutionalizing gender planning in agricultural technology generation and transfer process: Why is gender a development issue? Proceeding of the national workshop from 225-27 Oct., 1999. EARO, Addis Abeba.

Fan S, Rao N (2003), 'Public Spending in Developing Countries: Trends, Determination, and Impact', EPTD Discussion Paper, No. 99, IFPRI, Washington D.C. http://ideas.repec.org/p/fpr/eptddp/99.html

FAO (2003). 'Statement at the Ministerial Conference of the WTO - Circulated by H.E. Mr. Hartwig de Haen, Assistant Director-General', Fifth Session, Cancun, 10-14 September,Doc No: WT/MIN(03)/ST/61.

Feder G, Murgai R, Quizon JB (2004b), 'The Acquisition and Diffusion of Knowledge: The Case of Pest Management Training in Farmer Field Schools, Indonesia', J. Agric. Econ., 55 (2): 221-243.

Kabeer N (2003). Gender Mainstreaming in Poverty Eradication and the Millennium Development Goals: A Handbook for Policy Makers and Other Stakeholders. Ottawa: International development Research Centre.

Kotile DG, Martin RA (1998). Farmers' Perspective on Sustainable Farming Systems: A Case Study. Proceeding on the Fourteenth Annual Conference of the Association for International Agriculture and Extension Education, Tucson, Arizona

Odame HH, Hafkin N, Wesseler G, Boto I (2002). Gender and Agriculture in the Information Society. International Service for National Agricultural Research Briefing Paper No.55. The Hague, The Netherlands: ISNAR.

Rajak TA (1990). Improving the relevance and effectiveness of agricultural extension activities. Available at: www.fao.org/docrep/v4805e/v4805e07.htm.

REAC report (2010): East Showa quarter annual meeting. Research Extension Advisory council. Adama.

United Nations (2008). Trends in Sustainable Development – Agriculture, rural development, land, desertification and drought', Department of Economic and Social Affairs of the United Nations,

New York.
http://www.un.org/esa/sustdev/publications/trends2008/fullreport.pdf
Welch CJ, Alemu B, Msaki T, Sengendo M, Kigutha H, Wolff A (2000). Improving Household Food Security: Institutions, Gender, and Integrated Approaches. U.S.A: BASIS Management Entity.
World Bank (2001). Engendering Development through gender equality in Rights, Resources and Voice. World Bank Policy Research Report 21776. Washington, D.C. and London: World Bank and Oxford University Press.
World Bank (2007). World Development Report 2008: Agriculture for Development, World Bank, Washington, D.C. http://siteresources.worldbank.org/INTWDR2008/Resources/2795087

5

The model of family empowerment program for community development in West Java, Indonesia

Pudji Muljono

Department of Science of Communication and Community Development, Faculty of Human Ecology, Bogor Agricultural University, Jl Raya Pajajaran, Kampus IPB Baranangsiang Bogor, Indonesia. E-mail: pudji1962@yahoo.co.id.

Family empowerment is very important, as through empowerment programs the information and knowledge which are key determinants of progress of a society, may be passed on to the next generation. Community education, especially community empowerment which is considered a national investment, is expected to increase human skills, and knowledge of the society. It means that the entire population of Indonesia, regardless of their socio-economic backgrounds, is encouraged to learn and make productive use of knowledge all the time. The successful implementation of family empowerment would thus require intensive research and information gathering. A study on "the model of family empowerment program for community development" was designed by Bogor Agricultural University to response such needs. The objectives of the study were to evaluate models of the community development program which was implemented through systematic steps. The location of this research is all family empowerment programs of Bogor Agricultural University in Bogor, Cianjur, Sukabumi Regencies and Bogor City. Based on performance analysis and identification of problems in the management of family empowerment programs, it can be arranged various of action plans for family empowerment as follows: (1) training courses to improve the quality of management of the family empowerment program; (2) resocialization vertically and horizontally to all parties; (3) the establishment of prospective business networks to trigger the economic development of target communities; (4) comparative studies and bench marking for board members and personnel of family empowerment programs to stimulate learning environment; (5) the establishment of co-operative (economic enterprises) as the platform for the economic development of the community.

Key words: Family empowerment, development program, community development, poverty alleviation.

INTRODUCTION

Indonesia is a country comprising of 13.667 islands straddled out along the equator between the Indian and Pacific Oceans, and sandwiched between the continents of Asia and Australia. A country rich in natural resources, Indonesia's a population stood at of more than 202 (two hundred two) million people in the year 2003 (Statistics Indonesia of the Republic Indonesia, 2009), with more than 50% inhabiting the island of Java and the remainder distributed over Sumatra, Kalimantan, Sulawesi, Papua and other islands. The population of Indonesia is projected to be 258.2 million by 2025, thus, the need for food will increase dramatically. Thus, the successful development of agriculture in Indonesia should increase agricultural productivity. This enables the country to achieve self-sufficiency in food and improve employment and welfare levels in the rural areas. However, global and technological changes, the consequence of the new dynamics of industrial development, have greater repercussions on the development in Indonesia.

During Indonesia's most recent economic crisis (1997 to 1999), poverty and unemployment has prompted a renewed interest in the social networks that link rural and urban regions of the country. Moreover, the last crisis has impacted that In the year 2009, approximately 30 to 40 million labor forces are unemployed or working irregularly. Finally report Central Statistics Agency (Statistics Indonesia of the Republic Indonesia, 2009) in December 2009 showed that 37.4% of the citizens of Indonesia is experiencing absolute poverty (below the poverty line) and as many as 20% were very vulnerable to fall down the poverty line. All the evidence is clear that economic development in Indonesia considerably distorted (distorted development).

To respond to those changes, Indonesia has formulated macro-level strategies and initiatives in order to sustain and accelerate their economic growth. Improving the quality of human resources for people's welfare is prerequisite. The main objective of millennia of development or the Millennium Development Goals (MDGs) in Indonesia, hand in hand with poverty allevia-tion priorities, is determining the reduction of the proportion of poverty in 2015 into half portion or 8.2% of the total population. In 2004 to 2009 Indonesia's development planning, the poverty alleviation program will be accelerated into the 2009 program. The acceleration was the determination and government policies that require supports of all agencies and develop-ment institutions. What becomes more important is the growing connection between technology and development, and between the development potential of Indonesia and the intellectual skills of the Indonesians. Thus, technological change is an issue which require special attention for human resource development (HRD). While the policy implications of this shift in the economic process are somewhat complex, one such implication is that the universities must now become tools of development, and absolutely necessary for HRD. Within the multifaceted objectives of the university, viz. teachings, scholarships, researches, community services, etc.., there is an enormous scope for university to respond to the element of change, and, in particular to the need of the HRD. The universities, especially Bogor Agricultural University, can provide program to meet the emerging profile of the future workforce. The center for human resources development (CHRD) will be reponsible for the proposed program. New approaches and models were being explored to establish new forms of collaboration and closer interaction. Ensuing issues are those of contractual term, exploitation and commercialization of research results, and intellectual property rights.

There was now a growing awareness among academics of social-political disciplines, even among bureaucrats, that the community development programs cannot be managed in a partial way. Various poverty alleviation programs implemented by the government so far had been conceptually using the paradigm of empowerment. With referrence to the result of the various poverty alleviation programs that did not function as expected, it is therefore imperative to implement the study on "The model of family empowerment program for community development in West Java, Indonesia" to evaluate the community empowerment through to the family empowerment posts (posdaya) to poor families in town and regency of Bogor, Cianjur and Sukabumi Regencies respectively. The project will help the government to promote socio-economic development and to reduce poverty by implementing community development programs.

OVERVIEW OF COMMUNITY DEVELOPMENT IN INDONESIA

The concept of development and empowerment

According to Goldenberg (1993) "development is a process of self-reliant growth, achieved through the participation of the people acting in their own interests as they see them, and under their own control". The development programs are planned to end poverty, provide productive employment, and satisfy the basic needs of all the people. This implies that basic goods and services such as food and shelter, basic education and health facilities, and clean water must be accessible well. Compliment to the above statement. The new encyclopedia britanica (1978), describes that economic development is the process by which countries with low per capita national incomes may increase their production of goods and services so as to raise their own. Further, Mabogunje (1995) mentioned that development is about "wealth creation". Every society succeeds best in wealth creation when it adapts and transform its own institutions, its customs and the general attitude of its people towards the achievement of this goal.

Suharto (2005) stated that empowerment refers to the ability of people / groups / communities who are vulnerable and weak, so they have the strength or the ability to: (a) meet their basic needs so that they have the freedom, not only in the sense of free expression, but free from hunger, ignorance and pain; (b) reach out to productive resources that enable them to increase their income and obtain goods and services they need; (c) participate in development processes and decisions that affect them.

According to Sumarjo dan Saharuddin (2004) the characteristics of people who have been empowered by are as follows: (a) they are able to know and understand their potential; (b) able to plan (to anticipate changes in future conditions), and directs themselves; (c) they have the power to negotiate and cooperate mutually beneficial to the "bargaining power" adequate; (d) be responsible for their own actions.

Community empowerment is high on the current policy agenda of the Indonesian government. Some of Bogor Agricultural University policies relevant to the targets of rural development are: (a) promoting quality job creation by stimulating non-agriculture economic activities (rural industries and supporting services) and by enhancing the linkage between rural and urban areas; (b) increasing promotion and marketing of agriculture and non agriculture products; (c) improving peoples' access, in particular women's, to productive resources such as land, social-economy infrastructure, capital, information, technology and innovation, and to public services and market; (d) empowering rural communities by improving the peoples' capabilities, institutional and social capital strengthening including improving their networking, business partnerships and bargaining position; (e) improving the welfare by fulfilling the peoples' basic needs in education and health services and minimizing the risk of being vulnerable by improving unfair market structure condition; (f) developing of agriculture and non agriculture good practices which are environmentally friendly and meet the sustainable development principles.

When thinking about community empowerment we draw on decades of theory and practice concerned with developing and understanding community development. Enabling people to work towards their own development is part of the process of community empowerment. Defining what we mean by the "community" is never easy. Groups of people living in the same geographical area do not always have same needs or goals in life. But, in working for poverty alleviation, it is vital that people gain from the synergy of working together and organizing for community development. People who live in poverty and social disadvantage to not feel that they have any ability to get their voices heard or their needs met. Using participatory approaches can help them identify new ways of finding solutions to their problems. As groups of people begin to realize that there are ways in which their efforts can make a difference, they feel more powerful and are better able to contribute to the development process

Alleviating poverty persists as among the greatest problems facing the nation. The Indonesian's economic growth is not able to accommodate the millions of new people entering the job market each year and although some improvements have been achieved over the past years. Indonesia was able to make a leap out of an agriculture-based economy to a more diversified economy that includes labor-intensive manufacturing, and self-employment in urban services. The agricultural sector, however, prevail to be the biggest employer, absorbing 45% of the 99 million labor force in 1999, followed by service sector with 39% and industries 16%. Despite the economic growth, the economy has not yet recovered sufficiently to start generating anything close to the number of jobs urgently needed. Thus, the shift in the definition of quality, and the role of quality in productivity and competitiveness means that community development as source of future labor force must be for promoting quality in production and quality management skills.

Pos pemberdayaan keluarga (posdaya)

In the framework of the implementation of the MDGs (Figure 1), the development of family functioning is directed to five priority objectives, namely the commitment of leaders and elders of the village and hamlets, sub-district and district, the development of a religious function, the function of family planning and health, educational function, the function of entrepreneurship and environmental functions give meaning to a happy family life and prosperous.Posdaya is formulated between and inter family, thus it will stresses in posdaya personal basis, base groups, such as mosque-based posdaya, posdaya plant-based, or based posdaya education, and more. According to Suyono and Haryanto (2007) pos pemberdayaan keluarga (pos family empowerment – which was shortened into posdaya) was created as a forum for advocacy, communication, information, education to strengthen the coordination of activities of family functions. Regarding the main program is divided into four posdaya essential that educational programs, health, economic, and environmental.

Posdaya was created to respond the government's suggestion to build human resources through active participation in the family. The process of empowerment that prioritize of improving the family's ability to work hard eliminating ignorance, laziness and poverty in the broadest sense. The target activities are implementing joint effort so that every family has the ability to perform eight functions of families.

The aims of the study

The realization of HRD programs and the effectiveness of strategies to improve the productivity of the target audiences lie not only on the availability of methods, materials, infrastructures facilitated and government's guidance and control, through the center for human resources development (CHRD) – Bogor Agricultural University, but also on the ability of the audiences to adopt and implement these innovations, which depends on the level of the villagers education. In the context of HRD, especially the community empowerment, there have been various arguments for the contribution of community empowerment to economic development.

The main issue in this research is concerned with the impact of the community development model on the community livelihoods in West Java, especially those who involved in the community development program held by Bogor Agricultural University. This study aims not only to give insight on the development of people in Southeast Asia and their changing economic situation but also to understand whether or not the people gain benefit from the projects, especially in terms of getting access to primary products after the decentralization was implemented. Further aim of the study is to investigate other factors, which contribute to the changing position of people in relation to the community development during different periods, especially in the New order government period and until recent period.

THEORETICAL FRAMEWORK

Since 1960, birth of a concept called community

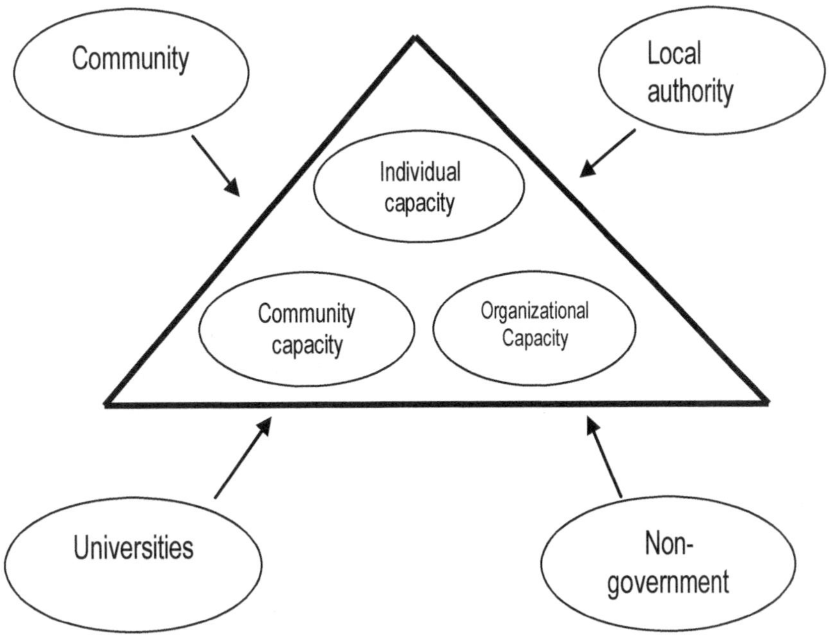

Figure 1. Framework of the community empowerment for the poverty alleviation through the posdaya program (Muljono, 2010).

empowerment and community development (hereinafter referred to as CD). CD is a process of interaction networks in order to improve the capacity of a community, to support sustainable development, and development of quality of life of the community (United States Department of Agriculture, 2005). CD does not aim to seek and establish a solution, or problem-solving structure to bring services to the community. CD is working with communities so that they can define and handle problems, and open to express their own interests in decision-making process (standing conference for community development, 2001).

The community empowerment program for poverty alleviation utilized the dualistic approaches such as the changing of client and the environment. These schematic approaches were described as follows (Muljono, 2010).

The concept of empowerment can be observed in individuals, groups or communities who control their own lives and seek the future in accordance with their wishes. The poor are vulnerable community groups and the weak do not have the strength and the ability to power. Through community empowerment, they may have the ability and strength to meet their basic needs and also that of their their families, so as to alleviate them from poverty (the condition of ignorance, hunger and pain).

Through their empowerment efforts are also expected to reach the productive resources that enable them to increase revenue, improve knowledge and skills, as well as participate in the development process.

Based on the concept of empowerment the individuals, groups or communities will control their own lives and seek their own future. Through their empowerment efforts are also expected to be the productive resources that enable them to increase revenue, improve knowledge and skills, as well as participate in the development process.

RESEARCH METHODS

The present study adopts a descriptive research approach, primarily because it is exploratory study, to describe the performance of posdaya, to identify problems in the management posdaya, and to formulate action programs for the development posdaya. Fieldwork was carried out in Bogor, Cianjur, Sukabumi regencies and Bogor City respectively for about 8 months, on April to November, 2009.

Data on institutional aspects and posdaya profile obtained through questionnaires, interviews, observation, key informant who is the manager of each posdaya. Additional information obtained from community leaders, religious leaders and cadres and members posdaya randomly selected based on their educational background, health and economy. Data on potential posdaya obtained through structured interviews and the perceptions and aspirations obtained through focussed group discussion (FGD). Topics of discussions in the FGDs are matters relating to the focus of study, including posdaya strengths and weaknesses, opportunities and challenges faced by posdaya. FGD participants were the members of each posdaya that represent the diversity of members posdaya based on aspects of education, health and economy.

To find information on the involvement of the target people in the community development programs, 70 informants consisting of community leaders, officials of the local governments, and members of the non-government organizations who represent the target people were interviewed.

The process of data collection involved identifying all bibligraphic records of both local and international authors on communty empowerment especially for West Java. Bernard's discussion on the use of unstructured and semi structured interviewing was very useful (1995). Geertz's research methods, such as his conceptualization of fieldwork as 'thick description, were also very important for this research by Geertz (1973).Data analysis were carried out with comparative method, which compares the ideal and the real conditions in the field, explore the opinions of the various elements involved in the implementation of activities posdaya based on empirical methods.

FINDINGS AND DISCUSSION

Overview posdaya

General description and profile of the seven posdaya research object which were carried out in Bogor city, and Bogor, Cianjur, Sukabumi regencies respectively, were presented in Table 1. Based on the results of FGDs with posdaya administrators note that the determination of criteria for poor families from each posdaya were different from each other. However, in general it can be concluded that the criteria of poor families by posdaya used are as follows: (1) their economy status are low; (2) they eat once a day; (3) they earned their living by working as coolies; (4) they live in semi-permanent house; (5) uncertainty income; (6) poor housing condition do not fit to the minimum of housing requirements; (7) have not any assets; (8) no opportunities to send their children to school.

In efforts to alleviate poverty in their communities, posdaya has set criterion of the less fortunate families based on a survey of poor families. The public considered that poor families were those who currently in the list of recipients of direct cash aid (BLT), and poor families rice (raskin) as well. The number of poor families as recorded by posdaya, the number of poor families posdaya has recorded by the board, the list of poor families as recipients of direct cash assistance (BLT), and the list of recipients of rice for poor families (raskin) are presented in Table 1.

Performance posdaya

Performance analysis posdaya was intended to evaluate the impact of the posdaya program as community empowerment for the socio-economic life of the community. In general the performance posdaya was considered well-implemented, because posdaya has produced some changes as follows:

1. Posdaya able to influence the way people view the forms of development interventions. At first they perceive any external intervention on the community to an aid, particularly material support / funding. But once they know posdaya, who brought the concept of self-reliance, mutual cooperation and independence, they began to understand that any external intervention into the community does not necessarily mean an aid. Interventions can include social activities, intervention ideas, values, empowerment and how to work on. Posdaya was able to increase their self reliance on the ability of citizen`s role in national development program. The trend of the citizens acted as a development target, but once engaged in posdaya, more people acting as designers, implementers and the evaluation of the development programs in the region.
2. Posdaya has aroused the dynamical community life through increased participation and commitment in the community development program. Prior to the the establishment of the posdaya project, the number of people involved in community development, both as beneficiaries / targets of the program as well as a relatively small cadre. After the formation of posdaya, more people perfomed services and posdaya provide services to the community. In addition, there were more people who wanted to be posdaya cadres. Citizens who had been less active and tend not to care about the environment, felt "disturbed" by the presence of posdaya. We have more public participation in the form of participation in energy and time, not in the form of funds or materials. This is understandable because the economic conditions of everyday relatively low.
3. The quality of poor families who were in the working areas of posdaya project, experienced significant changes after their involvement in posdaya project. Indicators of quality changes included: (1) their ability to change their perspective on the education as an important instrument for development; (2) their ability to express ideas in the official meeting/forum; (3) their more healthy and their regular visit to the posyandu (clinic integrated service), posbindu as part of posdaya activities; (4) the decreased number of malnourished infants.
4. Emerging economic activity in society, such as the emergence of small businesses in the field of food, handicrafts and services. As an example of business red guava juice, variety of chips, mushroom cultivation, mushroom chips, salted eggs, souvenirs and others. The business initially did not exist, after the formation of posdaya, community were urged to be creative to seek additional income.
5. Improving perception of the community of the importance of cleanliness and environmental sustainability by initiating efforts to process domestic waste into compost.

Problems in the management of posdaya

Problems in the management of an outline posdaya can be classified into two: (1) physical constraints; and (2) non-physical constraints. Physical constraints tend to be less exposed than the non-physical problems. Included in the physical constraints is the location of the secretariat of the posdaya, as they move from one place to another, and even they are located in the productive areas (for

Table 1. Profile of posdaya as the research object.

No.	Name of posdaya	Location	Date of establishment	Competitive program	Number of families			
					Target of posdaya	Poor family (gakin)	Recipients of raskin (rice for poor family)	BLT recipients
1	Bina sejahtera	Kampung Bojong Menteng village, Pasir Mulya, West Bogor sub-district, Bogor	8 May, 2007	Integrated health center (posyandu) for elderly people, microfinance institutions (mfis), integrated agricultural demplots, community library, productive economy business.	236	75	75	65
2	Mandiri	Tegal Gundil village, Northern Bogor sub-district, Bogor city	1 August, 2007	Integrated health center, library, training programs, MFI, garbage recycling, apotek hidup (green drugstore).	425	95	95	42
3	Benteng harapan	Benteng village, Ciampea sub-district, Bogor regency	11 January, 2008	Early childhood education, library, IHC, sewing course, empowerment traders, plant breeding.	400	150	300	140
4	Kenanga	Girimulya village, Cibungbulang sub-district, Bogor regency	2 May, 2007	Merchants development, eradication of illiteracy, integrated health center, elderly people development, reproductive health program, medicinal and ornamental plants.	159	88	27	27
5	Sirnagalih	Sirnagalih village Cilaku village, Cianjur regency	5 February, 2008	Early children library, reading park, IHC, sewing course, compost production, empowerment trader.	654	117	117	109
6	An-Nuur	Galudra village, Cugenang sub-district, Cianjur Regency	5 February, 2008	Early childhood education, library development for children, integrated health center for balita (children under five years old), field work for syster, mushrooms cultivation, cooperative.	136	136	136	136
7	Melati	Nagraksari village, Western Jampang sub-district, Sukabumi regency	7 February 2008	Credit unions, compost, package a youth training, *aloe vera* cultivation and its process.	56	56	56	56

example, hall / workshop). In addition, there is no class room for early childhood students available. At the moment the posdaya secretariat share together with the other institutions such as the assembly ta'lim, mosque, the housed of the management staff, or in farmers groups' stall. If permanent secretariat location is set for posdaya, the project will be able to do more activities and will be more managed.

For some community members, posdaya was regarded as a government program that would contribute something for them, and they think that the community were the target of the said project as labor. Although not most community have such perceptions, but it would affect the morale of committee members of posdaya, especially for the posdaya project that was not well developed.

Some comittee members of posdaya have been busy with their own routine activities that cause the difficulty for them to share thair spare time for posdaya activities. Some even fed up with the monotonous activities of posdaya, such as micro finance business which do not provide economic impacts for them. No honorarium are provided for the management of posdaya, and even did not get economic benefits while they are doing the bookkeeping and servicing the community for savings and loans. In addition, too often the posdaya was become the object for outside parties. This is because the pattern of visits that have been understood by posdaya and preparation programs of visits and time-consuming.

The availability of cadres on certain posdaya have become another contraints. Empowerment with self-reliance philosophy of human resources is really required volunteers of high social spirit. Posdaya is community empowerment activities with the principle of, by, and for the community. Therefore, there is no particular party that will bear the management fee, while the manager/cadre needed to spend time and energy to manage community empowerment through posdaya, and some board / cadre hopes of honorariums from posdaya.

Constraints in the quality of human resources has caused the lack of creativities of posdaya management for their innovative ideas, as well as their communication skill to conduct consultation with outside parties to develop their ideas for posdaya development. In addition, external support was also one critical success for posdaya. In some posdaya the outsiders did not provide support as expected. For example, some chief RT has not shown interest to encourage and assist the development posdaya and even some of them do not understand the posdaya program. For local authorties who understand the existance of posdaya tend to monitore from afar, posdaya support in the form of presence in Posdaya activities encouraged citizens, and not synergize the various rural development programs with potential posdaya. Funding support from the community for the development of many posdaya activities are really reqired to enhance the posdaya on its educational, health, economy and environment program.

Action plan of development program of posdaya

Based on the analysis of the potential and performance of posdaya, there were contraints encountered by the project and and how to solve such constraints. Alternatives action plans for posdaya to increase community empowerment and poverty alleviation have been identified. Identification of problem analysis, the factors causing, potential, alternative solutions and action plans are presented in Table 2.

Referring to the performance and the constraints which seventh object of posdaya research, as well as ideas and posdaya development plan submitted by the manager of posdaya, the program of action that needs to be formulated for the development of posdaya are as follows: (1) refresher management and training posdaya cadre; (2) resocialization vertically and horizontally of the said programs to all parties; (3) the establishment of prospective business networks to trigger the economic development of target communities; (4) training programs; (5) implementing comparative studies and bench marking for board members and personnel of family empowerment program that offers a stimulating learning environment; (6) the development of co-operative (economic enterprises) as the platform for the economic development of the community.

With community needs in mind, the village network component will establish soft infrastructure to network villages according to their information-generating activities and interests. In the village network environment, information sought by a village contact is provided, conditionally, to every other relevant village in the block, the district, or the country. Such a network makes every village a peer to other villages, with access to what other rural communities are learning and experiencing daily. The principal sources of information, news, opinion, and discussion within the network would be the villages themselves, through the experiences of their peers.

Conclusions

For further development of posdaya, there is need a reaffirmation of the goals of posdaya, refresher of its committee members and cadres posdaya, intensify the resocialization of posdaya to all parties from the public, community leaders, village, district officials and local government and build productive business networks in an attempt community empowerment. In addition, there is need for the development and strengthening posdaya activities on education, health, economy and environment. Posdaya as a participatory model of community development has proved to be successful performance as the program has produced positive changes in both

Table 2. Identifying problems and alternatives management of posdaya.

No.	Problems encountered	Factors	Potentials of the villagers	Alternative methods	Action plans
1.	The low support of local government officials.	Lack of socialization program on the understanding of the concept of posdaya.	Posdaya has been legalized by village chief / head of village.	Continuous socialization to the community and community leaders.	Posdaya tour. Local staff workshops
2	The caderization of the project officials is not going well.	The attitude of apathy and indifference of society. Lack of the involvement of the community.	The sufficient number of potential villagers/residents.	Designation as a cadre of officials directly.	Meeting program for cadres of the projects.
3.	The character of the community that used to instant and lazy.culture, they are reluctant for the development.	Patterns of development that makes people as objects of development.	Many critical s and innovators among villagers who are apathetic.	The requirement of co-workers who have adequate skills for posdaya. Accomplishments made by posdaya will acknowladge public for its success.	Partnership program by university' students. Intensifying activities programs needed by the communities.
4.	Lack of equipment / facilities as a center of activity for posdaya.	The limitated land and funds	Having good *Links* with the village's authorities.	Efforts to propose funding supports to potential donors.	The development of facilities and equipment for Posdaya.
5.	Limited support for the development of posdaya.	Lack of socialization program. Public assumptions that contributions for activities of non mosque activities is not rewarding.	Legality of posdaya through villages decisions letters and neighbourhoods as well.	Periodic special meeting for the potentials donors. Partnership development with other agencies.	Special meeting for the potentials donors Establish partnerships with other agencies.
6.	Stagnation of the village business development involving the mothers / women.	No cummunity development program. Lack of companionships. Limited capital funds.	Most of villager women are having business skills. High revenue from villagers.	Community development programs	Partnership programs for UKM posdaya. Training program for productive business.
7.	Lower level of economic status and the educational background of the villagers	Limitated management knowledge of the villagers on business resources. Limited cheap education facilities.	The untapped potentials of the villagers. Officials and cadres who are responsible.	The development of education facilities. Recruitment of tutors. The establishment of the networks.	The development of PAUD, and madrasah diniyah. The training program for tutors of PAUD. The coordination plan with local authorities.

Table 2. Contd.

8.	Not all villagers know and understand posdaya program.	The socialization programs have not been well implemented. The new posdaya is still in infancy and has not run well.	Socialization program of posdaya. Publication on the mass media by partner universities / local authorities.	Posdaya tour. Formalize posdaya as the government program. Social services of posdaya.	Posdaya tour. Reporting and audiency program to the higher level of the authorities. Social services of posdaya.

physical and non physical of the society. These changes include the public opinion of the program and activities in education, health, economy and environmental development. The Management of posdaya encountered both physical as well as non-physical contraints. Physical constraints which were related mainly to the limitations or lack of facilities and supporting facilities to run posdaya activties. While non-physical constraints were related to improper understanding about posdaya, the management capabilities of the board members are still weak, fed-up status of for the posdaya board, the low quality of human resources, and the lack of the external support. The project potential generates economic benefits for the country by creating business opportunities and employment. The main economic driver of posdaya is the potential to affect the livelihood strategies of small-scale enterprises and local entrepreneurs in the areas of: (1) natural capital (opportunities for gaining access to national government policies); (2) financial capital (communication with micro-credit and other lending organizations); (3) human capital (increased knowledge of new skills through distance learning and processes required for certification); (4) social capital (cultivation of contacts beyond the immediate community; and (5) physical capital. With improved rural connectivity, rural entrepreneurs will be able to upgrade their capital, which can be translated into better business opportunities.

RECOMMENDATIONS

Considering that the vast majority of the poor live in the rural areas and derive their livelihood directly or indirectly from agriculture, support for farming is a high priority for rural development. The project can deliver useful information and data to farmers on crop care, animal husbandry, fertilizer and feedstock inputs, drought mitigation, pest control, irrigation, weather forecasting, seed sourcing, and market demand and prices. Research results suggest that increasing agricultural productivity benefits the poor and landless through increased employment opportunities. The project will generate employment opportunities in three ways: (1) the unemployed can use the improved accessibility through ICT to find job opportunities; (2) the unemployed can become employed in new jobs created through the expansion of the businesses of rural entrepreneurs; and (3) even the employed can acquire new skills through specific education programs that could be facilitated by posdaya.

REFERENCES

Bernard HR (1995). Research methods in anthropology: Qualitative and quantitative approaches, 2nd. Eds. Altamira Press, a Division of Sage Publications, Inc. Walnut Creek, London and New Delhi. P 803.

Geertz C (1973). The interpretation of cultures: Selected essays, Basic Books, New York. P 476.

Goldenberg J (1993). The "leap - frogging" approach to national development. In: Facing the challenge: responses to the report of the South Commission.

Mabogunje AL (1995). A concept of development. Ibadan, DPC. P 258.

Muljono P (2010). The model of family empowerment for community empowerment. J. Commun. Culture Politics 23(1): 9-16.

The New Encyclopaedia Britanica (1978). Helen Hemingway Benton. Vol. 3, Chicago. P 1023.

Standing Conference for Community Development (2001). Strategic framework for community development. http://www.sccd.org.uk.

Statistics Indonesia of the Republic Indonesia (2009). Statistics Indonesia, SIRI, Jakarta. P 630.

Suharto E (2005). Community development for people empowerment. Refika Aditama, Bandung. 324p.

Sumarjo dan Saharuddin (2004). Participatory methods for community development. Department of Socio-economic, Faculty of Agriculture, Bogor Agricultural University, Bogor. P 215.

Suyono H, Rohadi H (2007). Handbook of posdaya development. Balai Pustaka, Jakarta. P 197.

United States Department of Agriculture (2005). Community development technical assistance: Handbook. http://ocdi.usda.gov.

6

Spatial relationship between human well-being and community capital in the Black Belt region of Alabama

Buddhi Raj Gyawali[1]*, Rory Fraser[2], James Bukenya[1] and Swagata "Ban" Banerjee[1]

[1]Department of Agribusiness, P. O. Box 323, Alabama A&M University, Normal, AL 35762, Alabama, USA.
[2]Center for Forestry and Ecology, P.O. Box 323, Alabama A&M University, Normal, AL 35762, Alabama, USA.

This paper examines the relationship between human well-being and forms of community capital in the eight Black Belt counties in the west-central region of Alabama. This region is one of the poorest in the United States with high proportion of African-American populations. Cross-sectional spatial regression models were estimated using the data from the U.S. Population and Economic Census, Geographic Information System, and satellite imageries of 2000. The results indicate that geographic space is highly segregated in these counties and African Americans are less likely to be found in areas high in built, natural, and political capital. Service-providing entities such as financial, industrial, and social capital are located more in urbanized centers. The results suggest that social capital is strongly correlated with human well-being. The findings provide spatially-explicit empirical insights and suggest targeting rural development policies to create more social capital and address specific needs of the region, especially of African-American populations.

Key words: African Americans, census data, land cover, proximity, rural, social capital, spatial.

INTRODUCTION

The west-central Black Belt region of Alabama has not only undergone transformation in land use and land cover, but has also experienced changes in different forms of community capital. Community capital includes tangible and intangible assets or resources that exist in a community, which are capable of producing new resources (Flora and Flora, 2004; Hancock, 2001). They exist in the form of social and cultural organizations, government or financial agencies, infrastructures and other entities to strengthen or retain community capability and provide resilience for various socioeconomic, political, and infrastructural developments (Emery and Flora, 2006; Ostrum, 1990). These institutions are grouped into the major forms of community capital that include cultural, human, social, political, financial, built, industrial and natural (Emery and Flora, 2006; Flora and Flora, 2004).These forms of capital are involved with creating and controlling natural resources and goods and services, while helping to enhance the relationships between people and the land, as well as other natural resources (Bullard, 1990; Myrdal, 1998; Schulman, 1991; Zabawa, 1991).

Recent studies have used a community capital framework to study community resilience, rural development and ecosystem management (Adger, 2000; Costanza et al., 1998; Emery and Flora, 2006; Flora and Flora, 2004; Putnam, 2000). Formation and retention of community capital is a prerequisite for the socioeconomic development of a community, its resilience and sustainable ecosystem management (Costanza, 2004; Deller et al., 2001; Flora and Flora, 2004). The resiliency of the community to respond to changes in the larger environment depends in part on the resources available in a community. Communities that have higher levels of community capital have more resilience and more economic, environmental, and cultural stability (Harris et al., 2000).

Researchers such as Donoghue and Sturtevant (2007) and Emery and Flora (2006) have advocated for a balance of foundational and mobilizing capital in a community. Foundational capital includes primary resources such as physical, natural, and financial capital. Mobilizing capital activates and mobilizes foundational

*Corresponding author. E-mail: buddhi.gyawali@aamu.edu.

capital into productive use by the community. Such capital includes human, social, and political capital. As Appleton (1999: 61) posits, "what brings about transformation is the people who can create innovative and creative ways," an indicator of human capital. Rupasingha and Goetz (2003) in their study of rural communities, found evidence of human capital or education, becoming increasingly important for economic growth. Investment in human capital enhances the community's attractiveness for more businesses (Johnson, 2002).

Utilization of cultural and social capital for maximum utilization of human capital and for economic development in rural regions is widely supported by empirical evidence (Robinson and Siles, 1999; Terluin, 2003). Communities that invest in such policies and practices empower citizens' values, attitudes, and beliefs about their potential and future opportunities and earn higher economic well-being. Cultural capital creates a favorable and an inclusive climate in which all of the citizens, irrespective of their race, gender, or religion become able to maximize their potentials and benefit from their labor (Florida and Gates, 2001).

The distinct socio-cultural, economic and natural resource characteristics and prevalence of persistent poverty in the Black Belt region justify that it is a good hotspot to study the relationship between community capital and poverty. The historical eras of slavery, reconstruction, the civil rights movement, and affirmative action created various institutions in the west-central Black Belt region of Alabama. Evidence of the formation of cultural, social, and political capital in the region arose in response to reconstruction efforts, the lack of voting rights (Myrdal, 1998; Schulman, 1991), race and class-based discrimination (Mitchell, 2001), and the loss of land ownership (Gilbert et al., 2001; Zabawa, 1991). The organization of political, cultural and social capital (civic institutions and informal networks) in communities was initiated in response to government agencies' non-inclusion of African Americans in the political processes and granting ownership rights to land (Bliss et al., 1998; Mitchell, 2001; Schelhas and Zabawa, 2000). The Black Belt communities' tradition of communal action through local churches and clubs, family gatherings, town meetings, community organizations, and cooperatives is legendary. However, many of these institutions, organizations and networks have disappeared under the continuation of biased policies and efforts from the late 19th and the early to mid-20th century, as well as a result of the degradation of civic engagement in community activities in the latter stages of the 20th century (Ayers, 1992).

Political, economic, and social institutions have continued to have a role in the formation of community capital in specific locations. This in turn has improved these communities' well-being. Yet, the creation and development of community capital have rarely been used as a policy tool in economic and social development (Fey et al., 2006). This is a shortcoming of both past and current research in the South (Durlauf, 2002; Glaeser, 2001; Rupasingha et al., 2006). Studies by Robinson and Siles (1999), Flora and Flora (2004), Putnam (1996), and others have shown the important role community capital plays in community development. However, they have not yet provided adequate insights as to why community capital is concentrated in specific places and why there are significant cross-sectional variations in community capital across a region (Durlauf, 2002).

Despite major efforts to alleviate poverty and bring about economic development, persistent poverty exists among many African Americans in the west-central Black Belt communities of Alabama. Social, economic, political, and natural resource factors have been used to explain this phenomenon (Bliss and Bailey, 2005; Joshi et al., 2000). However, these explanations have not examined the role of spatial patterns of the distribution of community capital and how it is connected to the well-being of people. Community capital (primarily cultural, political, and social) has helped African Americans withstand slavery, segregation and Jim Crow law; by reinforcing community resilience and cohesiveness in Alabama (Schweninger, 1997). Now, many of these modes of capital have been dissipated or isolated. At the same time, new forms of capital (such as financial, infrastructural, and built) have been formed in response to changes in population dynamics, land cover, land ownership and these have had differential impacts on rural development (such as changing patterns of urbanization and quality of life) across different ethnic groups and locations in the Black Belt region of Alabama.

The objectives of this paper are: (1) Identify the specific forms of community capital in the west-central Black Belt region of Alabama; (2) Identify their spatial patterns of distributions across the region, and (3) Examine the relationship between human well-being and different forms of community capital available in the region.

The main null hypothesis of this study is that there is no significant relationship between the human well-being index and the spatial pattern of distribution of community capital. Alternatively, this study hypothesizes that socioeconomic development issues in the study area may be better understood if the spatial context of the distribution of different forms of community capital and its relationship with human well-being index are examined.*

MATERIALS AND METHODS

Study area

The west-central Black Belt region of Alabama lies within the Gulf

* Explanation of the hypothesized relationship is provided in the *Empirical model* section.

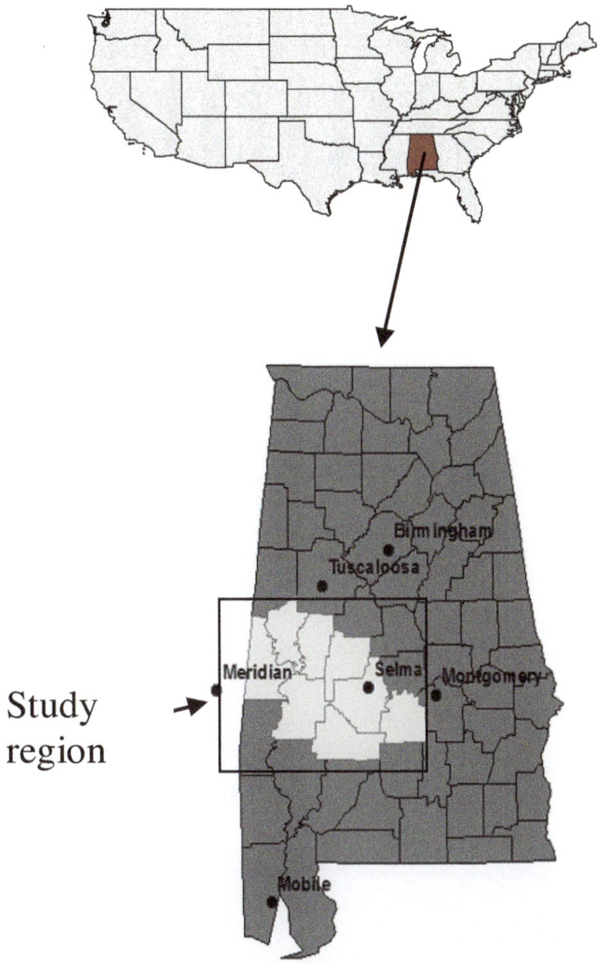

Figure 1. Study area showing eight black belt countries in the west-central region of Alabama in the United States.

south's coastal plane, 25-30 miles wide and stretching 300 miles from Eastern south-central Alabama into Northwestern Mississippi (Figure 1). The study site (-86.4° to -88.4° East, 31.13° to 33° North) consists of eight counties (Dallas, Green, Hale, Lowndes, Marengo, Perry, Sumter, and Wilcox). The area covers 6,479 mi^2 (16,781 km^2). The total population of the region is 149,378, of which 65% are African Americans (U.S. Department of Commerce, 2007). The population density is 22 people per square mile. Thirty-two percent of the people live with incomes below the national poverty line (Bukenya and Fraser, 2004).

This region contains vast amounts of forest resources and fertile agricultural land. The physical landscape characteristics are highly differentiated by soil type, forests, locations of industries, and demographic attributes, such as income, race and land-based income-earning activities. The region is mostly affected by a sub-tropical climate and is generally warm and humid. The mean annual temperature is 66°F (19°C), with a winter average of 49°F (9°C) and summer average of 81°F (27°C). The amount of precipitation is about 52 inches (1,320 mm). The African-American population is concentrated in the agricultural lands and in small towns. The socio-economic and land cover maps of the region clearly depict clustered patterns in terms of forests, ethnic population distributions, incomes, and poverty.

Data preparation

Data preparation and analysis were conducted using data from the U.S. Population Census 2000, the U.S. Economic Census 2002, U.S. TIGER Road and urban data, and land cover data derived from the classification of Landsat ETM+ for 2000 at the Census Block Group (CBG) level. There are 161 Census Block Groups (CBGs) in the eight Black Belt counties in the west-central region of Alabama (U.S. Department of Commerce, 2007). The CBG is the smallest unit for which the U.S. Census Bureau makes Census data available to the public in the form of average values of various attributes of the population residing in an area. Typically, CBGs range between 600 and 3,000 people, with an average size of 1,500 people (U.S. Department of Commerce, 2007). CBGs tend to be relatively homogeneous with respect to demographics, economic status, and living conditions. A geographically-tiered spatial database of the demographic and socioeconomic indicators, land cover types, and different forms of community capital - including layers representing the biophysical, infrastructural, and administrative boundaries - were created at the CBG level for 2000. The selection of the variables was based largely on earlier studies and the availability of data at the CBG level.

Demographic and socioeconomic data were obtained from the Time-Series Research Package released by Geolytics Inc. (2004a). These data are available for each Census Block Group (Geolytics, 2004a). The Census data provide the aggregated average value for each variable for each CBG (Geolytics, 2004b). The raw data were standardized by converting them into percentages (raw data of a variable divided by its total in a CBG). The percentage values were then considered as an average value for a CBG.

A human well-being index (HWBI) was created using income, education, and employment (UNDP, 2005). These measures incorporate the basic material for a good life and health - two factors considered important in the millennium ecosystem assessment (MEA) concept of human well-being (MEA, 2003).[§] The measure of education is the percentage of persons over 24 years who graduated from high school and the percentage of the same population who graduated from college. The measure of income is the per capita income in dollar value, and the employment measure is represented by the percentage of people, 16 years and older who are employed. Indexes of income (I_{in}), high school (I_{hs}), graduate education (I_{ge}), and employment (I_{em}) were created as per Bukenya and Fraser (2004). For instance, the per-capita income index was created using the following equation:

$$I_{in} = \frac{(X_{in} - MinimumX_{in})}{(MaximumX_{in} - MinimumX_{in})} \qquad \text{..................(1)}$$

where X_{in} is the average per-capita income for a CBG, and $MaximumX_{in}$ and $MinimumX_{in}$ are the highest and lowest average per-capita income, respectively, for all CBGs. These index values are computed in order to normalize the values of the variables that are included in the human well-being index (HWBI), so that all values fall between 0 and 1. Then, the four indexes were averaged to obtain the HWBI for each CBG (Bukenya and Fraser, 2004; UNDP, 2005; Vemori and Costanza, 2006).

§ Well-being is at the opposite end of a continuum from poverty, which has been defined as a "pronounced deprivation in well-being. The constituents of well-being, as experienced and perceived by people, are situation dependent, reflecting local geography, culture, and ecological circumstances (MEA 2003, p. 3)

Table 1. Definitions and indicators of different forms of community capital*.

Capital	Definition and indicators
Human	Intellectual, professional, and technical preparation of population that helps to make a community energetic
Financial	Financial resources available to invest in the community for business or infrastructure development, civic and social enterprise and wealth accumulation
Cultural	Entities that represent how people perceive the world including heritage, values, history, and identity, and entities such as art, language, symbols, and customs
Political	Resources and tools, local governments and other institutions that engage citizens and businesses in community development
Social	Collective norms, networks of reciprocity, and mutual trust that are utilized for mutual benefits, building trust and fulfilling shared needs
Built	Physical infrastructure that enhances other forms of community capital, such as roads, airports, railways, office buildings, schools, utilities, sewers and water systems
Natural	Assets occurring naturally for a long time in a particular community from which much of the economic wealth is derived. Examples are water, soil, minerals, plants, animals, biodiversity, aesthetic or scenic resources, mines and quarries, parks, and natural amenities
Industrial	Manufacturing base in the community that provides employment, food, fiber, and shelter for the people in a community. Examples include factories and facilities such as paper mills, sawmills, automobile plants, catfish and meat processing plants, lumber and cotton gin factories

*Definitions and indicators adopted from Flora and Flora (2004).

$$HWBI_X = \left(\frac{I_{in} + I_{hs} + I_{ge} + I_{em}}{4} \right) \qquad (2)$$

Land cover (forest cover) data were derived from the Landsat ETM+ imageries for 2000. These terrain-corrected (geographic, radiometric, and topographic correction) data with less than 10% cloud cover were acquired from USGS/EROS data center, Sioux Falls, South Dakota. A hybrid classification approach using both unsupervised and supervised techniques was employed to derive accurate and reliable land cover classes (Campbell, 1996). Anderson's Level II classification scheme was primarily employed to classify the study region into the nine major land cover classes using ERDAS 8.7 Image processing software (Anderson et al., 1976; Leica Geosystem, 2005). The ArcGIS 9.1 (ESRI, 2005) software was used for extracting area covered by each land cover type in each CBG in 2000. The three subgroups of forests (evergreen, deciduous, and mixed) were combined into a single "forest-cover" class (Gyawali et al., 2009).

Community capital data were obtained from the U.S. Economic Census 2002, compiled as a directory of all credit-approved establishments in Alabama. Information on different forms of businesses, industrial establishments, government offices, built infrastructures, and private and non-governmental organizations (NGOs), and cultural entities in each CBG were obtained. There were 5,308 of such establishments in the study region in 2002. North American industrial codes (NAIC) were used to categorize these entities into the proxies to represent eight different forms of community capital. The methodology adopted by Rupasingha et al. (2006) for creating proxies was followed. The identification of the proxies was based on literature (Flora and Flora, 2004). A total of 1,048 establishments met the criteria developed by Rupasingha et al. (2006) to represent the eight different forms of community capital (Table 1). Twenty-five CBGs did not have any such establishments; therefore, these CBGs were excluded from the analysis keeping 136 CBGs with at least one establishment to represent one of the eight forms of community capital. The data were standardized by computing the number of community capital establishments per 1,000 people in a CBG, (that is, the number of establishments for each type of capital available in a CBG was divided by the total population of a CBG and multiplied by 1,000). CBGs that contained at least one count of any community capital in 2000 were used in the analysis.

Empirical model

The major objective of this paper was examined by factoring the HWBI in a CBG with the proxies of the different forms of community capital (human, cultural, social, political, built, industrial, financial, and natural) and other control variables as suggested by related literature (Flora and Flora, 2004; Green and Haines, 2002; Johnson 2002; Rupasingha et al., 2002) (Table 2). Communities may have different forms of capital in different amounts and such differences may relate to the well-being of those communities. Different forms

Table 2. Variables description for the relationship between human well-being and forms of community capital in 2000.

Variables	Measures of the variables	Expected sign
Human well-being	**A composite index of education, income, and employment indices (between 0 and 1)**	**Dependent variable**
A. Community capital	No. of establishments per 1000 population in a CBG	
Human capital	Professional service sectors, such as schools, libraries, lawyers, doctors	+
Financial capital	Banks, insurance, finance, investments	+
Built capital	Electric, water, sewers, transportation, natural gas	+
Political capital	Federal agencies, conservation, legislative bodies/political offices	+
Social capital	Civic and social organizations, business clubs, NGOs, advocacy groups	+
Natural capital	Natural amenities such as parks, mining, historic sites	+
Industrial capital	Manufacturing and Processing plants such as pulp and paper mills, sawmills, cotton gin	+
Cultural capital	Cultural entities such as barber shops/beauty salon, church	+
B. Demographic (control)		
Population density	Population per acre in a CBG	+
African Americans	Percentage of AA population in a CBG	-
Age-dependency ratio	Ratio of % of persons <15 years and > 64 years to % of persons in the economically productive (16-64 years) years (ranges from 0 to 1) in a CBG	-
Homeownership	% of owner-occupied housing units (out of total occupied housing units)	+
Urban areas	Binary 0 or 1 (1 = Urban CBGs, with ≥ 1 person in 2 acres of land)	+
C. Landscape variables (control)		
Total forestlands	Percent of total forests in a CBG	+
Roads	Density of major roads (meters/acre) in a CBG	+
D. Spatial variables		
Spatial Lag	Neighborhood and autocorrelation effect	+

CBG = Census block group; AA = African Americans; NGOs= Non-government organizations.

of community capital will have additive effects on community well-being. The increasing proportion of community capital may generate significant positive effects on the human well-being of a community and help to reduce poverty. Also, a balance among these forms of capital is necessary for the sustenance of the desirable level of community well-being (Coleman, 1988; Flora and Flora, 2004). A cross-sectional spatial regression model was specified to examine the relationship between human well-being and community capitals in the year 2000. The counts (per 1000 population in a CBG) of establishment of each of the eight forms of community capital in a CBG were used as independent variables. The regression model includes other socioeconomic, landscape, and spatial variables to control for confounding factors (Equation 3). These variables are grouped into: (1) Demographics (population density, race, age structure, rural or urban areas, and homeownership); (2) Landscape variables (amount of forest land

and road networks), and (3) Spatial factors (neighborhood effects). The following spatial regression model explains the functional relationship between human well-being and vectors of both independent and control parameters:

$$HWBI_{2000} = \alpha + \beta_0 CC_{2000} + \beta_1 DV_{2000} + \beta_2 LV_{2000} + \rho WHWBI_{2000} + \varepsilon \quad (3)$$

where *HWBI* is the human well-being index in a CBG in 2000, *CC* is a vector of different forms of community capital, and *DV* and *LV* are vectors of demographic and landscape variables in 2000, respectively. $WHWBI_{2000}$ is the spatially lagged dependent variable for a spatial weights matrix for 2000 (Anselin, 2003), Scalar ρ is the spatial lag to be estimated and ε is a vector of error terms with normality and homoscedasticity. The spatial factor *W* for $HWBI_{2000}$ in a CBG is the spatially-weighted average of all other adjacent CBGs' HWBI in 2000. The weight matrix *W* is constructed using first-order binary "rook contiguity" by assigning a weight of 1 to all adjacent CBGs that are contiguous to the CBG or share a common border, and a weight of 0 is assigned to all other CBGs which are not contiguous to the CBG (Rey and Montouri, 1999). The spatially-lagged endogenous variable ($\rho WHWBI_{2000}$) for the dependent variable accounts for all forms of spillover effects from neighboring CBGs (Anselin, 1988, 2003; Janikas and Rey. 2005; Lim, 2003).

The control variables are used to account for their effects on human well-being. Those effects are either their direct contribution to the dependent variable or their indirect contribution through their interaction with community capital (Flora and Flora, 2004). These variables may significantly affect the formation, disappearance or stability of community capital, which ultimately will have effects on sustaining the human well-being indices of the communities.

Demographic Variables (DV) such as population density and characteristics are important factors that influence the relationship between human well-being and community capital. This variable may have direct or indirect (through interaction with different forms of community capital) effects or human well-being. For instance, as population increases, more demands for community capital will be created. Such demands are for roads, parks, jobs, schools, shopping centers, and open space for a growing population (Reynolds, 2001). On the other hand, cheap or uneducated or untrained labor force may be in high demand for newly opened manufacturing capital that may not require skilled labor. The behavior of people (profit-maximizing activities, or ambition for social recognition, community service or political positions) and a person's social qualities to associate with other community members may be related to the formation of more social or political capital (Flora and Flora, 2004; Putnam, 1996).

A larger population of a community helps to create or maintain collective voice, action, and community unity. These strength help to ensure the sustainability of community capital (such as built, financial, social, and political) and efficient and high quality services (such as equal access to income-earning activities, government subsidies, loans, training, and tax breaks), which are related to the community's well-being. It is found that the loss of community capital (or lower human well-being) is usually evident more in periphery and rural areas (where a loss of human population is most likely to happen) than in urban areas (Flora and Flora, 2004). Glaeser et al. (2002) found that migration reduces social capital levels in a community as it weakens local networks and associations.

Poor communities, composed mostly of minorities, tend to engage more with local resources (e.g., volunteers, social and cultural clubs, and cooperatives) than more affluent communities (Chaskin et al., 2001). However, Putnam (1995) found that racial differences have contributed to a decline in social capital in the United States. Alesina and La Ferrara's (2000) study supports this contention. The study found that participation in associational activities is significantly lower in ethnically-fragmented localities. In such a scenario, when minority residents feel alienated, participation in local affairs declines and collective action is fragmented (Israel and Beaulieu, 2004). Also, Glaeser et al. (2002) found that social capital is higher among homeowners. Homeowners have an incentive to improve the community they live in. Social capital first rises and then falls with changes in age groups. Also, educational level determines people's attitudes and behaviors while they choose whether or not to be a member of civic organizations (Putnam, 1995; Glaeser, 2001). Higher educational level may lead to higher levels of social capital in a community.

Landscape variables (LV), such as agricultural or forest lands, water, and good quality soil are important raw materials for manufacturing capital. They are a means for creation of more jobs. Higher amounts of raw materials in a community ensure sustenance of industries and jobs without a fear of closing industries or loss of jobs (Ahn et al., 2001; Alig, 1986; Platinga and Miller, 2001; Wears and Gries, 2002). Usually, higher quality land is used for agricultural farming so it increases the chances for the presence of agricultural processing industries. Similarly, the presence of high-density forestlands increases the chances for forestry-based industries.

Physical proximity increases opportunities for interactions among local residents that build community bonds (Israel and Beaulieu, 2004). Increasing distance to the major highways and shipping points (waterways) may result in a lower likelihood of the presence of industry capital, as it will increase the cost of hauling and the commuting distance for workers. Similarly, urban areas increase the likelihood of the presence of financial and service-oriented capital as several government and county or state agency offices are located in urban areas.

Table 2 provides information on the dependent and independent variables and control variables. All eight forms of community capital are expected to have a positive relationship with human well-being. Because these forms of capital assist communities to build their community capability, effectively manage local resources, and increase community resilience, they should thereby increase human well-being (Flora and Flora, 2004).

RESULTS

Descriptive statistics

Out of the total 1,048 counts of capital, the largest share was natural (249 establishments) capital, followed by political (203), social (140), built (113), human (106), financial (72), and cultural (78) capital, respectively. These counts of capital were not evenly distributed across the study region (Figure 2). For a better interpretation of the counts, these forms of capital were standardized per 1,000 CBG populations. The four columns on the right side of Table 3 represent the central tendency measures (minimum, maximum, mean, and median) of the standardized counts of community capitals per 1,000 population in a CBG and central tendency measures of other control variables. These measures suggest that distribution of the major forms of community capital are positively skewed because the majority of the forms of capital are concentrated in the CBGs that are relatively densely populated and are parts of towns or

Table 3. Descriptive statistics of the dependent and independent variables in 2000*.

Variables	Total counts of capitals in 136 CBGs	Min.	Max.	Mean	Median
Well-being 2000		0.24	0.82	0.52	0.51
Cultural capital	78	0.00	9.55	0.68	0.00
Financial capital	72	0.00	15.15	0.60	0.00
Human capital	106	0.00	9.09	0.94	0.00
Built capital	113	0.00	7.52	0.93	0.00
Industrial capital	87	0.00	8.95	0.75	0.00
Natural capital	249	0.00	16.47	2.00	1.22
Political capital	203	0.00	19.70	1.63	0.00
Social capital	140	0.00	13.64	1.24	0.68
Total	1048				
Population density	-	0.01	8.61	0.68	0.05
African Americans%	-	0.91	100.00	64.34	71.26
Dependent population ratio	-	0.16	1.03	0.67	0.66
Home ownership%	-	20.55	97.88	73.72	80.56
Urban area (dummy)	-	0.00	1.00	0.24	0.00
Total forests%	-	21.28	100.00	72.74	77.15
Road density (meter per acre)	-	0.00	22.74	3.94	2.29

*N = 136 CBGs. CBGs with at least one form of capital are included in the analysis. In the analysis, the capital represents the number of establishments per 1,000 population.

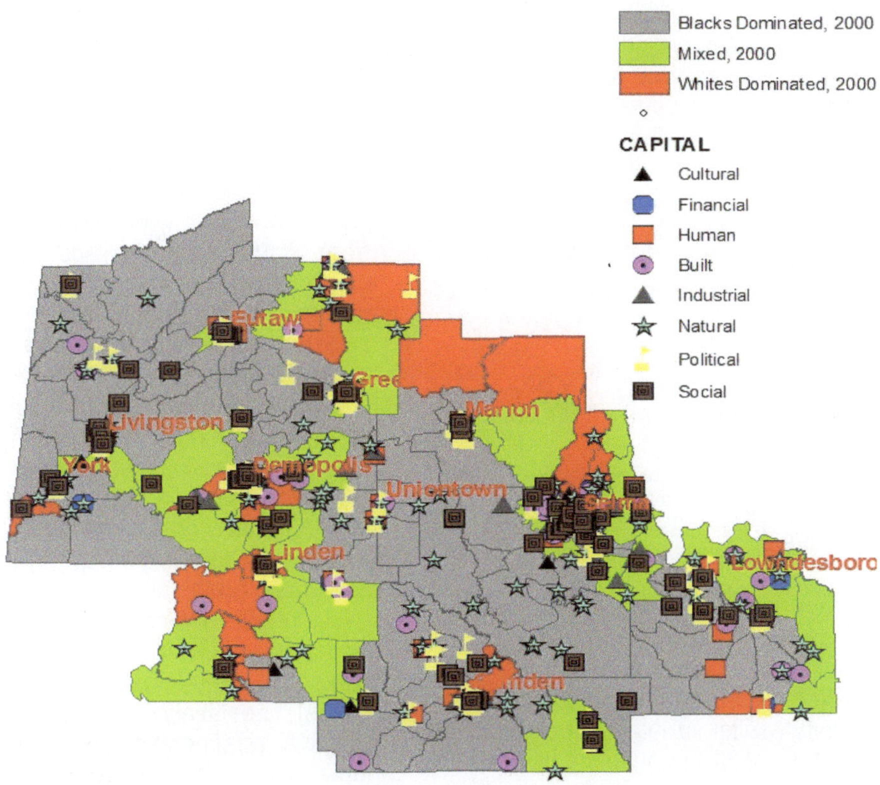

Figure 2. Spatial distribution of eight forms of community capital in the study region in 2000.

Table 4. Univariate Moran's I values for human well-being, community capitals and control variables in 2000.

Variables	Moran's *i* value
Human well-being	0.33
Population density	0.63
African Americans	0.35
Dependent population	-0.02
Homeownership	0.36
Roads network	0.41
Total forests	0.37
Cultural capital	0.07
Financial capital	0.02
Human capital	0.21
Built capital	-0.03
Industrial capital	0.09
Natural capital	-0.06
Political capital	-0.04
Social capital	0.08

nearby towns, parts of major road intersections, business hubs, or industrial zones, such as Selma, Demopolis, Camden, Marion, Livingston and Greensville.

Spatial data exploration

The spatial pattern of the distribution of the HWBI, community capital and the control variables were explored using univariate Moran's *I* values to examine the magnitude of autocorrelation among the variables (Table 4). The values of the HWBI, population density, the percentage of African Americans, roads, homeownership, and forests were highly correlated to each other. The CBGs that are clustered together (or have shared a border) have high likelihood of having similar values of these variables. However, among the community capital variables, only human capital had a high univariate Moran's *I* value. Other capital had very low Moran's *I* value suggesting a random distribution of these other forms of capital across the study area.

Spatial regression analysis

The ordinary least squares (OLS) model was highly significant overall (F = 18.18, P ≤ 0.001). The OLS model indicated that 65% of the total variation in human well-being was explained by the explanatory variables (Table 5). Among the community capital variables, social capital was significant (β = 0.24, t = 2.36), suggesting its positive relationship with well-being. Among the control variables, African-American population (β = -0.78, t = 12.43) was significant at the 1% level, the age-dependency ratio (β = -0.12, t = 2.24) was significant at the 5% level, and urban area (β = 0.34, t =1.71) was significant at the 10% level.

Due to the presence of significant spatial clustering effects in the OLS model, as observed by the presence of a significant Moran's error value (Moran's *I* = 0.14, P ≤ 0.001) and Lagrange multiplier for lag (7.70, P ≤ 0.001), a spatial lag model based on the maximum likelihood (ML) was estimated to obtain unbiased estimates of the relationship between human well-being and community capital (Anselin, 2003; Lim, 2003; Rey and Montouri, 1999).

The spatial lag model improved significantly and provided better explanations of the relationship than the OLS model. This was evident by the increase in R^2 value (from 0.65 to 0.71) and a decline in the Akaike Information Criterion (from 255.70 to 249.43). The model was significant (likelihood ratio test value = 8.27, P ≤ 0.001) and was free of multicollinearity (the multicollinearity condition number was < 30) and heteroskedasticity (a Breusch-Pagan test was not significant at the P ≤ 0.10 level). However, the model did not add any new significant variables. The variables that were significant in the OLS model, e.g., social capital, African Americans, age-dependency ratio, and urban areas, were also significant in the spatial lag model with the same signs.

The results indicate that the CBGs with more social capital are likely to have a higher level of human well-being. Similarly, the presence of micropolitan[**] areas (towns) positively affects human well-being.[**] The negative relationship between African Americans and human well-being is suggested by the lower HWBI in the

[**] A micropolitan area contains an urban core of at least 10,000, but less than 50,000, population

Table 5. Results of the ordinary least square (OLS) and spatial lag model for the relationship between human well-being and forms of community capitals in 2000.

Variables	OLS model		Spatial Lag model	
	β-coef.	t-statistics	β-coef.	*z*-value
Constant	0.08	1.17	0.07	1.12
A. Community capital				
Cultural	-0.10	1.19	-0.12	1.51
Financial	0.05	0.60	0.05	0.79
Human	-0.09	1.09	-0.108	1.37
Built	-0.04	0.71	-0.02	0.46
Industrial	0.01	0.09	0.01	0.01
Natural	-0.03	0.52	-0.04	0.69
Political	-0.03	0.37	-0.01	0.04
Social**	0.24	2.36	0.27	2.84
B. Demographic variables				
Population density	-0.09	1.11	-0.07	-0.93
African Americans***	-0.78	12.43	-0.68	10.84
Age dependency ratio**	-0.12	2.25	-0.11	2.34
Homeownership	0.03	0.37	0.05	0.78
Urban area*	0.34	1.71	0.31	1.71
C. Landscape variables				
Forestlands	-0.01	0.01	-0.05	-1.11
Roads density	-0.07	1.09	-0.05	-0.86
Spatial lag *(Rho)*			0.24	
R^2	0.65		0.71	
F value	18.18***			
AIC	255.70		249.42	
Multicollinearity condition number	6.24			
Breausch-Pagan test	14.61 (ns)		13.37 (ns)	
Moran's I value	0.14***			
Lagrange multiplier (Lag)	7.70***			
Robust LM (lag)	2.58*			
Lagrange multiplier (error)	5.51**			
Robust LM (error)	0.39 (ns)			
Likelihood ratio test			8.27***	

*** ≤ 1%, ** ≤ 5%, * ≤ 10% significance levels; z-values are absolute values.; ns = not significant at ≤ 10% level; AIC = Akaike information criterion; LM = Lagrange multiplier.

CBGs where a higher number of African Americans live. The spatial lag (*rho*) estimate was positively significant (ρ = 0.24, *z* = 4.625) suggesting a one-unit change in the HWBI in the adjacent CBG causes the human well-being of the CBG to go up by 0.24 unit. This suggests a positive neighborhood effect on the HWBI.

DISCUSSION

The results suggest that human well-being and various forms of community capital have a weak relationship. Among the eight community capital variables, only social capital has a positive significant relationship with human well-being. The results reveals that the increasing amount of social capital – such as social networks, social and cultural clubs, farmers' associations, and non-government organizations have had positive effects on the HWBI of the studied region. The results of this study, consistent with the findings by Flora and Flora (2004), Parent and Lewis (2003), and Svendsen et al. (2007), suggested that social capital is one of the strongest assets of a community, which not only helps to bond communities but also plays a role in the socioeconomic development of rural areas. Social capital helps to increase the bargaining power of communities (Krishna, 2002; Putnam, 2000).

The control factors such as urban areas indicated a positive relationship with human well-being. However, the African-American population and the ratio of the dependent population were negatively correlated with the human well-being index suggesting the likelihood of lower HWBI in the African-American and dependent-dominant CBGs. The study also found significant neighborhood or clustering effects on human well-being, suggesting a high likelihood of the presence of well-off communities around the communities that have high records of human well-being, which is also evident by the results of Gyawali et al. (2009).

A closer examination of the distribution of community capital results indicates an uneven spatial distribution of different forms of community capital across the study region. Cultural, financial, human, industrial, and social capital differed significantly in their numbers and distribution between urban and rural areas. Service-providing entities such as roads, financial, political, industrial, and social capital were located in greater proportion in the urbanized centers. Such pattern suggests that community capital tends to be clustered around high-population areas (urban centers) and along the highways. It may be because urban growth brings more businesses and creates more revenue for the investment in urban infrastructure and human capital development (Callaghan and Colton, 2007; Flora and Flora, 2004; Terluin, 2003). Such growth increases the likelihood of creating more benefits to the people who are living nearby urban centers or the locations of those businesses. As suggested by Woodhouse (2006), the lower poverty rate of urban places in the study region may be the result of the concentration of major forms of capital in the urban areas.

The African-American population variable had a negative relationship with human well-being suggesting that this group of people (who are the majority in the region) lags behind in socioeconomic conditions. The results show that the situation in predominantly African American communities in Alabama's Black-Belt is no different from that in the rest of the country (Fraser et al., 2005; Gyawali et al., 2008, 2009). The uneven distribution of community capital and the lower proportion of the financial, built, natural, and political capital provide a gloomy condition of the status of African-American-dominant CBGs, and the results are consistent with previous findings (Alesina et al., 1999; Gilbert et al., 2001; Rupasingha et al., 2006). Such isolation from the major resources is counterproductive for their overall economic development.

As the study did not find a strong relationship between human well-being and all eight forms of community capital (except social capital), there might be many confounding factors that affect human well-being. Size and edge of the CBGs and exogenous effects such as state, local, and county policies may have an important role in human well-being or in the formation of community capital or in causing the concentration of community capital in a specific location.

The study suffered from limitations related to availability of data at the CBG level. The high correlations among the forms of community capital may have overestimated the prediction of human well-being. The proxies of community capital may not have represented the core definition or values of community capital. The human well-being index did not include life expectancy, one of the major constituents of quality of life, due to the unavailability of information at the CBG level. At the same time, all four factors (high school and bachelor-level education, income, and employment) were provided equal weights. These attributes may have contributed differently to the human well-being and using weighted values may have improved the analysis. Due to these data constraints, the regression models may be underspecified. Even though the spatial regression model explained over two-thirds of the total variations (R^2 = 71%), the explanatory power could have been stronger if these factors were included.

The question of whether the trends of human well-being measures such as education, income, and employment have improved over time as a function of community capital is a major concern and can be addressed by correctly selecting variables at multiple levels. A multi-scale approach using time-series data at micro-level (household), meso-level (communities) or macro-level (counties) could be employed to better understand the relationship between human well-being and community capital. A combination of diverse data collection methods (such as surveys and focus group interviews) could be the way to obtain or supplement data that truly represent human well-being and community capital.

Conclusion

This study provides empirical evidence of the major issues facing African-American communities in Alabama's Black Belt. This region is highly racially segregated with an uneven (some would suggest inequitable) distribution of the resources needed to foster economic development (Bliss and Bailey, 2005). This highly segregated area has pockets of well-developed communities with high income growth and higher level of social capital. However, for large areas in this region, there is limited capital, especially in the areas where African-American communities are in majority. The observed limited linkage of human well-being to major forms of community capital may have been the evidence of discrimination and exclusions of African Americans from local social networks, government institutions and economic opportunities. African Americans were less likely to be found in areas high in natural, infrastructural and social capital such as forests, roads, and industrial facilities, market and financial centers. Such hindrances may have contributed to the formation of community capacity and perpetuation of poverty in the region. These are the important issues for further research. Policymakers interested in addressing poverty issues and bringing real economic development to persistently poor African-American communities may need to rethink their development policies and strategies. Such policies aimed at improving conditions in this region need to recognize as this study did, the importance of different forms of social capital and target efforts to address their specific needs, such as developing financial, built, social, and human capital and engaging citizens in these capitals.

ACKNOWLEDGMENTS

The authors would like to thank the anonymous reviewers of the manuscript for their suggestions and comments. The authors would also like to thank USDA Forest Service, the Alabama Agricultural Land Grant Alliance (AALGA), and National Sci. Foundation (NSF-HRD-0420541) for providing the financial support for this study.

REFERENCES

Adger NW (2000). Social and ecological resilience: Are they related? Progress Hum, Geogr., 24: 347-364.

Ahn S, Platinga AJ, Alig RJ (2001). Historical trends and projections of land use for the south-central United States. Research Paper PNW-RP-530. Portland, OR: US Forest Service, Pacific Northwest Research Station.

Alesina A, la Ferrara E (2000). Participation in heterogeneous communities. Q. J. Econ., 115: 847-904.

Alesina A, Baqir R, Easterly WR (1999). Public goods and ethnic divisions. Q. J. Econ., 114 (4):1243-1284.

Alig RJ (1986). Econometric analysis of the factors influencing forest acreage trends in the southeast. For. Sci., 32 (1): 119-134.

Anderson JR, Hardy EE, Roach JT, Witmer RE (1976). A land use and land cover classification system for use with remote sensing data. USGS Professional Paper 964. Reston, Virginia: U.S. Geological Survey.

Anselin L (1988). Spatial econometrics: Methods and models. Boston: Kluwer Academic Publishers.

Anselin L (2003). Geoda 0.9.5-I5. User's Guide. Urbana, IL: Department of Agricultural and Consumer Economics, University of Illinois.

Appleton S (1999). Education and health at the household level in Sub-Saharan Africa. Unpublished manuscript. Oxford University: Center for the Study of African Econ.

Ayers EL (1992). The promise of the new south life after reconstruction. Oxford University Press, New York.

Bliss JC, Sisock M, Birch T (1998). Ownership matters: forestland concentration in rural Alabama. Soc. Nat. Resour., 11 (4): 115-137.

Bliss JC, Bailey C (2005). Pulp, paper and poverty: Forest-based rural development in Alabama, 1950-2000. In Communities and forests: Where people meet the land, eds. RG Lee and DR Fields,. Corvallis, OR: Oregon State University Press, pp. 138-158.

Bukenya JO, Fraser R (2004). Estimation of the human development index for Alabama counties. In Southeast Decision Sci. Institute's Proceedings of the 33rd annual meeting proceedings, ed. E. Omojokum. Williamsburg, VA, pp. 45-50.

Bullard R (1990). Dumping in Dixie: Race, class and environmental quality. Boulder, CO: Westview Press, pp.138-158.

Callaghan EG, Colton J (2007). Building sustainable and resilient communities: a balancing of community capital. Earth, Dev. Sustain., 10: 931-942.

Campbell JB (1996). Introduction to remote sensing. New York: The Guilford Press.

Chaskin RJ, Brown P, Venkatesh S, Vidal A (2001). Building community capacity. New York: Aldine de Gruyter.

Coleman JS (1988). Social capital in the creation of human capital. Am. J. Sociol., 94: S95-S120.

Costanza R (2004). Estimates of the genuine progress indicator (GPI) for Vermont, Chittenden county and Burlington, from 1950 to 2000. Ecol. Econ., 51: 139-155.

Costanza R, d'Arge R, de Groot, R, Farber S, Grasso M, Hannon B (1998). The value of the world's ecosystem services and natural capital. Ecol. Econ., 25: 3-15.

Deller SC, Tsai T, Marcouiller DW, English DBK (2001). The role of amenities and qualities of life in rural economic growth. Am. J. Agric. Econ., 83 (2): 352-365.

Donoghue E, Sturtevant V (2007). Social science constructs in ecosystem assessments: Revisiting community capacity and community resiliency. Soc. Nat. Resour., 20 (10): 899-912.

Durlauf SN (2002). Bowling alone: A review essay. J. Econ. Behav. 47: 259-273.

Emery M, Flora C (2006). Spiraling-up: Mapping community transformation with community capitals framework. J. Community Dev. Soc., 37(1): 125-139.

ESRI (2005). ArcGIS 9.1. Redlands, CA: ESRI Inc.

Fey S, Bregendahl C, Flora C (2006). The measurement of community capitals through research. A study conducted for the Claude Worthington Benedum Foundation by the North Central Regional Center for Rural Development. Online J. Rural Res. Policy 1(January)

Flora C, Flora J (2004). Rural communities: Legacy and change. Boulder, CO: Westview Press.

Florida R, Gates G (2001). Technology and tolerance: The importance of diversity to high technology growth. Washington, DC: Brookings Institute.

Fraser RF, Gyawali B, Schelhas J (2005). Land Tenure and Well-being in Perry County, Alabama. Small-Scale For. Econ. Manage. Policy, 4(1): 21-33.

Geolytics (2004a). U. S. census CD 1980 long-form in 2000 boundaries. East Brunswick, NJ: Geolytics, Inc.

Geolytics (2004b). U.S. Census DVD Time Series Normalized Research Package Data. East Brunswick, NJ: Geolytics, Inc.

Gilbert J, Sharp G, Felin MS (2001). The Decline (and revival?) of black farmers and rural land owners: A review of the research literature. Working paper No. 44 North America series. Madison, Wisconsin: Land Tenure Center, University of Wisconsin-Madison.

Glaeser EL (2001). The formation of social capital. Canadian J. Public Policy, 2: 34-40.

Glaeser EL, Laibson D, Sacerdote B (2002). An economic approach to social capital. Econ. J., 112(483): 437-458.

Green GP, Haines A (2002). Asset building and community development. London: Sage Publications.

Gyawali BR, Fraser RF, Schelhas J, Tadesse W, Wang Y, Bukenya J (2009). Human Well-being and Land Cover Types in the Southeastern U.S.A. Int. J. Ecol. Dev. (IJED), 14 (F09): 81-94.

Gyawali BR, Fraser R., Bukenya J, Schelhas J (2008). Income Convergence in a Rural, Majority African America Region. Rev. Regional Stud., 38(1): 45-65.

Hancock T (2001). People, partnership and human progress: building community capital. Health Promotion Int., 16(3): 275-280.

Harris CC, McLaughlin W, Brown G, Becker DR (2000). Rural

communities in the inland northwest: An assessment of small communities in the interior and upper Columbia River basins. In Interior Columbia basin ecosystem management project: scientific assessment, ed. TM. Quigley, 120. Gen. Tech. Rep. PNW-GTR-477. Portland, OR: USDA/Forest Service, Pacific Northwest Research Station.

Israel GD, Beaulieu LJ (2004). Laying the foundation for employment: The role of social capital in educational achievement. Rev. Regional Stud., 34 (3): 260-287.

Janikas MV, Rey SJ (2005). Spatial clustering, inequality and income convergence. Region Dev. 21: 45-64.

Johnson JH (2002). A Conceptual model for enhancing community competitiveness in the new economy. Urban Affairs Rev., 37(6): 763-779.

Joshi ML, Bliss J, Bailey C, Teeter L, Ward K (2000). Investing in industry, under-investing in human capital: forest-based rural development. In Alabama. Soc. Nat. Resour., 13: 291-319.

Krishna A (2002). Active social capital. tracing the roots of development and democracy. Columbia University Press, New York.

Leica G (2005). ERDAS IMAGINE 8.7. Atlanta Georgia: Leica Geosystems Inc.

Lim U (2003). A spatial analysis of regional income convergence. Plan. Forum, 9: 66-80.

Millennium Ecosystem Assessment (MEA) (2003). Ecosystems and Human Well-Being. World Resources Institute/Island Press, Washington DC.

Mitchell TW (2001). From reconstruction to deconstruction: Undermining black landownership, political independence, and community through partition sales of tenancies in common. Northwestern Univ. Law Rev., 95(2): 505-580.

Myrdal G (1998). An American dilemma, the Negro problems and modern democracy Vol II. Panthoen Books, New York.

Ostrum E (1990). Governing the commons: The evolution institutions for collective action. Cambridge University Press, New York.

Parent FD, Lewis BL (2003). The concept of social exclusion and rural development Southern Rural Sociology, 19: 153-175.

Platinga AJ, Miller DJ (2001). Agriculture land alues and values of rights to future land development. Land Econ., 77(1): 56-67.

Putnam RD (1995). Tuning in, tuning out: the strange disappearance of social capital in Am. Pol. Sci. Pol., 28: 664-683.

Putnam RD (1996). Who killed civic America? Prospect, 7: 24-72.

Putnam RD (2000). Bowling Alone: The collapse and revival of American community. Simon and Schuster, New York.

Rey SJ, Montouri B (1999). US regional income convergence: A Spatial Econometric Perspective. Regional Stud., 33: 143-156.

Reynolds JE (2001). Land use change and competition in the South. J. Agric. Appl. Econ., 33(2): 271-281.

Robinson L, Siles M (1999). Social capital and household income distributions in the United States: 1980, 1990. J. Socio-Econ., 28: 44-93.

Rupasingha A, Goetz SJ (2003). The causes of enduring poverty: an expanded spatial analysis of the structural determinants of poverty in the US. Rural development. paper #22. Pennsylvania, PA: The Northeast Center for Rural Development, Pennsylvania State University.

Rupasingha A, Goetz SJ, Freshwater D (2002). Social and institutional factors as determinants of economic growth: Evidence from the United States Counties. Papers in Regional Sci., 81: 139-155.

Rupasingha A, Goetz, SJ, Freshwater D (2006). The production of social capital in US Counties. J. Socio-Econ., 35: 83-101.

Schelhas J, Zabawa R (2000). Minority and limited resource land owners and forests in the South: Developing a research agenda. Tuskegee University, Al: USDA/Forest Service, Southern Research Station.

Schulman BJ (1991). From Cotton Belt to Sun Belt federal policy, economic. development. and the transformation of the South, 1938-1980. Oxford University Press, London.

Schweninger L (1997). Black property owners in the South, 1790-1915. University of Illinois Press, Urbana, IL.

Svendsen GLH, Sorensen JFL (2007). There's more to the picture than meets the eye: Measuring tangible and intangible capital in two marginal communities in rural Denmark. J. Rural Stud., 23: 453-471.

Terluin IJ (2003). Differences in economic development in rural regions of advanced countries: an overview and critical analysis of theories. J. Rural stud., 19(3): 237-344.

U. S. Department of Commerce (2007). Census of the United States. Bureau of the Census, Washington DC.

United Nations Development Program (UNDP) (2005). Human development report, 2005. Oxford University Press, New York, NY.

Vemuri AW, Costanza R (2006). The Role of human, social, built, and natural capital in explaining life satisfaction at the Country level: toward a national well-being index (NWI). Ecol. Econ., 58: 119-133.

Wear DN, Gries JG (2002). The southern forest assessment summary report. USDA Forest Service, Atlanta, GA.

Woodhouse A (2006). Social Capital and economic development in regional Australia: a case study. J. Rural Stud., 22: 83-94.

Zabawa R (1991). The black farmer and land in South-Central Alabama: strategies to preserve a scarce resource. Hum. Econ., 1: 61-81

7

Risk management strategies of cattle farmers in Namibia: Case study from Omaheke and Otjozondjupa regions

Mogos Y. Teweldemedhin and Lucia Kafidii

Department of Agriculture, Polytechnic of Namibia, Namibia.

A sample survey of 269 questionnaires was administered to commercial and communal cattle farmers in the two regions of Namibia, (Omaheke and Otjozondjupa) to identify factors which affect the adoption of livestock insurance. About 205 questionnaires were collected and the 64 questionnaires were regarded as irregular responses or not returned. A computer software programme was used to generate a logit model. This model was used to test the alternative risk management strategies used by farmers in the two regions. In addition, this research took into account the off-farm investment and farm enterprise diversification. The logit model produced results that are statistically significant and negative estimated coefficient of the household characteristics. This implies that the Namibian livestock industry growth can be achieved with improved education, experience and support from other income as way of diversifying risk strategy. However, the positive relationship of variables FTHEFT and PROD implies the sector is suffering from continuous risk of theft and requires quality production to get market access. This necessitates the need for policy makers and insurers to design programme to educate farmers so that they can adopt proper risk management tools and thereby increase their participation in insurance. The low level of education of many farmers in the study area may have negatively influenced the decision to purchase livestock insurance in addition to other factors.

Key words: Risk management strategies, logit model, Namibia livestock industry

INTRODUCTION

Natural disasters such as floods and bushfires can have a major impact on the management and financial viability of rural properties, as well as major implications for animal welfare. Landholders have a responsibility to ensure that management and property development plans recognise the risks and incorporate the strategies that are necessary to ensure the safety of all persons, livestock and any residential dwellings, and the security of the plants and equipments in the event of such a disaster. The cyclical nature of production is a characteristic of livestock farming which is caused primarily by the climatic conditions such as flooding and drought. Moreover, human-made disaster such as theft and starting of fires are other risks for losing livestock. To avoid such kind of loss of livestock productivity, it is important to increase insurance awareness campaigns among the farmers. Insurance awareness enable farmers to realise the need for livestock insurance, therefore insure their livestock, which can then be considered as a way of risk management and management skills. Risk assessment and risk reduction strategies need to be considered by Namibian livestock farmers regardless of the size of the property. These are the basic elements that assist in developing an awareness of the range of risks and issues (Bowler, 2007). In Namibia, cattle farming are the main agricultural production sector with annual estimated value of N$900 million, and of which about N$400 million is being contributed by exports of weaner. The total number of cattle in 2006 was estimated to be 2.3 million (Meat Board of Namibia, 2007). In this context, the objective of this study is to assess the appropriateness of insurance as a risk management tool for farmers in communal and commercial areas. The

*Corresponding author. E-mail: tmogos@polytechnic.edu.na.

research includes a literature study of the principles, opportunities and problems of risk-sharing strategies in Namibia. Secondly, an empirical study of farmers' perceptions of risk and risk management were carried out by administering questionnaires among a large sample of livestock farmers. Since livestock production is the backbone of the economy of Namibia, it is vital for farmers to gain awareness about the importance of insurance. Therefore, this study makes a contribution for improvement of the know-how of farmers regarding risk assessment in their farming activities. Therefore, the main objectives of this paper are to examine factors that influence livestock insurance adoption by livestock farmers in Namibia's Omaheke and Otjozondjupa regions and draw some policy implications from the results.

METHODOLOGY AND DATA

General

A structured questionnaire was used to identify factors that affect the use of insurance. Semi-structured interviews were conducted with farmers at their respective farms. Information on livestock production characteristics in the smallholder farming systems was obtained from communal and commercial farms in Omaheke and Otjozondjupa. The Omaheke and Otjozondjupa regions are the two major producers of livestock in Namibia. Hanagara (2009) on his study shows that communal and commercial farmer in Omaheke registered with Meat board of Namibia estimated to be 5727 and 390 respectively. Whereas number of farmers at Otjozondjupa were unable to registered, when an inquiry made to the farmer from that area estimation from his knowledge indicated it should be less than Omaheke. A total of 269 questionnaires were distributed to the farmers in these regions. From this 42 of questionnaires were collected from commercial farmers and about 163 were collected from communal farmers. The remaining 64 questionnaires were regarded as irregular response or not returned. With the assumption of Otjozondjupa the same population as Omakeke, the surveyed questionnaire is fairly representative; this accounts for 5 and 1% communal and commercial farmers respectively.

Household characteristics

The following household characteristics were included: sex; age; level of training/education attained; farming experience; and size of the household.

Study area

The study was conducted in Omaheke and Otjozondjupa regions and focussed on both communal and commercial farming systems using 205 respondents, which is fairly representative of the study area. The respondents were chosen for particular purposes on the basis that they were involved in livestock farming in these two regions, and on basis that they were "typical" of a group or represented diverse perspectives of livestock farming (Leedy and Armrod, 2000). District interviews were conducted using students residing in these regions. This proved to be cost effective way of colleting data. The estimated population of farmers was supplied by Agricultural Extension Officers or village heads as the sampling frames.

Review of current literature

Jarvie and Nieuwoudt (1988) define insurance as "... the elimination of the uncertain risk of loss for the individual through the combination of a large number of similarly exposed individuals who each contribute ...premium payments sufficient to make good the loss caused to any one individual." Thus, the idea behind insurance is that of risk pooling, which involves combining the risks faced by a large number of individuals who contribute through premium payments to a common fund that is used to cover the losses incurred by any individual in the pool (Hardaker et al., 1997). Insurance, in general, can provide protection against adverse economic losses experienced by individuals and firms, and caused by natural phenomena such as fire, hail and floods. The decision to buy insurance against risk in agriculture should be an economic one. In making that decision, requires the consideration of the following two critical factors:

1) How much loss can the manager (farmer) withstand without insurance?
2) What are the trade-offs between insurance costs and potential losses (Casavant and Infanger, 1984)?

Therefore, insurance is more attractive to risk-averse farmers and in situations where risks warrant paying a premium significantly higher than the expected loss without insurance (Hardaker et al., 1997). However, in some countries government subsidises premiums, making the purchase of insurance more attractive (Eidman, 1990). There are two basic issues that affect both the insurer and insured. These are asymmetric information and systemic risk. Asymmetric information relates to the problem that the insurer and the insured may not have the same information regarding the probability of losses occurring. The problem could arise due to either adverse selection or moral hazard. Adverse selection occurs if those more at risk purchase more insurance than others, without the insurer being aware. "As a result, the insurer's expected indemnity outlays exceed total premium income, and, in the long run, the insurance operation loses money" (Nieuwoudt, 2000).

RESULTS AND DISCUSSION

Descriptive result analyses

Table 1 shows the diversification of risk, non- farm income and insurance awareness. The research revealed that 95.2% commercial and 98.2% communal farmers in Omaheke and Otjozondjupa regions have no insurance cover for their livestock. They use non- farm income and diversify their farm activities as means of risk management strategy. Furthermore, the study reveals that 71.4% of commercial farmers and 92% of the communal farmers have off farm income to support their farming. From the survey conducted, the study finding implies that farmers in Namibia might not insure their livestock. 26.2% of the commercial farmers revealed that they cannot afford to pay insurance premium, and 27% revealed that they do not see the importance of insuring livestock. For similar research question, results from communal farmers show 17.2 and 82.8%, respectively (Table 2). This indicates that the livestock farmers' awareness to insure is very poor in Omaheke and Otjozondjupa regions of Namibia.

Table 1. Percentage of response to different questions.

	Commercial farmers			Communal farmers		
Description	**Yes**	**No**	**No response**	**Yes**	**No**	**No response**
Do you have other non- farm income?	71.4	23.8	4.8	92	6.7	1.2
Do you think insurance is important?	57.1	42.9		93.2	6.8	
Do you have insurance?		95.2	4.8		98.2	1.8
Do you diversify your risk?	33.3	54.8	11.9			

Table 2. Summary statistics of reasons for not insuring livestock.

		Commercial farmers			Communal farmers		
		Frequency	**Percent**	**Valid percent**	**Frequency**	**Percent**	**Valid percent**
Valid	I could not afford	11	26.2	28.9	28	17.2	17.2
	I do not see the importance	27	64.3	71.1	135	82.8	82.8
	Total	38	90.5	100.0	163	100.0	100.0
Missing	System	4	9.5				
Total		42	100.0				

Table 3. Summary description of the study areas and sample sizes used for the diagnostic study.

		Commercial farmers			Communal farmers		
		Frequency	**Percent**	**Valid percent**	**Frequency**	**Percent**	**Valid percent**
Valid	Male	39	92.9	92.9	127	77.9	77.9
	Female	3	7.1	7.1	36	22.1	22.1
	Total	42	100.0	100.0	163	100.0	100.0

Table 4. Summary description of type of farming.

		Commercial farmers			Communal farmers		
		Frequency	**Percent**	**Valid percent**	**Frequency**	**Percent**	**Valid percent**
Valid	Livestock	29	69.1	76.2	82	50.3	50.3
	Mixed farming	13	30.9	23.8	81	49.7	49.7
	Total	42	100.0	100.0	163	100.0	100.0

Table 3 summarises the gender distribution data in the survey, in both farming sectors, the largest portion of the respondents were male and this shows that there is gender imbalance in this sector. This could imply that government and other non-governmental stakeholders in gender equality need to encourage female farmers' participation.

Table 4 shows that commercial farmers who practise livestock farming, account for about 69% of the respondents. This shows that commercial farmers focus mainly on increasing productivity by specialising on the specific enterprise. On the other hand, communal farmers responded that they are engaged in both livestock and mixed enterprise on equal basis (50:50%). This implies that the communal farmers try to diversify their farming activities. This could be due to poor resource ownership preferred to diversify risk and getting advantages of multiple income generation.

The responses on diversifying enterprise indicates that about 16.7% commercial farmers regard profit as the main motive for farming, 9.5% of farmers indicated that they diversify in order to avoid risk and 4.8% indicated that they diversify to create jobs for their families. On the other hand, communal farmers reveal that about 14% diversify with the expectation of creating jobs, 12.3% to avoid risk and 11.7% for profit reasons (Table 5).

Tables 6 and 7 illustrate the perception of insurance by communal and commercial farmers. The research found that the livestock farmers' perception to insurance is good

Table 5. Summary of reasons for diversifying farming enterprises.

		Commercial farmers			Communal farmers		
		Frequency	Percent	Valid Percent	Frequency	Percent	Valid Percent
Valid	To get better profit	7	16.7	53.8	19	11.7	23.5
	To avoid risk	4	9.5	30.8	20	12.3	24.7
	To create a job	2	4.8	15.4	23	14.1	28.4
	Total	13	31.0	100.0	19	11.7	23.5
Missing	System	29	69.0		81	49.7	100.0
Total		42	100.0		82	50.3	
					163	100.0	

Table 6. Summary result of importance of livestock insurance.

		Commercial farmers		Communal farmers	
		Frequency	Percent	Frequency	Percent
Valid	Yes	24	57.1	150	92.0
	No	18	42.9	11	6.7
	Total	42	100.0	161	98.8
Missing	System			2	1.2
Total				163	100.0

Table 7. Summary result of degree livestock insurance importance.

		Commercial farmers		Communal farmers	
		Frequency	Percent	Frequency	Percent
Valid	Very important	16	38.1	139	85.3
	Important	7	16.7	11	6.7
	Not important	1	2.4	3	1.8
	Total	24	57.1	153	93.9
Missing	System	18	42.9	10	6.1
Total		42	100.0	163	100.0

good. These accounts for 92 and 57% of communal and commercial farmers, respectively (Table 6). When the respondents were asked how important insurance is to their livestock, 85% of communal farmers and 38% commercial farmers said it was very important (Table 6). This implies that the insurance companies in Namibia did not give much weight to advocate livestock insurance; this might be due to the nature of the farms or complicity of the enterprise to handle the sector in the insurance system.

Model

In explaining a model the dependent variable (Y_i), (where one represents the farm enterprise diversified as risk minimisation and zero represents not diversified), different regression methods can be used. Some of the methods would make use of discriminant analysis, linear probability model, logit and probit. In this study the following independent variables were considered:

Gender of the head of household; age of the head of household; year of farming experience; education level; family size; off farm income; awareness to insurance; frequency of theft; safety; productivity; frequency of sales; value of sale; credit, character of a person to risk (captured by asking his/her prescription in starting new technique).

This study follows the general modelling of Mohammed and Ortmann (2005) to test the relationship between dependent and exploratory variables. The model is as follows:

$$\ln(pi/(1\text{-}pi)) = \beta o + \beta_1 Gender + \beta_2 Age + \beta_3 EXPLVi + \beta_4 EDUi + \beta_5 FSIZE + \beta_6 OFFi + \beta_7 INFOi + \beta_8 FTHIEFT + \beta_9 SAFETY + \beta_{10} PROD + \beta_{11} FSale + \beta_{12} Vsales + \beta_{13} CREDIT + \beta_{14} Ntech + \beta_{15} Dummy$$

Where:

Table 8. Variables that influence the adoption of risk management and expected sign.

Variables	Description	Expected sign
Gender	Gender of the household,	+/-
Age	Age of the household	+/-
EXPLV	Year of farming experience	+/-
EDU	Education level head of the household	+/-
FSIZE	Number of people in the house	+/-
OINCOME	Off farm income outside the farm	-
INFO	Awareness to insurance	+
FTHIEFT	Frequency of theft	+
SAFETY	A need to increase safety of production	-
PROD	A need to increase productivity	-
FSALE	Frequency of sales	-
Vsales	Value of sales last season	-
CREDIT	Amount of loan outstanding	-
Ntech	Character of a person to risk	-
Dummy	Dummy variable for type of farming, 1 for commercial farmer, otherwise zero	+/-

(pi/1-pi) is the probability of insurance awareness, gender of the household, age of the household, EXPi is the farming experience, EDU is the educational level, Fsize is the number of family members in the house, OFFi is the off farm income, INFOi is the awareness of insurance, Ftheft is the frequency of theft, SAFETY is a need for safe production, PROD is a need to increase productivity, Fsale is frequency of sales, Vsale is a value of sales last season, CREDIT is the amount of an outstanding loan.

Consideration of the model variables

As indicated earlier, livestock insurance adoption in Omaheke and Otjozondjupa regions of Namibia is not well established. This may be due to a number of factors, such as lack of information on the livestock insurance, lack of awareness of the insurance scheme, low level of educational attainment among farmers, poor rural infrastructure (making communication difficult and limiting access to insurance), affordability, degree of farmers' risk aversion and diversification of farm enterprise. These could be among the factors that affect the insurance of livestock. The definitions of the most important variables expected to influence the adoption of livestock insurance are presented in Table 8.

The expected sign for risk management between male and female farmers could be positive or negative; male farmers are most likely to be risk takers and expected to invest in one enterprise for higher return. On the other hand, since men have relative higher power and time to spend at the farm, they could do mixed farming to keep themselves busy at the farm. Age and experience holds similar impact to the adoption of risk management.

The expected sign for educational qualification could be positive or negative. Education may promote an understanding of the effects of risk and hence may increase the demand for diversification of the enterprise as risk management. On the other hand, increasing education levels are associated with an increase in transferable human capital, facilitating greater risk taking by individuals with lower risk aversion (Szipiro and Outreville, 2003 cited in Mohammed and Ortmann, 2005). Family size was expected to have a positive effect as the numbers of family members were dependent on the farm. As farmers participate in off-farm investments this will be take as a risk management strategy, the probability of diversifying enterprises decrease, either due to other secured income or they do not have time to involve in additional farm enterprises. Therefore, the expected sign for off-farm income will be negative. As farmers' attitude towards insurance increases, it is most likely that farmers will see need to subscribe to insurance. Further, as the frequency of theft increases, it is necessary for farmers to diversify farming to avoid potential losses.

Although diversification is an alternative risk management strategy, it does not necessarily mean that diversification and insurance always can closely substitute each other (Blank and McDonald, 1996).

LOGIT MODEL RESULTS

A chi-square test was used to test the equality of the standard deviation of a population to a specified value. This test can be either a two-sided test or a one-sided test, the two-sided version tests against the alternative that the true standard deviation is either less than or greater than the specified value. The one-sided version only tests in one direction. The choice of a two-sided or one-sided test is determined by the problem (Snedecor

Table 9. Logit model results for livestock risk management adoption in Namibia (n = 200).

Variables	Coefficient estimate	Standard error	Wald	t-stat	Significance
Gender	7.667	3.808	4.073	0.162**	0.0436
Age	-0.0654	0.0291	5.074	-0.197**	0.0243
EXPLV	-0.0212	0.0536	0.157	0.692***	0.0001
EDU	-1.003	0.598	2.809	-0.101*	0.093
FSIZE	0.594	0.268	4.931	0.192**	0.0264
OINCOME	3.77	1.511	6.22	0.231**	0.0126
INFO	0.883	1.58	0.312	0.001*	0.5763
FTHIEFT	1.338	0.838	2.553	0.083*	0.1001
SAFETY	0.0314	0.893	0.0012	0.0001	0.971
PROD	4.015	1.929	4.33	0.172**	0.0374
FSALE	-3.792	1.485	6.519	-0.239*	0.0107
Vsales	0.0005	0.0003	3.469	0.1358*	0.0629
CREDIT	-4.256	1.74	5.989	-0.224***	0.0629
Ntech	-1.725	1.409	1.498	-0.001	0.2209
Dummy	-12.833	6.72	3.65	-0.1442*	0.0562
Model chi-square		36.49*** on 20 degree of freedom			
Correct prediction (%)					
Overall			93		
Diversification			60		
Non diversified			98		

Note: *, **, and *** indicates statistical significance at 10, 5 and 1% respectively.

and Cochran, 1983).

The chi-square, which tests the joint significant of the explanatory variables in this study, is statistically significant at the level 5% (Table 9). The estimated model correctly classified 93% of the respondents. The success rates for predication of adoption of diversification and non diversified enterprises are 60 and 98%, respectively.

On the basis of results obtained and shown in Table 9, the techniques described in the methodology section were applied and factors affecting diversification as a means of risk management are reported in this section. Variables NSTOCK, EXPLV, SAFETY, HOWIMP and NTECH are not significant (Table 9), all other variables were found to be statistically significant at the specified level of significance.

The positive estimated coefficient sign to the variables (Dummy, GENDER, PROD, OINCOME, VSALE, FTHEFT, and WSALE) indicate that the greater the values of these variables, the higher the tendency for farmers to diversify their enterprises. The negative sign for the remaining variables indicate the opposite implication of the previous explanation, that is, the greater the value of these variables, the lower the probability for diversified enterprise.

The positive sign of the FSIZE was as hypothesised. It indicates that family members are dependents on the farm. Responsibility and creativity increases as the farmer wants either to avoid risk or to obtain better income for the family. This makes it necessary for the family to diversify its enterprise, especially communal farmers.

The formal education level (EDU) has a negative coefficient estimate indicating that, *ceteris paribus*, the probability of diversifying risk deceases as the level of formal education of the farmers' increases; this implies that farmers preferred to specialise on the specific enterprise to maximise output. Bullock et al. (1994) and Vandeveer (2001) found education was negatively related to a farmer's willingness to take risk. However, Mohammed and Ortmann (2005) found that education was positively related to farmers' willingness to take risk.

The gender of the farmer was found to be positive and significant at 5% level, whereas, age was found to be negative and significant. This implies that female farmers are risk averse. Age indicates that specialisation as the farmer matures is most likely to be more commercialised. The negative estimate coefficient for age implied the decision of diversification. It appears, therefore, that older and more experienced farmers are less willing to diversify their enterprise. Farmers with such characteristics might have acquired enough knowledge through time to deal with income and risk without diversification. Results of the studies by Jarvie and Nieuwoundt (1988) and Vandeveer

(2001), however, indicate that younger farmers, or those with less experience, were less likely to diversify their enterprise.

Off-farm income (OINCOME) has a negative coefficient estimate implying that the more farmers engage in off-farm activities the less probability to diversify the enterprise. This may be due to the fact that time constraints to allow stretching to many enterprises. The off- farm income is also a good cash injection to the livestock farming especially for the emerging commercial farmers. They support their farms through their other income outside farm. The off-farm income helps many farm households as a means of diversification of risk. This negative estimate coefficient for this variable implies that farmers engage in off farm activities.

FSALES shows significant and negative estimated coefficient. This implies that there is a high frequency of sales in the livestock. This is a good proxy for good market opportunity for the livestock farmers. On the other hand, frequency of theft (FTHEFT) shows positive estimated coefficient. This implies that as the risk of theft gets high, farmers prefer to diversify. CREDIT indicates that negative and significant values show that farmers focus on one enterprise to get more return in order to increase the payment/debt commitment to the debtors.

Lastly, the puzzling result of the estimated coefficients on PROD and VSALE, shows significant levels, and are positive to influence diversification. Perhaps insured and non-insured category data were not found for this study. This might jeopardize the result of this variable, or this might suggest that a specific study is needed on the impact of productivity and value of livestock relationship to risk management strategy to obtain a more reasonable estimate.

CONCLUSIONS AND POLICY IMPLICATIONS

It was very surprising that none of the surveyed livestock farmers insured their livestock. As indicated in Table 1 farmers were aware of the importance of insurance. However, when they were asked why didn't they insure their livestock, 64.3% of commercial farmers and 82.8% communal farmers said, they did not see the importance of insuring. About 26.27% of commercial farmers and 17.2% communal farmers said that the insurance premium was unaffordable. This implies that the insurers need to increase campaign on increasing awareness on the role of livestock insurance to farmers. It requires the combined efforts of all stakeholders namely, governments, civic society organisations and the private sector. Furthermore, it is important to ask question; how can livestock farming in Namibia make a significant contribution to the viability of farming systems in the country to achieve the Millennium Development Goals without considering how to minimise risk in the farming? Is it impossible to support the farming industry? Therefore, it is important for good risk assessment and risk reduction strategies to be considered by Namibian livestock farmers in Omaheke and Otjozondjupa regions, regardless of the size of the property.

The statistically significant and negative estimated coefficient of the household characteristics at the specified significance level (such as family size, educational qualification, age, and off farm-income) implies that Namibian livestock industry growth was achieved with improved education, experience and support from other income as a strategy for diversifying risk. However, the positive relationship of FTHEFT and PROD implies that the sector is suffering from continuous risk of theft and requires quality production to get market access. Therefore, farmers need to follow effective risk management systems to achieve the required profit rate.

The negative statistical estimation of education attainment plays a role in creating insurance awareness and implies that farmers prefer to diversify their enterprise as a risk management tool, rather than buying insurance. On the other hand, the positive correlation of information toward insurance (which is captured by asking "How is insurance important to you?") among the livestock farmers in Omaheke and Otjozondjupa regions. The answer to this question reveals that there is a need for policy makers and insurers to design a programme to better educate farmers so that they can assess risk management tools and thereby increase their participation in insurance. The low level of education of many farmers in the study area may have negatively influenced the decision to purchase livestock insurance in addition to other factors such as not knowing about the provision of insurances, the premium being too expensive and so forth. The insurance companies in Namibia should intensify their advertising efforts and inform farmers of the importance of insurance in the industry. Currently, livestock farmers seem to follow diversification activities as alternative risk management strategies.

While the results of this study are encouraging, there remains considerable scope for refining and deepening the research.

REFERENCES

Blank SC, Mcdonald J (1996). Preferences for crop insurance when farmers are diversified. Agribusiness. 12(6): 583-592.

Bowler J (2007). Risk management for stock owners in times of fire and flood. http://www.dpi.nsw.gov.au/__data/assets/pdf_file/0004/104278/risk-management-for-stock-owners-in-times-of-fire-and-flood.pdf date accessed on May 20, 2009.

Casavant KL, Infanger CL (1984). Economics and Agricultural Management. An Introduction. A Prentice-Hall Company, Reston, Virginia, USA. pp. 144-147.

Eidman VT (1990). Quantifying and managing risk in agriculture. Agrekon. 29(1): 11-23.

Hardaker JB, Huirne RBM, Anderson JR (1997). Coping with risk in Agriculture. CAB International, Wallingford, Oxon, UK. pp. 106-111.

Jarvie EM, Nieuwoudt WL, (1988). Factors influencing crop insurance participation in maize farming. Agrekon. 28(2): 11-16.

Leedy P, Armrod JE (2000). Research Planning and Design. Merrill Prentiss Hall, USA. Pp. 49-50.

Meat Board of Namibia (2007). Cattle review: production and marketing

trends. http://www.nammic.com.na date accessed April 10, 2009.
Mohammed MA, Ortmann GF (2005). Factors influencing adoption of livestock Insurance by commercial dairy farmers in Three Zobatat of Eritrea. Agrekon. 44(2): 172-186.
Nieuwoudt WL (2000). An economic evaluation of a crop insurance programme for small-scale commercial farmers in South Africa. Agrekon. 39(3):269-291.
Snedecor S, George W, William G (1989). Statistical Methods, Eighth Edition, Iowa State University Press. pp. 89-93.
Vandeveer ML (2001). Demand for area crop insurance among litchi producers in northern Vietnam. Agric. Econs., 26(2):173-184.

Validation of socio-economic status indicators for heads of farm families in Delta North agricultural zone, Delta State, Nigeria

B. O. Ovwigho

Department of Agricultural Economics and Extension, Delta State University, Asaba Campus, Asaba, Nigeria. E-mail: bishopvwigho@yahoo.com.

The specific objectives were to collate a universe of socio-economic status indicators; carry out item analysis to validate socio-economic status indicators; select valid socio-economic status indicators for heads of farm families; and discuss the sociological implication of the valid items. Stratified and simple random sampling designs were used in selecting the towns and villages. Twelve percent of heads of farm families in selected Ibo speaking Local Government Areas of the state comprising Aniocha South (41), Ika South (47) Ndokwa West (55) and Oshimili North (31) were selected to give a sample size of 174 heads of farm families. Structured interview schedules were used in data collection. Data were analysed using point-biserial correlation coefficient and t-test. A universe of 106 items were collated from a pre-research survey. Sixty nine (69) items were found as valid indicators of socio-economic status. The number of relatives trained by the individual, axe, yam barn, motor cycle, motor car, bicycles and water cistern toilet had very important social and economic implications among the people. The indexes are recommended to development agencies for evaluating changes in socio-economic status of heads of farm families particularly after an agricultural development intervention programme.

Key words: Indexes, item analysis, valid items, socio-economic status indicators, farm families.

INTRODUCTION

Item analysis is an empirical procedure for selecting valid items which could be used in construction of socio-economic status scale as well as test and non test instrument. It involves selecting valid items from a universe of items based on a pre-determined criterion. Ebel (1972) and Gronlund (1976) stated that item analysis was a procedure used to find out items that might be too easy or difficult for the purpose of test construction. The process helps to identify items which failed to discriminate between the better and poor examinees. Several items analysis procedures have been used in the social and behavioural sciences. Grounlund (1976), Ebel (1972) and Ali et al. (1988) described the discrimination index procedure for selecting objective test items. The procedures are as follows: (1) Arrange the test scores or answer sheets in an order ranging from high to low; (2) Separate two groups of test papers, an upper consisting of approximately 27% of the total group who receive highest scores on the test and a lower group consisting of an equal paper from those who receive lowest scores; (3) Count the number of times each possible response to each item was chosen by the upper group. Do the same separately for the papers of the lower group; (4) Record the response counts opposite the responses they refer to on a copy of the test.; (5) Add the counts from the upper and lower groups to the keyed correct response; (6) Subtract this sum from the maximum possible sum, that is, the sum of the number of papers in upper and lower groups and divide the differences by the maximum possible. Express this as a percentage, that is, multiply the decimal fraction by one hundred (100), the result is an item difficulty; (7) Subtract the lower group count of correct responses from the upper group count of correct responses. Divide this difference by the maximum possible difference that is the number of papers in the upper (or lower) groups. This quotient expressed as a decimal fraction is called index of discrimination.

This could be expressed mathematically as follows (Gronlund, 1976);

$$DI = \frac{RU - RL}{½T}$$

Where:

DI = Discrimination index
RU = Number of persons in the upper group who got the item right.
RL = Number of persons in the lower group who got the item right
T = Total number of persons in both upper and lower group.

The above procedure showed that the simple discrimination index procedure did not take into account the performance of an individual on total items in the test. An item analysis procedure such as the point-biserial correlation coefficient takes into cognizance the total test scores and an item response. Karabinus (1975), Henrysson (1976) and Kraemer (1981) remarked that the point-biserial correlation coefficient was used when an association existed between random variables X and Y, where X is a dichotomy and Y is measured on a continuum. The procedure involved simple tabular presentation of each test item to give the following information: (1) Number choosing the correct answer; (2) Number choosing the wrong answer; (3) Frequency distribution of criterion scores for the group choosing either of the options. The resulting Table would show that students with high criterion scores have a much stronger tendency to choose the right answer. They maintained that the relationship could be quantified by the use of the point-biserial correlation coefficient. Other statistical tools which could be used for item analysis included the tetrachoric correlation coefficient, t-test, Pearson r, multiple correlation and contingency Table. Divgi (1979) and Zalinski et al. (1979) explained that the tetrachoric correlation generally denoted by "rt", was used to measure the degree of linear association between two variables X and Y where both were dichotomized and the true underlying distribution were assumed to be normal. If the underlying psychological traits of the dichotomous variables were normally distributed then the tetrachoric correlation coefficient is the product-moment correlation between the traits. It was difficult and time consuming because the formula was based on series of approximation. Akinola and Patel (1987) made use of the t-test in validating socio-economic status indicators. Total score on the scale was calculated by adding 2 or 1 for possession and non-possession, respectively for 75 items. The scores were ranked from the highest to the lowest. The highest and lowest 25% of the farmers were then compared to see if there existed any significant difference in the proportion of farmers possessing the item at 0.01 level of significance. Gullickson and Hopkins (1976) maintained that the Pearson r was an extremely useful means of estimating the discrimination index of linear relationship between two variables for a given population. The Pearson r is a consistent and efficient estimator of p only under the conditions that r was obtained from a random sample of the population. Anderson et al. (1976) found that it was possible to relate one characteristic at the same time to more than two through multiple correlation coefficients. By this definition, the multiple correlation was the product – moment correlation between the predicted independent and best linear composite of the predictor variables. Most researchers especially in developing countries find it difficult to ascertain items that should be used to measure socio-economic status of farmers. Bills et al. (1985) observed that empirical research on the process of social stratification in developing nations has been hampered by the lack of adequate instrumentation and measurement procedures. This was not the case in industrialised nations where such instruments have been developed to a substantial level. This study, therefore, attempt at developing indexes of socio-economic status indicators for heads of farm families in the zone. The objectives of the study were to collate a universe of socio-economic status indicators; carry out item analysis to validate socio-economic status indicators; select valid socio-economic status indicators for heads of farm families; and examine the social and economic importance of some valid socio-economic status indicators.

MATERIALS AND METHODS

Sampling procedure and sample size

Stratified and simple random sampling designs done on multi stage basis were used in selecting the respondents as well as the towns and villages. Delta North agricultural zone is inhabited by the Ibo speaking people. The 9 local government areas in Delta North Agricultural Zone were grouped into 4 on the basis of the old local government map and sub-ethnic groups. The selection was done in three stages. In the first stage, one local government area namely Aniocha South, Ika South, Ndokwa West and Oshimili North were randomly selected from each group. In the second stage, 40% of the towns Aniocha South (6), Ika South (6), Ndokwa west (6) and Oshimili North (4) were randomly selected. The third stage involved random selection of 12% of the heads of farm families and this corresponded to Aniocha South (41), Ika South (47), Ndokwa West (55) and Oshimili North (31). This gave a sample size of 174 respondents.

Data collection

Oral interview and structured interview schedule were used in data collection. The interview schedule contained items which enhance individual's socio-economic status in the agricultural zone. This was done by asking a few farmers in each of the local government areas to mention items that could enhance the socio-economic status of farm families. One hundred and twelve items were collated from the pre-research survey. These items were built into an interview

Table 1. Evaluation of discrimination indexes after item analysis.

Index of discrimination	Item evaluation
0.40 and above	Very good item, accept
0.30 to 0.39	Reasonably good but subject to possible improvement
0.20 to 0.29	Marginal items, usually need and subject to improvement
Below 0.19	Poor items to be rejected or improved by revision

Source: Ebel (1972). Essentials of educational measurement.

schedule.

The respondents were required to indicate possession, non-possession or tick the number of items possessed in the case of quantitative items. The instrument was reliable. The reliability test showed a coefficient of (r = 0.88). Oral interview was used to elicit information related sociological implication of valid items.

Data analyses

A uniform scoring method which assigned a value of 1 for possession and 0 for non-possession was used for the measurement of dichotomous items. For quantitatively measured items, possession scores ranged from 0 to 6 depending on the number of items listed against the particular indicator.

Gronlund (1976) suggested that items with average discriminating power usually fall around 0.58. Ebel (1972) stated various values of discriminating indexes and how to evaluate them (Table 1).

A score was obtained for each individual following the above scoring procedure. The scores for all the respondents were arranged from low to high in order to form the criterion scores and the total for each criterion score were tabulated. The possession scores of the upper and lower 25% were compared by the use of t-test at 0.01 level of significance for quantitative items (Table 2). Significant items were selected as valid. The point-biserial correlation coefficient was used in the item analysis of the dichotomous items. The criterion scores, number of possession, number of non-possession and total for each criterion score were tabulated. Items with rpbis 0.55 and above were selected as valid items (Table 3).

RESULTS AND DISCUSSION

Indexes of valid socio-economic status indicators in the Ibo ethnic group of Delta State Indexes of valid socio-economic status indicators in the Ibo ethnic group of Delta State is given in Table 4. From the universe of socio-economic status indicators, 112 items were collated from the pre-research survey. These items were pre-tested and ambiguous and localised items were removed. The items dropped were elephant tusk, mud house, pit toilet, granaries/silos, outside broom and cooking retort stand. These items were either possessed by very few persons or a particular socio-economic status group. After the pre-test, the universe of the socio-economic status indicators became 106 items. The pre-test was used to show items which did not discriminate or have zero discrimination.

Valid socio-economic status indicators

The results of the item analysis showed that 69 items were valid indicators of socio-economic status in the ethnic group (Table 2). The items were children in primary school, secondary school, tertiary school, number of relatives trained by the sponsor, cement house in the village, cement house outside the village, traditional hats, traditional attires, pair of shoes, George wrappers, single wrappers, room with cemented floor, chieftaincy title, cutlass, spades/shovel, water cistern toilet, wash hand basins, cabinets beds, wall hanger, framed photograph of oneself, axe, farm size, poultry, fish ponds, goats, hired labourers, yam barn, plots of land owned in the village, plots of land owned outside the village, personal borehole, motor cycle, motor car, compact disc player, radio/cassette player, television, ceiling/table fans, executive chairs, stove, GSM handset, personal generator, wheelbarrow, floor carpet, floor rug, wardrobe, rain coat, umbrella, book shelve, dinning table, metal buckets, plastic buckets, blender, frying pan, tumblers, kettle, bicycle, electric/coal iron, metal, spoons, suitcases/traveling bags, wash hand basins, glass plates, wrist watch, ability to read in English, ability to read native dialect, membership of social clubs, official in a Christian organization and membership of cooperative societies. These items could be standardized into a socio-economic status scale to measure the socio-economic status of heads of farm families in the Ibo ethnic group of Delta State, Nigeria.

Social and economic implication of some valid items

The valid items were cherished by the people socially, culturally and economically. A few of the items namely cutlasses, axes, yam barn, motor car, bicycle and number of relatives trained by a farmer have far reaching social and economic implication in the study area.

Cutlasses and axes

These two items were valid because they were the major tools used for farm labour. Beside the economic use, the

Table 2. Item analysis procedure for quantitatively measured item using t-test

	Item: Number of children in primary school						
Criterion scores	**0**	**1**	**2**	**3**	**4**	**>4**	**Total**
186						1	1
176				1		2	3
171				1		1	2
166		1				2	3
161				1			1
156				1			1
151			1				1
146						3	3
141						1	1
131			1				1
126	1		2				3
121			5		1		6
116			6	2	2		10
111			4	1			5
106		1	2			Upper (3) 25%	
101		1	5	1			7
96			8				8
91		1	3	3	1		8
86			2	2			4
81			1	1	1		3
76			2	1	1		4
71		1		2			3
61		1	1	1	1		4
56	1		2		1		4
51		1	2	1	1		5
46	1	10	8	1	2		22
41	3	3	3	1	1		11
36	1	2	6			Lower (9) 25%	
31	5	10	1	1			17
26	4	13	3				20

Scores of upper 25% group who Possess the item = 5,5,5,5,5,5,5,5,5,5,4,4,4,3,3,3,3,3,3,3,2,2,2,2,2,2,2,2,2,2,2,2,2,2,2,2,2,2,2,2,2,2,1. Where N = 43, x = 2.98.
Scores of lower 25% group who possess the term = 1,1,1,1,1,1,1,1,1,1,1,1,1,1,1,1,1,1,1,1,1,1,1,1,1,1,2,2,2,2,2,2,2,2,2,2,3. Where N = 35, x = 1.34.

$$t\text{ - ratio} = \frac{2.98 - 1.34}{0.2280} = 7.19$$

But from Table 2, t = 2.660 at df= 76, p = 0.01. Decision = item is valid in the Ibo ethnic group since the observed t is higher than critical t.

axe and cutlass were used during the Ikenga festival in Delta Ibo area of Delta State. Men come out to brandish cutlass ostensibly to demonstrate their physical prowess during the Ikenga festival which precedes the planting season around February/ March.

Yam barn

The yam barn is a local storage structure for yam. The Ibo ethnic group produces and consumes a lot of yam compared to other ethnic groups in Delta State. The number of yam barn is often used to assess the socio-economic status and performance of a farmer in the preceding farming year. The new yam festival is performed by the people in September every year before the sales and consumption of the new crop. Igbokwe (2005) stated that no member of the Ibo community was allowed to consume and plant yams until the new yam festival and rituals for planting yams were performed.

Table 3. Item analysis procedure for dichotomous item using point-biserial correlation.

	Item: Ownership of cement house(s) in the village		
Criterion scores	**Yes (possession)**	**No (non-possession)**	**Total**
186	1		1
176	3		3
171	3		3
166	2	1	3
161	11		11
156	1		1
151	3		3
146	1		1
141	1		1
131	3		3
126	6		6
121	8		8
116	3	2	5
111	1	2	3
106	5	2	7
96	6	2	8
91	5	2	7
86	2	3	5
81	1	2	4
76	1	2	3
71	1	2	3
61		2	2
56	1	4	5
51	1	3	4
46	1	4	5
41		21	21
36		11	11
31		9	9
26		17	17
21		20	2
Total	(62)	(112)	

$$rpbis = \frac{MP - MN}{St} \cdot \sqrt{P(1-P)}$$

Where:

rpbis = symbol for point- biserial correlation
MP = Mean score for heads of farm families who possessed the item was obtained by multiplying the scores on the yes column by the corresponding criterion score divided by N
MN = Mean criterion score for heads of farm families who did not possess the item was obtained by multiplying the scores on the no column by the corresponding criterion score divided by N.
ST = Standard deviation of the criterion scores
P = Proportion of heads of farm families who possessed the item
(Source: Adapted from Henrysson (1976).

$St = 43.42$, $MP = 117.69$, $MN = 50.06$, $P = {}^{62}/_{174} = 0.36$, $1 - p = 0.64$

$$rpbis = \frac{117.69 - 50.06}{43.42}\sqrt{0.36(0.64)}$$

$= 1.56 \times 0.48 = 0.75$

Decision = the item is valid since the calculated rpbis is above 0.55

Table 4. Indexes of valid socio-economic status indicators in the Ibo ethnic group of Delta State.

	Item	Statistical tool	Discrimination index
1.	Children in primary school	t	7.19
2.	Children in higher school	t	3.53
3.	Children in secondary school	t	2.73
4.	Number of relatives trained by you up to secondary school	t	2.79
5.	Ownership of cement house in the village	rpbis	0.75
6.	Ownership of cement house outside the village	rpbis	0.65
7.	Traditional hats	t	5.54
8.	Traditional attires	t	11.70
9.	Pair of shoes	t	5.23
10.	George wrappers	t	6.63
11.	Single wrappers	t	2.72
12.	Rooms with cemented floor	t	29.56
13.	Chieftaincy title	rpbis	0.56
14.	Cutlasses	t	6.80
15.	Spade/shovel	t	6.80
16.	Water cistern toilet	rpbis	0.64
17.	Wash hand basins	rpbis	0.70
18.	Cabinet beds	t	5.18
19.	Wall hanger	rpbis	0.65
20.	Framed photographs of yourself	t	18.86
21.	Axe	rpbis	0.64
22.	Farm size	t	7.94
23.	Poultry	t	4.48
24.	Fish ponds	t	5.55
25.	Goats	t	5.36
26.	Hired labourers	rpbis	0.69
27.	Yam barn	"	0.55
28.	Plots of land owned in the village	t	6.64
29.	Plots of land owned outside the village	rpbis	0.62
30.	Personal bore- hole	"	0.74
31.	Motor cycle	"	0.60
32.	Motor cars	"	0.73
33.	Turn table/speakers	"	0.70
34.	CD Player	"	0.57
35.	Television	"	0.65
36.	Ceiling / Table fans	"	5.89
37.	Executive chairs	"	0.74
38.	Lantern	t	14.25
39.	Stove	rpbis	0.55
40.	Personal generator	"	0.59
41.	Wheel barrow	"	0.75
42.	Floor carpet	"	0.61
43.	Rug	"	0.64
44.	Wardrobe	"	0.58
45.	Rain coat	"	0.63
46.	Umbrella	t	6,53
47.	Book shelves	rpbis	0.73
48.	Refrigerator	"	0.58
49.	Standing mirror	"	0.71
50.	Dining table	"	0.78
51.	Metal buckets	t	10.67

Table 4. Contd.

	Item	Statistical tool	Discrimination index
52.	Plastic buckets	t	11.92
53.	Electric blender	rpbis	0.68
54.	Frying pan	"	0.72
55.	Tumblers	t	16.08
56.	Kettle	"	6.94
57.	Bicycles	rpbis	0.69
58	Electric/coal Iron	"	0.76
59	Metal spoons	t	21.50
60.	Suitcase/travelling bags	t	3.15
61.	GSM handset	rpbis	0.53
62.	Glass plates	t	8.93
63.	Wrist watch	rpbis	0.69
64.	Can you read English	rpbis	0.55
65.	Can you write English	"	0.55
66.	Can you read your native dialect	"	0.67
67.	Membership of social clubs	"	0.61
68.	Official in a Christian organisation	"	0.59
69.	Membership of co-operative societies	t	9.73

Motor cycle, cars and bicycles

These three items were found valid in the Ibo ethnic group because they were the most popular means of transporting farm produce. Women and girls were often seen in this areas riding motor cycles and bicycles to farm. It was favoured means of transportation in the area because of its land-locked nature.

Number of relatives educated or trained by the farmer

The number of relatives trained by a farmer was a measure of socio-economic status and altruism in the Ibo ethnic group of Delta State. If an individual was wealthy and his relatives were not educated he is not usually accorded a pride of place in the Ibo speaking society of Delta State, In some cases, the individual might be given a chieftaincy title if he is able to train many of his relatives. In essence such an individual is respected for contributing his quota to the preservation of the African extended family system.

Conclusion

The empirical indicants for the measurement and construction of socio-economic status scale for farm families in the Ibo ethnic group of Delta State were established from the study. The 69 valid items play significant role in socio-economic lives of the farm families. Agricultural production like every other business venture is geared towards increase standard of living.

Indexes of socio-economic status provide the basis for evaluating changes in socio-economic of farm families. On the whole, the use of the indexes generated in this study might reduce the arbitrary assignment of socio-economic status to rural farmers in the study areas.

REFERENCES

Akinola CA, Patel AU (1987). "Construction and Stardardisation of a Scale of Measure Socio-economic Status of Heads of Rural Household (Gandu) in the Funtua Zone of Kaduna State, Nigeria" J. Rural Dev. 10: 85-94.

Ali A, Ezeadi S, Ogbazi JN (1988). Introduction to Educational Measurement. Awka(Nigeria) : Meks Unique Publishers Pp. 17-19.

Anderson SB, Ball S, Murphy RT (1976). Encyclopedia of Educational Evaluation. San Francisco : Jossey – Bass publishers pp.32-379.

Bills DB, Godfrey DS, Haller AO (1985). "A Scale to Measure Socio-economic Status of Occupations in Brazil." Rural Socio. 50(2): 225 – 250.

Divgi DR (1979). "Calculation of the Tetrachoric Correlation Coefficient" Psychometrica 44(2): 169 – 172.

Ebel RL (1972). Essentials of Educational Measurement. London: Prentice Hall. Pp. 361-435.

Gronlund NE (1976). Measurement and Evaluation in Teaching . New York: Macmillan Publication co., Inc. Extension in Nigeria. Ilorin: Agricultural Extension Society of Nigeria. pp. 265-278.

Gullickson A, Hopkins K (1976). "Internal Estimation of Correlation Coefficients Corrected for Restriction of Range." J. Educ. Psychol. Meas., 36: 9-25.

Henrysson S (1976). Gathering, Analyzing and Using Data on Test Items". In: Thornike RD (ed). Educational Measurement. Washington DC : American Council on Education. pp.130-139.

Igbokwe EM (2005). Concepts in Rural and Agricultural Sociology' 'In: Adedoyin, S.F.(ed). Agricultural . pp. 91-100

Karabinus RA (1975.) "The r –Point Biserial Limitation". J. Educ. Psychol. Meas., 35.277 – 282.

Kraemer HC (1981). "Modified Biserial Correlation Coefficient". Pyschometrica. 46 (3):275.

Zalinski J, Abraham NM, Alf E (1979). "Computing Tables for the

Tetrachoric Correlation Coefficient." J. Educ. Psychol. Meas., 39:267 – 268.

Scope of ericulture in Assam: A micro-econometric analysis

Utpal Kumar De[1]* and Manjit Das[2]

[1]Department of Economics, North-Eastern Hill University, Shillong, 793022, Meghalaya, India.
[2]Department of Economics, M. C. College, Barpeta, 781301, Assam, India.

The paper analyses the scope of ericulture in Assam, India by theoretically and empirically examining its profitability and the excess capacity lies at the family farm level. Ericulture is a traditional agro-based sericulture activity that has been playing a significant role in generating employment and income in a slowly progressing rural economy of Assam. Among all the major silk varieties, it occupies the prime position in the sericulture sector of Assam. Though it generates substantial income and employment each year especially for the rural poor masses, its growth has not been noticeably high as the potential. The analysis proved the existence of substantial scope for enhancing the ericulture activities even at the existing level of technology and with little effort in the food leaves production and financial assistance, it may be expanded significantly that would generate further income and employment in rural Assam.

Key words: Sericulture, ericulture, informal sector, excess capacity, Assam.

INTRODUCTION

Ericulture is a traditional agro-based eco-friendly economic activity, the end product of which is "silk". In terms of contribution to employment and income by all the four major sericulture activities of Assam (viz eri, muga, mulberry and tasar), ericulture occupies the prime position among the poor rural masses (De and Das, 2009). Moreover, because of its utility (especially as winter cloth) and comparative lower price than the other silk varieties; it gained much popularity among the middle and lower middle income population across the state and other parts of India (Das, 2008). In Assam, more than 1.35 lakh families were engaged in ericulture activities during 2005 - 2006 (about 70% of the then total sericulture practising families in the state) produced about 700 MT of eri cocoon (Directorate of Sericulture, Government of Assam, 2007). Through the export of ericultural products under the initiative of Assam Apex Weavers and Artisans Co-operative Federation Limited (ARTFED), significant amount of foreign exchange have been added to government treasury every year (De and Das, 2007).

Though ericulture output has been growing significantly at 5.9% annual exponential rate (during 1980-83 to 2002-05), that has been much lower than the capacity it has for the solution of unemployment problem in rural Assam. It has been mostly practised by the poor tribal communities like Misings, Kacharis, Bodos, Mikirs, Rabhas, Karbis and Garos and in most cases their women folk practice the same to supplement their family income and also provide nutritional support through the delicious by-product, pupae. A large number of families were found to come out of poverty or acute poverty due to their engagement in eri rearing and weaving activities (De and Das, 2009). Therefore, if undertaken properly, there is ample scope for sustainable employment and income of a wider section of rural population.

Eri-fabric is called "Poor man's Silk" as it is cheaper than muga and mulberry silk (Benchamin and Jolly, 1987). Eri fabric is highly durable and has a specific thermal property, which makes it an alternate fibre to wool. Erstwhile undivided Assam is known to be the

*Corresponding author. E-mail: de_u@yahoo.com.

original home of eri and muga silk in the world and thus, they are popularly known as Assam Silk. Though ericulture is practised in almost all the districts of Assam, it is highly concentrated in the districts of Karbi Anglong, North Cachar Hills, North Lakhimpur, Dhemaji, Barpeta, Kokrajhar, Sibsagar and Darrang.

Like other sericulture activities, ericulture can be broadly divided into two parts: ericulture proper and endi-textile industry. Ericulture proper consists of sowing of seed; plantation of host plant (castor, kesseru, bar kesseru etc) on which eri worms are fed; maintenance of host plants; plucking of leaves from the planted trees or from wildly grown trees or from Eri Concentration Centres[1] (ECC); feeding and rearing of pupae up to cocoon stage; while endi textile comprises spinning of yarn and weaving of fabrics. As most of the rearers are poor, they carry out rearing activities within the premises of their dwelling houses. Here spinning and weaving are carried out by a section of eri-worm rearers, mostly at home on a very small scale with simple traditional tools (like Takli, Charkha in case of spinning and Throw-shuttle or Fly-shuttle loom in case of weaving). Of course, all the cocoon rearers are not weavers and a large section of them owing to their ignorance and poverty sell their produced cocoon to the few professional weavers working in the area or to the middlemen who collect from them on their terms and supply to the weaving mills. Therefore, the rearers receive comparatively lower price and their earning is much lower than those of eri-rearers cum endi-entrepreneurs. Besides its utility and rising popularity among the middle class population, the activities suffer from several economic and non-economic limitations that limit its growth and thus raise the questions of prospects of ericulture activities in Assam.

Although there are some studies on the scientific aspects of ericulture till now, no serious study has been made on the economic aspects of silk-culture (sericulture) especially on ericulture of Assam. An earlier survey was conducted by the Sericulture and Weaving Department of Government of Assam during 1975 - 1976 in 598 Gaon Panchayats of the plain districts of Assam to assess the position of the silk industry in terms of production and employment. It was reported that the majority of population pursuing the silk culture as leisure time occupation[2]. A few other studies in Assam and other parts of India conducted by Dutta (1983, 1988), Choudhury (1984), Ratnala et al. (1990), who tried to analyse the problems, prospects and economic conditions of sericulture sectors. However, most of them stressed on mulberry, mugaculture and related cottage industries except Dookia (1984) and Das (2007) who tried to throw some light on ericulture and eri-silk industry as a source of income and employment in India.

Dutta (1983) tried to assess the income generated by silkworm rearers using primary data despite the fact that the sample used in his study was very inadequate. He (1988) also analysed the problems and future prospects of various sericulture activities in Assam particularly in the Sibsagar district. His study however suffered from the limitation that the transport cost associated with the collection of wildly grown food leaves in case of eri and imputed labour cost was not taken into consideration for the estimation of income and profit while explaining the prospects. Dookia (1984) tried to analyse the position of eri silk industry in Indian economy as a cottage industry and discussed its role in the creation of employment and income in the rural economy. Choudhury (1984) also attempted to assess the economic importance of each variety of silk by estimating gross and net returns per hectare of land under host plant and per family of silkworm rearers. This assessment seems to be based on experiment rather than field study.

Ratnala et al. (1990) studied the employment of human labour in sericulture across different size of the farms especially of the mulberry farms in Andhra Pradesh. In their study, they found that utilisation of human labour was more in smaller farms where the attention on each activity was more than the bigger farms. They also observed a significant positive relationship between hired labour use and the size of land holding. The conclusion of the study was in favour of the utilisation of the abundant and disguisedly unemployed human resources in rural India.

Thereafter "A Study of Mugaculture with Reference to Income and Employment Generation in Kamrup District" by Paresh Chandra Das in 2002 may be mentioned. Analysing primary data collected from a sample of 736 families in Kamrup district during January 1999 to December 2002, he tried to estimate the share of mugaculture in the generation of employment. Out of 736 families, 26.87% of the respondent households were found to adopt mugaculture as a source of livelihood in his study. However, he did not estimate the scope of mugaculture to generate employment per unit of output or area under the growth of host plant.

Though there are a few studies on sericulture and particularly on ericulture in Assam and other parts of the country, none of them tried to estimate the scope and excess capacity remain in ericulture at the existing technological set up through microeconomic analysis.

With this background the present paper is an attempt to examine the prospect of ericulture in Assam. It is examined through the analysis of scope of increasing activities and thereby the revenue and profit. This paper successfully shows the existence of excess capacity of the rearers and that shows the scope for expansion of such activities in the presence of widespread unemployment, poverty and absence of smooth industrial progress in

[1]Eri Concentration Centres are set up to grow eri feed plants in order to supply leaves regularly to the rearers. At present, there are 94 ECCs in Assam covering 733.49 hectares of land (Directorate of Sericulture, Government of Assam).

[2] As there is disguised unemployment in agriculture, the rural people, especially the women folk have been engaged in such occupation. Also the male people during the lean agricultural season take part in such activities for raising a part of their family earning.

the last few decades.

MATERIALS AND METHODS

Collection of data

For the purpose of analyses, primary data are collected from 180 families chosen by multistage sampling procedure from the district of Barpeta,[3] the fourth populous district of Assam with 16.42 lakhs population staying in 1073 villages under 12 Community Development (CD) blocks. Sericulture is mostly practised in 140 villages, which are mainly concentrated in the CD blocks of Gobardhana, Jalah and Sarukhetry. At the first stage, thus those three CD blocks out of total twelve blocks are chosen purposively on the basis of the concentration of ericulture activities. Thereafter, three villages have been chosen from each of the chosen blocks (total nine villages) also purposively depending upon the concentration of such activities. Though the sample villages were chosen purposively, it represents the state of ericulture in rural Assam due to the fact that (1) it is one of the major ericulture practicing district and (2) the technology used by the farms in the chosen area are similar to those in the other districts except the concentration of activities.

From the nine selected villages, total 180 sample families are selected (23, 10 and 17 from Gahia, Agdia and Garartari villages under Sarukhetry block; 14, 21 and 35 from Salbari, Hahchara and Bhuyapara villages under Jalah block and 18, 32 and 10 from Bashbari, Nimua and Khurabari villages under Gobardhana block respectively) on the basis of the number of families engaged in such activities. From each village, the final sample families are chosen by simple random sampling without replacement from all the families practising ericulture.

From each family, information regarding the number of broods reared in the previous year, production of cocoon during the year, number of people engaged in this occupation, cost of production of cocoon, depreciation of instruments, interest paid on hired capital if any, selling price of cocoons, family income etc are collected by direct interview through a structured schedule during June 2005 to July 2006. The collected data are compiled, re-grouped and reclassified for the purpose of analysis. Moreover, cost, revenue and profit per kilogram of eri cocoon production through the plantation of host plant during 2005 - 2006 has been collected from Central Silk Board (North-Eastern Region), Guwahati. Those are considered to estimate the profitability after the revision of cost due to cultivation of castor instead of collection of food leaves from the wildly grown castor. That would provide the justification of adoption of ericulture even if there is scarcity of castor leaves in the wild.

Methods

First of all, average number of broods of cocoon harvested by an average sample family has been determined. Thereafter, correlation between the annual number of broods reared and the annual family income of the sample households has been calculated. Average revenue and cost functions are estimated and also plotted graphically for different level of broods practised by the sample rearing families. The average cost/revenue function is estimated by using the following stochastic regression equation.

After the trial of several specifications, the quadratic form of the type $Y_i = \beta_1 + \beta_2 X_i + \beta_3 X_i^2 + U_i$ is found to be the best possible fit in this case. Here, Y_i represents average cost/revenue of ith farm family, X_i the level of output of the ith family and U_i the random disturbance term with the classical linear regression assumptions. The intercept represents the efficiency factor as all the farms use almost identical traditional technology. Also the possibility of heteroscedastic random disturbance is checked by Goldfeld-Quandt test and it is found that the random disturbance term is homoscedastic (Ramanathan, 2002).

From those estimated average cost and revenue functions the profit function is also examined and the profit maximising level of output is estimated by simple optimisation technique. Also, the level of output corresponding to the minimum average cost is estimated by minimising the best fitted average cost curve with respect to the level of output of eri cocoon, which provides the maximum scale of an average farm at the existing level of technology. Profit maximising level of output is estimated by maximising the estimated profit function with respect to the level of output (here cocoon) while the output corresponding to minimum average cost is obtained by minimising the estimated average cost function respectively. Those are also compared with the existing level of actual output to see how much expansion of activities would be possible in order to maximise profit or minimise the average production cost. That provides a clear indication of the scope for the expansion of ericulture activity at the existing level of technology.

RESULTS AND DISCUSSION

Ericulture is multi-voltine. In a year maximum six broods of eri cocoon can be harvested consecutively if there is no constraint on the supply of food leaves. Here it is observed from the actual observation that the number of broods of the sample families practising ericulture varies from one to four in a year. A sample family on an average harvest 2.43 broods of cocoon in a year (Table 1). The short fall of broods harvested in most cases is because of shortage of castor leaves which are collected by the rearers from the surrounding wildly grown areas. The correlation between the annual number of broods reared and the annual family income of the sample households during 2005 - 2006 is observed to be –0.492, which is significant at 5% level of significance by two tailed test. It indicates that relatively poorer families are more interested in practising ericulture than the relatively well-off families in the study area. In other words, the poorer families in the area are more dependent on ericulture for their survival. Also, the correlation between the revenue generated annually from ericulture proper and the annual family income of the rearers is to be significantly negative (–0.283), which also indicates that the high income families are less dependent on ericulture for their income and vice versa. The regression of proportional contribution of ericulture to family income (Y_i) on the average family income (X_i) across the sample farms also yields

$$LnY_i = 13.154 - \underset{(0.098)}{1.086}\, LnX_i^{*} \ldots \quad \ldots \quad R^2 = 0.946 \ldots (1)$$

While the regression of percentage of families adopted

---

[3] Though contribution to total ericulture output of the state and number of families engaged in ericulture is much larger in Karbi Anlong, North Lakhimpur, North Cachar Hills and Dhimaji district, growth rate of number of families engaged in ericulture and area under host plant is the second highest in the district of Barpeta. Moreover, growth of production per unit of land is the highest in Barpeta during the last decade (De and Das, 2007).

Table 1. Number of broods operated per family in the sample households during 2005 - 2006.

Block	Village	Household (Number)	Total number of broods	Broods per family	Total income per family (Rs)
Sarukhetri	Gohia	23	51	2.22	51869.56
	Agdia	10	32	3.20	47700.00
	Garartari	17	37	2.17	51294.11
	Sub-total	**50**	**120**	**2.40**	**50840.00**
	Bashbari	18	36	2.00	50555.55
	Nimua	32	86	2.68	48500.00
Gobardhana	Khusrabari	10	25	2.50	50800.00
	Sub-total	**60**	**147**	**2.45**	**49500.00**
	Salbari	14	34	2.43	160500.00
	Hahchara	21	53	2.53	53000.00
Jalah	Bhuyapara	35	83	2.37	56200.00
	Sub-total	**70**	**170**	**2.43**	**76100.00**
Grand Total		**180**	**437**	**2.43**	**60216.66**

Source: compiled from the field survey.

spinning and weaving (Y_i) on the average family income (X_i) across sample farms yields

$$Y_i = 31.13 + \underset{(7.58E^{05})}{0.0002}\, X_i^{*} \ldots \quad \ldots\ R^2 = 0.512\ \ldots\ (2)$$

Here figures in the parentheses represent standard error of the coefficients and slope coefficients are significant at 1% level of significance by two tailed test. The first equation showed that the poorer across all categories are more dependent on ericulture while the second one reveals that relatively richer village adopted more spinning and weaving activities along with rearing. In other words, due to lack of capital the poorer section have little chance of adopting weaving activities simultaneously along with rearing (Das, 2008).

Though rearing of cocoon alone does not contribute much to the family income (Appendix 1), the spinning and weaving activities along with the rearing contribute significantly to the employment and income of those families (De and Das, 2007). Not only that, it also helps a significant portion of those extremely poor families to come out of poverty (Appendix 2).

Total revenue generated and cost incurred in the production of eri cocoon by the sample families has been illustrated in Table 2. The table showed that 180 sample families together generated a sum of Rs 368064.5 total revenue from the production of 1267.63 kilogram of eri cocoon (revenue from cocoon and pupae) during 2005 - 2006. A sum of Rs 19947.00 was spent by the rearing families to make up the annual depreciation of rearing appliances, cost of seed, and cost of transportation for the collection of leaves etc. Thus, the rearers together earned a gross profit of Rs 348117.50 over the explicit cost of production. Gross profit from a kilogram of cocoon production was thus Rs 274.62. Of course, this gross profit included the imputed cost of self-employed labour. Imputed labour cost is much higher than the explicit cost of production of cocoon. Taking into account the wage of similar other activities (existing agricultural wage), a sum of Rs 185592.50 would be the approximate opportunity cost of family labour used for the purpose. Therefore, total net profit earned by the rearers would be only Rs 162525. Net profit per kilogram of cocoon production was calculated at Rs 128.21 while estimated gross profit per kilogram was Rs 274.62, and that was associated with significant inter-village variation.

Gross profit as well as net profit per kilogram of cocoon production was the highest in Gahia village in Sarukhetry block with Rs 313.88 and 160.48, respectively, while it was the lowest in Agdia village with Rs. 228.61 and Rs. 41.77 within the same block. In the village Agdia, gap between gross and net profit per kilogram of eri cocoon was much higher due to relatively higher wage rate that these villagers could earn as agricultural labourer as compared to their counterpart in other villages.

From Figures 1 to 4, It is observed that the average revenue curves are almost horizontal but the average cost curves corresponding to different (one, two, three and four) broods harvested by different sample families have different shapes. The average revenue curve always lies above the average cost curve across all levels of activities and hence there is substantial positive profit generated over the cost per unit of output by various farms. From the pictures it appears that the rate of profit margin is more in case of smaller farm, rearing one or two broods in that year. However, the scope of profitability rises with the level of output of eri cocoon, as the average revenue curve is almost horizontal (as most

Table 2. Cost, revenue and profit in the production of eri-cocoon by the sample households during 2005 - 2006.

Block	Village	Production of cocoon (Kg)	Total revenue (Rs)	Average revenue (Rs)	Total explicit cost (Rs)	Gross profit over total explicit cost (Rs)	Implicit labour cost (Rs)	Net profit over total cost (Rs)	Gross profit per Kg (Rs)	Net profit per Kg (Rs)
Sarukhetri	Gohia	148.00	49090.00	331.68	2637	46453.00	22702.50	23750.50	313.88	160.48
	Agdia	79.00	19690.00	249.24	1630	18060.00	14760.00	3300.00	228.61	41.77
	Garartari	120.60	36446.75	302.22	1985	34461.75	16965.00	17496.75	285.76	145.08
	Sub-Total	347.60	105226.80	302.73	6252	98974.75	54427.50	44547.205	284.74	128.16
Gobardhana	Bashbari	106.73	29812.45	279.33	1495	28317.45	14480.00	13837.405	265.32	129.65
	Nimua	249.55	77245.50	309.54	4220	73025.50	34680.00	38345.50	292.63	153.66
	Khusrabari	76.45	20537.50	268.64	1010	19527.50	10479.00	9048.50	255.43	118.36
	Sub-Total	432.73	127595.00	294.86	6725	120870.00	59639.00	61230.905	279.32	141.50
Jalah	Salbari	93.50	25695.00	274.82	1365	24330.00	14238.00	10092.00	260.22	107.94
	Hahchara	150.50	41282.50	274.31	2100	39182.50	22407.00	16775.50	260.35	111.47
	Bhuyapara	243.30	68266.00	280.59	3505	64761.00	34881.00	29880.00	266.18	122.81
	Sub-Total	487.30	135243.50	277.54	6970	128273.50	71526.00	56747.5	263.24	116.45
Grand Total		1267.63	368064.50	290.36	19947	348117.50	185592.50	162525	274.62	128.21

Source: Compiled from the field survey.
Notes: (1) Total revenue includes the revenue from cocoons as well as pupae. (2) Total cost includes expenditure incurred on seeds, rearing house, annual appliances, cost and transport, cost of collection of leaves of feed plant.

of the rearers sell their cocoon at the same price) and average cost curve is downward slopping and convex to the origin and becomes asymptotic after certain level of output and that varies for different size of farms in terms of broods harvested.

The level of performance and the scope of enhancing production and earning by the different level of rearers at the existence level of technology are explained with the help of the estimated average cost and revenue functions. Figures 1 to 4 present the average revenue and average cost functions for different groups of rearers harvesting one to four broods of eri cocoon throughout the year. The average revenue and cost functions for different sample brood farms are estimated by regression method. Here the quadratic forms are found to be the best estimates. The estimated average revenue (AR) and average cost (AC) functions for different brood levels are presented in equations 3 to 10.

For the farms practicing one brood in a year

$$AR = 294.46 + 61.89X - 9.137X^2 \ldots R^2 = .253, \bar{R}^2 = .202, F = 4.93^*, N = 32$$
(9.42)* (2.96)* (-2.77)* ... ... (3)

$$AC = 453.378 - 136.45X + 12.732X^2 \ldots R^2 = .786, \bar{R}^2 = .771, F = 53.12^*, N = 32$$
(8.28)* (-3.73)* (2.202)** ... ... (4)

For the farms practicing two broods in a year

$$AR = 476.11 - 26.366X + 2.0075X^2 \ldots R^2 = .108, \bar{R}^2 = .16, F = 3.18^{**}, N = 69$$
(12.54)* (-2.03)** (1.85)** ... ... (5)

$$AC = 466.67 - 80.576X + 4.532X^2 \ldots R^2 = .934, \bar{R}^2 = .932, F = 463.97^*, N = 69$$
(20.413)* (-10.30)* (6.908)* ... ... (6)

For the farms practicing three broods in a year

$$AR = 101.317 + 54.172X - 3.49X^2 \ldots R^2 = .252, \bar{R}^2 = .202, F = 3.04^{**}, N = 37$$
(2.005)** (2.344)**(-2.40)** ... ... (7)

$$AC = 709.97 - 108.66X + 5.258X^2 \ldots R^2 = .929, \bar{R}^2 = .925, F = 223.95^*, N =$$
(11.76)* (-7.044)* (5.411)* ... ... (8)

For the farms practicing four broods in a year

$$AR = 416.72 - 3.025X + 0.008X^2 \ldots R^2 = .31, \bar{R}^2 = 26, F = 9.6^*, N = 39$$
(2.98)* (-2.28)* (.008) ... ... (9)

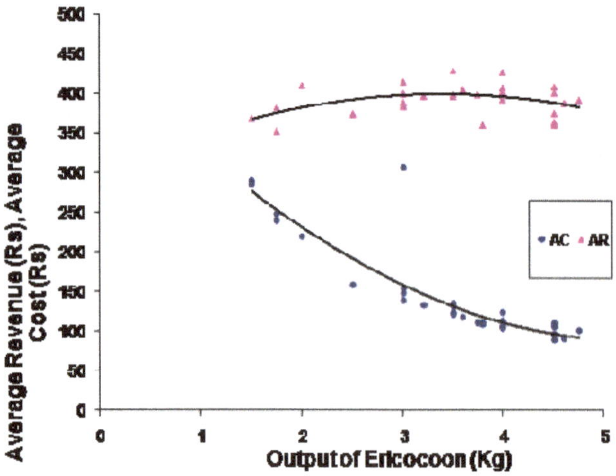

Figure 1. Average revenue and average cost curves of the rearers practicing one brood in a year.

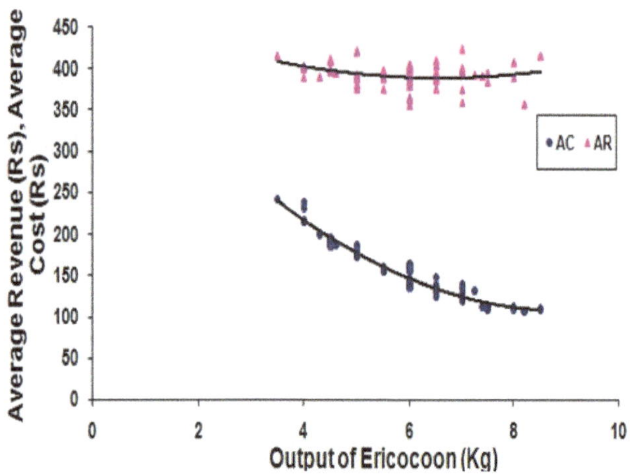

Figure 2. Average revenue and average cost curves of the rearers practicing two broods in a year.

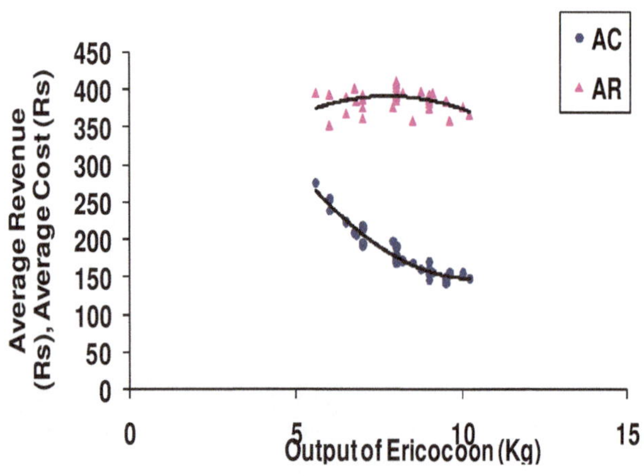

Figure 3. Average revenue and average cost curves of the rearers practicing three broods in a year.

$$AC = 463.13 - 35.001X + 0.85X^2 \ldots R^2 = .919, \bar{R}^2 = .915, F = 204.64^*, N = 39$$

(9.302)* (-4.14)* (2.422)** ... ... (10)

The figures in the parentheses represent t value. Here * and ** indicate that the test statistic is significant at 1 and 5% level of significance by two tailed test.

Table 3 shows that for various levels of broods actual eri-cocoon outputs are well below their efficient level of productions. In order to maximise profit, each producer with different brood level can increase production on an average from 30% to about 143% in a year at their existing technology and even without increasing the number of broods harvested by each of them. Thereby, the level of profit can be increased from about 32 to 168% approximately. Among all the sample farms, those who rear brood twice a year; are found to seriously under-utilise their capacity. The under-utilisation in some cases has been due to the shortage of caster leaves that may be grown under the initiative of the rearers or may be supplied by others. If the food leaves are cultivated then there is a chance of rise in cost of production (as shown in Table 4) of the extra unit of output and thus the average cost curve would shift upward. Yet there would be some profit as shown in Table 4 and hence total profit will certainly increase if not exactly to the extent as estimated here. The rate of profit may even be increased if modern equipments are used to reduce unit cost of production and the activity is undertaken on a larger scale that would generate economies of scale (as seen here an average rearer at each level is producing at the point on the downward falling portion of the average cost curve).

It is thus observed that instead of perusing ericulture as a part time occupation, there is enough scope of expansion of this culture. If ericulture is practised in a scientific and professional way and managed through the production of castor (food) leaves, cost of production undoubtedly increases as mentioned earlier. For example, if plantation of host plant is done by the rearers and they use separate rearing house instead of their residential compound, cost of production increases by several times. However, with regular supply of food leaves, production of cocoon as well as pupae also increases. Increase in production helps them to be organised and attract better price for cocoon and its by-product pupae. It is because of the fact that the traders will visit them regularly as they can get cocoon in bulks and quality of output will be better. Hence revenue earned is also found to be much higher if all the pupae are sold at the reasonable price, which may be even higher than the revenue from cocoon (as observed from the data obtained from Central Silk Board).

Table 4 reveals that profit per kg of cocoon in case of those in other areas who cultivate food leaves and maintain separate house for culturing ericocoon is slightly lower than that of rearers in Barpeta who collect leaves from nature. Although average profit per kg of cocoon is higher in case of naturally collected leaves, there is a

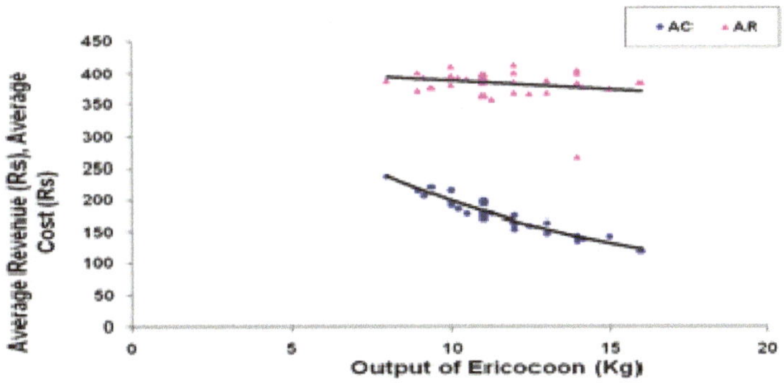

Figure 4. Average revenue and average cost curves of the rearers practicing four broods in a year.

Table 3. Brood-wise actual, profit maximising and lowest possible average cost output of eri-cocoon and scope of increasing output and profit as obtained from the estimated result by using sample data.

Number of broods	Average yearly output of cocoon per family (Kg)	Average yearly profit maximising output of cocoon per family (Kg)	Average yearly output of cocoon per family corresponding to lowest average cost (Kg)	Scope of percentage increase in	
				Output to maximise profit	Yearly total profit
1	3.428	5.615	5.359	63.80	64.56
2	5.920	14.400	8.890	143.24	168.06
3	8.038	10.488	10.330	30.48	31.60
4	11.597	24.460	20.500	110.92	130.01

Note: Output corresponding to lowest average cost and maximum profit is estimated by minimising best-fitted average cost function and maximising the best-fitted profit function respectively as obtained from the sample data.

Table 4. Cost, revenue and profit per Kg of Eri-cocoon production with and without plantation of host plant during 2005 - 2006.

	Cost, revenue and profit per kg of cocoon production	*With plantation (Rs)	Without plantation (Rs)
Items of cost	1. Plantation	172.76	00
	2. Fixed Cost	126.60	2.73
	(a) Rearing house	100.00	00
	(b) Appliances cost	26.60	2.73
	3. Wages	175.00	146.41
	4. Others	32.50	12.85 (Transport cost)
	Total cost	506.87	161.99
Revenue	Revenue from cocoon	280.00	209.26
	Revenue from Pupae	350.00	81.10
	Total revenue	630.00	290.36
	Net profit	123.13	128.37

Sources: (1) Compiled from Field Survey. (2) *Office of the Directorate, Central Silk Board (North-Eastern Region), Guwahati, India.

constraint in the expansion of their activity (average 2.43 broods only). If host plants are cultivated, more broods (maximum six in a year) can be harvested and thus more profit can be generated by the rearers. Hence cultivation of food leaves is not uneconomic if it is done scientifically as well as commercially.

It is also observed from Table 4 that even if castor is cultivated scientifically and marketing of main and by-products is done properly, profit rate would not reduce much, rather it enhances the scope for increasing overall profit through the rise in gross production. Therefore, to meet the shortages of food leaves, systematic plantation of host plant may be advocated. If plantation of host plant (castor) is done scientifically, the rearers can have regular and adequate supply of leaves and thus can practice and harvest maximum possible broods, successively every year as full time occupation and earn reasonable income.

If more people are engaged in such activities or the existing rearers expand their activities, then their marketing potential would certainly improve. However, most of the sample rearers are found not to avail institutional credit (Das, 2008). This bottleneck can be removed through the formation of self-help group or expansion of banking activities in the area.

Conclusions and policy implications

The present study reveals that there is substantial scope for the expansion of ericulture in Assam and particularly in Barpeta district. There is a scope to further increase earning through increasing activities due to the existence of both external and internal economies of scale. This would help increase in employment and income of those poor households who have been suffering from chronic unemployment and sustenance. Deficiency of eri food plants is one of the important limitations for the growth of ericulture in Assam. The rearers can be encouraged to cultivate feed plants as it is observed that even if castor is cultivated scientifically, there is still sufficient profit in this culture. Moreover, the Government may establish more ECCs and expand the existing ECCs in collaboration with the department of social forestry and encourage private entrepreneurs to establish ericulture farm. At the same time, indiscriminate cutting down of naturally grown silkworms' feed plants should be prevented by strictly enforcing the existing law. Moreover, to meet the shortage of silkworms' feed plants and to increase production of cocoons, the state government may acquire wasteland and allot the same to the local silk rearing co-operatives, self-help groups (if any) or diligent rearers for the growth of silk production. Extensive plantation of secondary feed plant like *Borkesseru*, Tapioca etc can also be encouraged to meet the deficiency of feed leaves for eri silkworms during the crisis of castor.

Apart from that, produced seeds of castor can be used for the production of highly expensive castor oil, which will definitely raise the income of the rearers. Though marketing of cocoon is found to be an important problem to the rearers, if the proposed spun mill in Assam is materialised (there is a proposal of large spun mill, which is yet to come) it will surely raise the demand for eri cocoon and help them to have remunerative price for their product in near future. Moreover, to eliminate the role of middleman traders who exploit the rearers by offering a low price for cocoon, co-operatives, self-help groups may be formed. This process will raise the bargaining strength of the rearers cum weavers and help them to have respectable price. It is observed that the rearers of Hahchara village within Jalah CD block who have formed self-help group have been able to get higher price. Similarly, Roje Eshanshali Co-operative Society Ltd of Kokrajhar has been successful in removing the plight of the rearers of the area to a certain extent (Das, 2008). Government marketing personnel (Cocoon Marketing Inspectors for cocoon) and agencies like ARTFED; Assam Government Marketing Corporation limited (AGMC), North-Eastern Handicrafts and Handloom Development Corporation Limited (NEHHDC), Assam Khadi and Village Industries Board (AKVIB) etc (for eri fabric marketing) can be re-activated. Finally co-operation of various sections like government officers, artisans, traders, rearers, weavers etc engaged in activities related to ericulture is necessary for the successful growth of the sector.

Limitations of the present analysis

Majority of rearers have provided some quantitative information such as figures of annual eri cocoon production, annual sales proceeds from ericulture, amount of investment in plantation and collection of leaves etc, from their memory other than from books of accounts. Some of them even were not aware of the modern weight and measures system. Most of the rearers have not maintained proper books of accounts. Appropriate care was however taken in filling up the schedules during the survey.

REFERENCES

Benchamin KV, Jolly MS (1987). Employment and Income Generation in the Rural Areas through Sericulture. Indian Silk. June: 9.

Chowdhury SN (1984). Sericulture in Assam Past, Present and Future. paper presented in the seminar on Sericulture and Handloom Industries. organised by Directorate of Sericulture. Government of Assam.

Das M (2008). Problems and Prospects of Ericulture in Assam with Special Reference to Barpeta District. PhD dissertation. North Eastern Hill University, Shillong, Meghalaya, India.

De UK, Das M (2007). Village–Based Informal Ericulture and Endi-Textiles in Assam: An Economic Inquiry. Icfai J. Entrepreneurship Devel., 4: 8-29.

De UK, Das M (2009). Ericulture as a Remedy of Rural Poverty in Assam: A Micro Level Study in Barpeta District. In Mukherjee D (ed) Indian Handloom and Handicrafts: Issues and Experiences. Icfai University Press. Hyderabad, pp. 170-192.

Dookia BR (1984). Studies on Ericulture for exploitation as a Cottage Industry. PhD dissertation. University of Jodhpur, Jodhpur. Rajasthan.

Dutta PC (1983). Economics of Silk Production in Assam. Agro-Economic Research Centre and Assam Agriculture University, Jorhat, Assam.

Dutta PC (1988). Problems and Prospects of Silk Production in Assam with Special Reference to Sibsagar District. PhD dissertation. Dibrugarh University, Dibrugarh, Assam.

Government of Assam (2007). Assam Apex Weavers and Artisans Co-operative Federation Limited (ARTFED). Official Statistical Records. Guwahati, Assam.
Government of Assam (2007). Directorate of Economics and Statistics. Statistical Handbook. Various issues. Guwahati, Assam.
Government of Assam, Directorate of Sericulture (2007). Official Statistical Records. Guwahati, Assam.
Government of India (2007). Office of the Directorate (North East). Central silk Board. Official Statistical Records, Guwahati, Assam.
Ramanathan R (2002). Introductory Econometrics (5e). Cengage Learning.
Ratnala GR, Mallikarjuna B, Datta RK (1990-92). Human Labour Employment in Sericulture- an empirical study in Andhra Pradesh. Indian J. Sericulture, 34(2).

Appendix 1. Contribution of Ericulture to the Income of Sample Households during 2005 - 2006.

Block	Village	House-hold (Number)	Total income of the sample families (Rs)	Income from ericulture proper (Rs)	Contribution of ericulture proper to total family income (%)	Total income per family (Rs)
Sarukhetri	Gohia	23	1193000	49090	4.11	51869.56
	Agdia	10	477000	19690	4.12	47700.00
	Garartari	17	872000	36446.75	4.17	51294.11
	Sub-total	50	2542000	105226.75	4.13	50840.00
Gobardhana	Bashbari	18	910000	29812.45	3.27	50555.55
	Nimua	32	1552000	77245.5	4.97	48500.00
	Khusrabari	10	508000	20537.5	4.04	50800.00
	Sub-total	60	2970000	127594.95	4.29	49500.00
Jalah	Salbari	14	2247000	25695	1.15	160500.00
	Hahchara	21	1113000	41282.5	3.70	53000.00
	Bhuyapara	35	1967000	68266	3.47	56200.00
	Sub-total	70	5327000	135243.5	2.53	76100.00
Grand Total		180	10839000	368064.45	3.39	60216.66

Source: Compiled from field survey.

Appendix-2. Percentage change in poverty due to ericulture.

Block	Village	House-hold (Number)	Number and percentage of households below poverty		Percentage decline in poverty due to ericulture activity	Percentage of families practicing spinning and weaving
			Without income from ericulture	With income from ericulture		
Sarukhetri	Gohia	23	7 (30.43)	6 (26.08)	14.29	26.09
	Agdia	10	6 (60.00)	6 (60.00)	0.00	50.00
	Garartari	17	6 (35.29)	6 (35.29)	0.00	35.29
	Sub-total	50	19 (38.00)	18 (36.00)	5.26	34.00
Gobardhana	Bashbari	18	6 (33.33)	5 (27.77)	16.68	44.44
	Nimua	32	15 (46.87)	12 (37.5)	19.99	43.75
	Khusrabari	10	7 (70.00)	6 (60.00)	14.28	40.00
	Sub-total	60	28 (46.66)	23 (38.33)	17.85	43.33
Jalah	Salbari	14	5 (35.71)	5 (35.71)	00	64.29
	Hahchara	21	12 (57.14)	12 (57.14)	00	47.62
	Bhuyapara	35	14 (40.00)	11 (31.42)	21.45	45.71
	Sub-total	70	31 (44.28)	28 (40.00)	9.66	50.00
Grand Total		180	69 (38.33)	78 (43.33)	11.54	43.33

Source: Compiled from field survey.
Note: Figures in the parentheses represent percentage of families below poverty line.

10

The savannah accelerated development authority: Factors to consider in modernising agriculture in Northern Ghana

Emelia Guo, Mamudu A. Akudugu and Seidu Al-Hassan

Institute for Continuing Education and Interdisciplinary Research (ICEIR) UDS, Tamale, Ghana.

The move by government through the Savannah Accelerated Development Authority (SADA) to modernise agriculture in the savannah areas of Ghana is in the right direction. However, the specific factors to consider in ensuring that farmers in the operational areas of SADA modernise their farming activities are not well known. This paper examines the factors that SADA must consider in its efforts to modernise agriculture because the success of the initiative depends on farmers' readiness to adopt the agricultural modernisation technologies to be rollout by the authority. Data were obtained from 150 farmers across the Bawku West District of Ghana and fitted into a logit model. The results showed that there is about 70% probability that farmers interviewed will adopt the SADA agricultural modernization initiative. Farm size, cost of technology, expected benefits, off-farm activities, education and access to information among others are the factors that will significantly influence farmers' decision to adopt modernisation of their farming activities. It is concluded that success of the SADA initiative depends on socio-economic circumstances of farmers and institutional effectiveness. It is recommended that SADA should take advantage of factors that positively influence farmers' adoption of its initiative by emphasising on them and take steps to mitigate the negative ones.

Key words: Agriculture, Farmers, logit model, Northern Ghana, savannah ecology.

INTRODUCTION

The economy of Ghana is basically agrarian. This is against the backdrop that agriculture accounts for about a third of the Gross Domestic Product (GDP) of the country (ISSER, 2010). Besides, agricultural activities constitute the main use to which Ghana's land resources are put. The agricultural sector is the major source of occupation for about 47% of the economically active age group of Ghanaians (Wayo, 2002). Despite the fact that the country covers an area of approximately 239,000 km^2 of which agricultural land forms about 57% of the total land area, only 20% of this agricultural land is under cultivation. This means that Ghana is yet to fully utilise its natural resource base, particularly land for agricultural production.

Agricultural production in Ghana is influenced by agro-ecological conditions which is said to be responsible for the differentials in agricultural productivity across the country. According to Wayo (2002), there are six distinct agro-ecological zones in Ghana. These are the high rain forest, the semi-deciduous rain forest, the forest-savannah transition, the Guinea savannah, the Sudan savannah and the coastal savannah. The conditions of these ecological zones limit the types of crops that can be successfully cultivated in them. In general, tree crops do better in the forest zones while food crops do well in the transitional and savannah zones. The transitional and savannah zones are therefore the food baskets of the country.

For food and nutrition security to be guaranteed in Ghana there is the need to formulate pragmatic policy frameworks that will enhance food productivity in the transitional and savannah zones. Though, a number of policies have been implemented over the years by successive governments in Ghana to enhance food production in the country, including subsidies on agricultural inputs, most of these policies have largely failed. There is therefore the need for paradigm shift that focuses on developing a holistic framework for the modernisation of agriculture in the areas of Ghana's food basket. In this regard, the establishment of the Savannah Accelerated Development Authority (SADA) by government with the main objective of bridging the chronic development gap that exists between Southern and Northern Ghana through agricultural modernisation is timely. This is more so following the discovery of oil in commercial quantities in Ghana which has the potential of attracting the attention of policy makers and implementers at the expense of agricultural development. For SADA to achieve its objective of modernising agriculture in its operational areas there is the need for it to know the specific economic, social and institutional factors likely to influence farmers' adoption of the initiative. This will enable SADA develop strategies to take advantage of the positive factors and to deal with the negative ones.

This paper therefore examines the different factors that could promote or otherwise of the adoption of the agricultural modernisation initiative of SADA by farmers in the savannah agro-ecological zone of Ghana using selected farmers in the Bawku West District of the Upper East Region of Ghana as a case.

Choice of factors for logit model estimation

The factors that influence adoption can be broadly put into three main categories. These are economic factors, social factors and institutional factors which are discussed following.

Economic factors

Farm size: Much empirical adoption literature focuses on farm size as the first and probably the most important determinant (Shakya and Flinn, 1985; Harper et al., 1990; Green and Ng'ong'ola, 1993; Adesiina and Baidu-Forson, 1995; Nkonya et al., 1997; Fernandez-Cornejo, 1998; Baidu-Forson, 1999; Boahene et al., 1999; Doss and Morris, 2001; Daku, 2002). This is because farm size can affect and in turn be affected by the other factors influencing adoption. The effect of farm size on adoption could be positive, negative or neutral. For instance, Feder et al. (1985), McNamara et al. (1991), Abara and Singh (1993), Fernandez-Cornejo (1996) and Kasenge (1998) found farm size to be positively related to adoption. On the other hand, Harper et al. (1990) and Yaron et al. (1992) found negative relationship between adoption and farm size. Interestingly, Mugisa-Mutetikka et al. (2000) found that the relationship between farm size and adoption is a neutral one.

Farm size affects adoption costs, risk perceptions, human capital, credit constraints, labour requirements, and tenure arrangements among others. With small farms, it has been argued that large fixed costs become a constraint to technology adoption (Abara and Singh, 1993) especially if the technology requires a substantial amount of initial set-up cost. In this regard, Feder et al. (1985) noted that only larger farms will adopt these kinds of innovations. With some technologies, the speed of adoption is different for small- and large- scale farmers which is critical for SADA in its pursuance of modernisation of agriculture. In Kenya for example, a study by Gabre-Madhin and Haggblade (2001) found that large commercial farmers adopted new high-yielding maize varieties more rapidly than smallholders.

Furthermore, access to funds including credit is expected to increase the probability of adoption. Yet to be eligible for credit from formal sources, the size of operation of the borrower is important. Farmers operating larger farms tend to have greater financial resources and chances of receiving credit are higher than those of smaller farms. In contrast, it has been argued by Yaron et al. (1992) that a small land area may provide an incentive to adopt a technology, especially in the case of an input-intensive innovation such as a labour-intensive or land-saving technology. In that study, the availability of land for agricultural production was low, consequently most agricultural farms were small. Hence, adoption of land-saving technologies seemed to be the only alternative to increased agricultural production.

Further, in the study by Fernandez-Cornejo (1996), farm size did not positively influence adoption. The majority of the studies mentioned previously consider total farm size and not crop acreage on which the new technology was practised. While total farm size has an effect on overall adoption, considering the crop acreage with the new technology may be a superior measure to predict the rate and extent of adoption of technology (Lowenberg-DeBoer, 2000). Therefore in regards to farm size, technology adoption may best be explained by measuring the proportion of total land area suitable to the new technology. This was measured in hectares for the logit model estimation.

Cost of technology: The decision to adopt is often an investment decision. According to Caswell et al. (2001), the decision to adopt presents a shift in farmers' investment options. Therefore, adoption can be expected to be dependent on cost of a technology and on whether farmers possess the required resources. Technologies that are capital-intensive are only affordable by wealthier farmers (El-Osta and Morehart, 1999). The adoption of such technologies is therefore limited to larger farmers who have the wealth (Khanna, 2001). Changes that cost little are adopted more quickly than those requiring large expenditures. SADA must therefore be conscious of the fact that both the extent and rate of adoption of its agricultural modernisation initiative is dependent on the cost. This is in line with the law of demand – the lower the price, the higher the demand. Those farmers who said the cost of modernisation does not matter were assigned a dummy value of zero (0). On the other hand, those who said SADA must come out with low-cost agricultural modernisation technologies were assigned a dummy value of one (1) for the logit model estimation.

Level of expected benefits: Development interventions that produce significant gains can motivate people to participate more fully in them. This is because people will not participate unless they believe it is in their best interest to do so. Farmers must see an advantage or expect to obtain greater utility in adopting a technology. In addition, farmers must perceive that there is a problem that warrants an alternative action to be taken. Without a significant difference in outcomes between two options, and in the returns from alternative and conventional practices, it is less likely that farmers, especially small-scale farmers will adopt the new practice (Abara and Singh, 1993). A higher percentage of total household income coming from the farm through increased yield tends to correlate positively with adoption of new technologies (McNamara et al., 1991; Fernandez-Cornejo, 1996). The farmers who did not expect more benefits from modernising their farming activities were assigned a dummy value of zero (0) and those who expected more benefits from the modernisation of their farming activities through the SADA initiative were assigned a dummy value of one (1) for the logit model estimation.

Off-farm activities: The availability of time is an important factor affecting technology adoption. It can influence adoption in either a negative or positive manner. Practices that heavily draw on farmers' leisure time may inhibit adoption (Mugisa-Mutetikka et al., 2000). However, practices that leave time for other sources of income accumulation may promote adoption. In such cases as well as in general terms, income from off-farm labour may provide financial resources required to adopt the new technology. The farmers who said they did not have off-farm income generating activities were assigned a dummy value of zero (0). On the other hand, those who said they had off-farm income generating activities were assigned a dummy value of one (1).

Social factors

Age of adopter: Age is an important factor that influences the probability of adoption of new technologies because it is said to be a primary latent characteristic in adoption decisions. However, there is contention on the direction of the effect of age on adoption. Age was found to positively influence chemical control of rice stink bug in Texas (Harper et al., 1990), adoption of IPM on peanuts in Georgia (McNamara et al., 1991) and sorghum in Burkina Faso (Adesiina and Baidu-Forson, 1995). The effect is thought to stem from accumulated knowledge and experience of farming systems obtained from years of observation and experimenting with various technologies. Since adoption pay-offs may occur over a long period of time, while costs occur in the earlier phases, age (time) of the farmer can have a profound effect on technology adoption.

Age has been found to be either negatively correlated with adoption or not significant in farmers' adoption decisions. In studies on adoption of IPM sweep nets in Texas (Harper et al., 1990), fertilizer in Malawi (Green and Ng'ong'ola, 1993), rice in Guinea (Adesiina and Baidu-Forson, 1995), land conservation practices in Niger (Baidu-Forson, 1999) and Hybrid Cocoa in Ghana (Boahene et al., 1999), age was either not significant or was negatively related to adoption. This implies that older farmers, after investing several years in a particular practice, may not want to jeopardize it by trying out a completely new method. In addition, farmers' perception that technology development and the subsequent benefits require a lot of time to realize, can reduce their interest in the new technology because of farmers' advanced age, and the possibility of not living long enough to enjoy it (Caswell et al., 2001; Khanna, 2001).

Elderly farmers often have different goals other than income maximization, in which case, they will not be expected to adopt an income –enhancing technology. As a matter of fact, it is expected that the old that do adopt a technology do so at a slow pace because of their tendency to adapt less swiftly to a new phenomenon. This was measured in years for the logit model estimation.

Education: A number of studies that sought to establish the effect of education on adoption in most cases relate it to years of formal schooling (Feder and Slade, 1984; Tjornhom, 1995). Generally, education is thought to create a favourable mental attitude for the acceptance of new practices, especially of information-intensive and management-intensive practices (Waller et al., 1998; Caswell et al., 2001). According to Rogers (1983), technology complexity has a negative effect on adoption and this could only be dealt with through education. According to Ehler and Bottrell (2000), one of the hindrances to widespread adoption of new technologies is because of the requirement for greater ecological understanding of the production system. The ability to read and understand sophisticated information that may be contained in a technological package is an important aspect of adoption. This was measured as number of years of formal schooling for the logit model estimation.

Gender concerns: Gender issues in agricultural production and technology adoption have been investigated for a long time. Most of such studies show mixed evidence regarding the different roles men and women play in technology adoption. Doss and Morris (2001) in their study on factors influencing improved maize technology adoption in Ghana and Overfield and Fleming (2001) studying coffee production in Papua New Guinea show insignificant effects of gender on adoption. Since adoption of a practice is guided by the utility expected from it, the effort put into adopting it is reflective of this anticipated utility. It might then be expected that the relative roles women and men play in both 'effort' and 'adoption' are similar, hence suggesting that males and females adopt practices equally. Women were assigned a dummy value of zero (0) and men assigned a dummy value of one (1).

Institutional factors

Information: Acquisition of information about a new technology demystifies it and makes it more available to farmers. Information reduces the uncertainty about a technology's performance and hence may change individual's assessment from purely subjective to objective over time (Caswell et al., 2001). Exposure to information about new technologies affects farmers' choices about it. Feder and Slade (1984) indicate how, provided a technology is profitable, increased information induces its adoption. However, in the case where experience within the general population about a specific technology is limited, more information induces negative attitudes towards its adoption, probably because more information exposes an even bigger information vacuum thereby increasing the risk associated with it. A good example is said to be the adoption of recombinant bovine Somatotropin Technology (rbST) in dairy production (McGuirk et al., 1992; Klotz et al., 1995). Information is acquired through informal sources like the media, extension personnel visits, meetings, and farm organizations and through formal education. It is important that this information be reliable, consistent and accurate. Thus, the right mix of information properties for a particular technology is needed for effectiveness in its impact on adoption. Respondents who said they were not aware of the SADA agricultural modernisation initiative were assigned a dummy value of zero (0) and those who said they were aware were assigned a dummy value of one (1).

Extension services: Good extension services and contacts with producers are a key aspect in technology dissemination and adoption. This assertion is consistent with the observation made by the International Food Policy Research Institute (2001), that "a new technology is only as good as the mechanism of its dissemination" to farmers. Most studies analyzing this variable in the context of agricultural technology show its strong positive influence on adoption. According to Yaron et al. (1992), access to extension services can counter balance the negative effect of lack of years of formal education in the overall decision to adopt some technologies. Farmers who said they did not have access to extension services were assigned a dummy value of zero (0) and those who had access to extension services were assigned a dummy value of one (1) for the logit model estimation.

METHODOLOGY

The survey

The methodology employed for the survey was divided into two parts. The first part was a desktop review of existing literature on the factors influencing adoption of modern agricultural production technologies. The second part was to gather data from selected individuals in the Bawku West District on agricultural development issues through face-to-face interviews using semi-structured questionnaires supplemented by key informant interviews and focus group discussions. The information elicited from respondents bordered on demographic characteristics of respondents, their awareness and understanding of SADA, and the factors that influence farmers' choice of agricultural modernisation strategies. The key informants interviewed included frontline officers of the District Ministry of Food and Agriculture (MoFA); officers of agricultural related Non-Governmental Organisations (NGOs) operating in the District; frontline officers of the District Assembly including secretaries of the Town and Area councils; and officers of the Rural Bank.

In all, one hundred and fifty (150) individuals including hundred (100) farmers, ten (10) input dealers, twenty (20) agricultural commodity traders, five (5) transporters and five (5) food processors were interviewed. These categories of people were sampled from five (5) out of the seven (7) Area and Town Councils in the District. The sampled Town and Area Councils are the Zebilla Town Council, the Sapeliga Area Council, the Gbantongo Area Council, the Tilli-Widnaba Area Council and the Binaba Town Council. In each of the selected Town and Area Councils, 20 farmers; two input dealers; four agricultural commodity traders; one transporter; and one food processor were selected for the individual interviews. In addition, focus group discussions and key informant interviews were held in each of the selected areas to validate the data from individuals. The pieces of information gathered from the interviews were the basic inputs for analyses. STATA (version 11) was the software used for the data analyses.

The analytical framework

Using the logit model, the factors that influence farmers' decision to adopt agricultural modernisation over traditional agriculture were estimated. The use of the logit model for this analysis is consistent with the literature on adoption (Griliches, 1957; Lionberger, 1960; Rogers, 1983; Alston et al., 1995) which describes the process of adoption as taking on a logistic nature. The study used the threshold decision-making theory proposed by Hill and Kau (1973) and Pindyck and Rubinfeld (1998) to analyse the factors that farmers consider in adopting one agricultural production strategy over others. The theory points out the fact that when farmers are faced with a decision to adopt or not to adopt an innovation, in this case agricultural modernisation, every farmer has a reaction threshold, which is dependent on a certain set of factors. As such, at a certain value of stimulus below the threshold, no adoption is observed while at the critical threshold value, a reaction is stimulated. Such phenomena are generally modelled using the relationship:

$$Y_i = \beta X_i + u_i \tag{1}$$

Where Y_i is equal to one (1) when a choice is made to adopt and zero (0) otherwise; this means: $Y_i = 1$ if X_i is greater than or equal to a critical value, X^* and $Y_i = 0$ if X_i is less than a critical value, X^*. X^* represents the combined effects of the independent variables (βX_i) at the threshold level.

Equation (1) represents a binary choice model involving the estimation of the probability of adoption of a given technology, innovation or intervention (Y) as a function of independent variables (X). Mathematically, this is represented as:

$$Prob(Y_i = 1) = F(\beta' X_i) \tag{2}$$

$$Prob(Y_i = 0) = 1 - F(\beta' X_i) \tag{3}$$

Y_i is the observed response for the ith observation of the response variable, Y. This means that $Y_i = 1$ for an adopter (that is, farmers who decide to adopt agricultural modernization by SADA) and $Y_i = 0$ for a non-adopter (that is, farmers who decided not to adopt agricultural modernization by SADA). X_i is a set of independent variables such as farm size among others, associated with the ith individual, which determine the probability of adoption (that is making the decision to modernize one's farming activities), (P). The function, F may take the form of a normal, logistic or probability function. The logit model uses a logistic cumulative distributive function to estimate, P as follows (Pindyck and Rubinfeld, 1998):

$$P(Y = 1) = \frac{e^{\beta' X}}{1 + e^{\beta' X}} \tag{4}$$

$$P(Y = 0) = 1 - \frac{e^{\beta' X}}{1 + e^{\beta' X}} = \frac{1}{1 + e^{\beta' X}} \tag{5}$$

According to Greene (2008), the probability model is a regression of the conditional expectation of Y on X giving:

$$E(Y/X) = 1[F(\beta' X)] + 0[1 - F(\beta' X)] = F(\beta' X) \tag{6}$$

Since the model is non-linear, the parameters are not necessarily the marginal effects of the various independent variables. The relative effect of each of the independent variables on the probability of adoption (that is making the decision to modernize farming activities) is obtained by differentiating Equation (6) with respect to X_{ij} resulting in equation (7) (Greene, 2008):

$$\frac{\partial P_i}{\partial x_{ij}} = \left[\frac{\lambda^{\beta' x}}{\left(1 + \lambda^{\beta' x}\right)^2}\right]\beta = F(\beta' x)[1 - F(\beta' x)]\beta \tag{7}$$

The maximum likelihood method was used to estimate the parameters. The implication for applying the logit model in this paper is that, the farmer would decide to adopt agricultural modernisation in his or her farming activities at a given point in time when the combined effects of certain factors exceed the inherent resistance in him/her to change. The preference for the probability model (logit) to the conventional linear regression models, in analysing the factors farmers will consider in adopting agricultural modernisation over the traditional agriculture is based on the fact that, the parameter estimates from the former are asymptotically consistent and efficient. The estimation procedure employed also resolves the problem of heteroscedasticity and constrains the conditional probability of making the decision to adopt agricultural modernisation to lie between zero (0) and one (1). The main reason for choosing the logit model over the probit model for this paper is because of its mathematical convenience and simplicity (Greene, 2008) and the fact that it has been applied in similar studies (Feder et al., 1985; Shakya and Flinn, 1985; Green and Ng'ong'ola, 1993; Rogers, 1995; Nkonya et al., 1997; Boahene et al., 1999).

The empirical model for the logit model estimation is specified as follows:

$$log \frac{P_i}{1 - P_i} = \alpha + \beta X_i + \varepsilon_i \tag{8}$$

Where X_i is the combined effects of X explanatory variables that promote or prevent farmers' decision to adopt SADA's agricultural modernization initiative. $log\frac{P_i}{1-P_i}$ = The log-odds in favour of farmers' decision to adopt agricultural modernization; $X_1 ... X_i$ are factors that promote or prevent farmers from adopting agricultural modernisation and are defined as follows: X_1 = Farm size in hectares; X_2 = Cost of technology/modernization; dummy (1 = Cost matter; 0 = Otherwise); X_3 = Level of expected benefits; dummy (1 = High expected benefits; 0 = Otherwise); X_4 = Has off-farm income generating activities, dummy (1 = Yes; 0 = Otherwise); X_5 = Age of farmer in years; X_6 = Age of farmer in years squared; X_7 = Level of education measured as years of formal schooling; X_8 = Gender; dummy (1 = Man; 0 = Otherwise); X_9 = Access to extension services; dummy (1 = Has access; 0 = Otherwise), and X_{10} = Access to information on SADA initiative; dummy (1 = Has access; 0 = Otherwise).

RESULTS AND DISCUSSION

Socio-demographic characteristics of respondents

The study was conducted in the Bawku West District of the Upper East Region of Ghana. In all, one hundred and fifty (150) individuals were interviewed. Out of the people interviewed, 2% of respondents were below the age of 18 years (Table 1). These people are below the economically active age group and are expected to be in school. About 91% of the respondents belonged to the economically active age group of between 18 and 60 years with the remaining 7% of respondents being above the economically active age group. The implication of the age distribution of respondents across the different age groups is that, the findings indicated in this paper are representative of the different generations.

Further, 54% of the respondents were men with the remaining 46% of them being women. This implies that, the outcome of the study represents the collective views, concerns and opinions of both men and women with regards to the adoption of agricultural modernisation being spearheaded by SADA. Majority of the respondents

Table 1. Socio-demographic characteristics of respondents.

Age	Frequency	Percentage (%)
<18	3	2
18-45	96	64
46-60	41	27
>60	10	7
Total	150	100
Gender		
Men	81	54
Women	69	46
Total	150	100
Education		
No formal	78	52
Basic	45	30
Post basic	27	18
Total	150	100

Source: Field Survey Data, 2011.

(52%) did not have any form of formal education. Besides, 30% of the respondents had basic education and 18% of them had post-basic education (Table 1). This means that the findings are representative of both literate and illiterate residents of the Bawku West District of the Upper East Region of Ghana.

Factors to consider in the promotion of agricultural modernisation

The estimated factors influencing adoption of technology (that is, agricultural modernization by SADA) using the logit model included the economic factors such as farm size, cost of adoption, expected benefits from the adoption and the off-farm income generation activities. The social factors included the age of farmers, the level of education and gender. The institutional factors included access to information regarding the specific strategies to be adopted by SADA in its efforts to modernize agriculture and access to extension services. The logit model estimation gave a Pseudo R^2 of 0.7153 which implies that the variables included in the model are able to explain about 72% of the probability of adoption of the agricultural modernization initiative by SADA (Table 2). The Log-likelihood Ratio (LR) was also found to be significant at the 1% level. This means that all the explanatory variables included in the model jointly influence farmers' probability of adoption of the initiative. The model results also gave a predicted probability of adoption of 0.7023. This means that there is about 70% probability that the people interviewed will adopt the SADA agricultural modernization initiative but this is contingent on a certain set of factors as mentioned earlier. Given the foregoing goodness of fit measures, it is concluded that the logit model employed had integrity and hence appropriate.

Farm size was found to have a positive relationship with the probability of adoption of agricultural modernization. It was found to be significant at the 1% level. This finding is consistent with the literature that large scale farmers are more inclined to adopt new technologies than small scale farmers (Feder et al., 1985; McNamara et al., 1991; Abara and Singh, 1993; Fernandez-Cornejo, 1996; Kasenge, 1998). It is however, important for SADA to come out with strategies for encouraging small scale farmers to agree to modernize their farming activities given that they form the majority in SADA's operational areas, particularly the Bawku West District and elsewhere in the Upper East Region of Ghana.

The cost of modernization was found to be negatively related to the probability of adoption. It was found to be significant at 1%. The finding is consistent with Caswell et al. (2001) who noted in their study that the decision to adopt a new technology presents a shift in farmers' investment options. This means that if the technology is costly to the farmer, there is low probability that he or she will adopt it. Besides, the fear of losing livelihoods is a social cost that farmers consider in their adoption decisions. For instance, a farmer whose main source of livelihood is maize farming will hesitate to replace this crop with improved groundnut cultivation for the fear that if the crop fails his or her livelihood will be greatly affected. In effect, as noted by El-Osta and Morehart (1999) and Khanna (2001), technologies that are capital-intensive are only affordable by wealthier farmers and hence the adoption of such technologies is limited to larger farmers who have the wealth. For SADA to achieve its objective of modernizing agriculture, it must consider coming out with technologies that are affordable, especially to poor rural dwellers about 90% of who depend on agriculture for their livelihoods.

The expected benefits from modernization of agriculture were found to be positively related to the probability of adoption. This was found to be significant at the 1% level. This implies that if farmers expect benefits from modernizing their farming activities to be higher than their current methods of farming, they are most likely to modernize. This is consistent with Abara and Singh (1993) who observed that without a significant difference in outcomes between two options, and in the returns from alternative and conventional practices, it is less likely that farmers, especially small-scale farmers will adopt the new practice. Farmers currently expect that the SADA agricultural modernization initiative will bring them higher returns which is a good sign for SADA to succeed in getting farmers adopt whichever technologies that the authority will come out with.

Off-farm activities were found to have negative relationship with the probability of adoption. This was

Table 2. Logit regression results of agriculture modernisation adoption (N = 150).

Variable	Coefficient	Std. Error	Z	P>|z|
Farm size	0.4241	0.1384	3.0643	0.0001
Modernisation cost	-0.5535	0.1221	-4.5332	0.0000
Modernisation benefits	0.5361	0.1318	4.0675	0.0012
Off-farm activities	-0.4859	0.1248	-3.8020	0.0027
Age of farmer	-0.4812	0.2278	-2.1124	0.0351
Age of farmer squared	0.0454	0.1002	0.4531	0.2832
Educational level of farmer	0.3224	0.1611	2.0012	0.0387
Gender	0.4158	0.1253	3.3184	0.0009
Access to information	0.4814	0.1301	3.7002	0.0034
Extension services	0.2392	0.1657	1.4436	0.1931
Constant	0.3806	0.7379	0.5158	0.2861
Goodness of fit measures				
LR Chi Square (10)		131.29		
Prob>Chi^2		0.0000		
Log likelihood		-51.8321		
Predicted probability of adoption		0.7023		
Pseudo R^2		0.7153		

Source: Field Survey Data, 2011.

found to be statistically significant at 1%. This implies that the higher the off-farm activities, the lower their probability of adopting agricultural modernisation technology. This is in line with the observation made by Mugisa-Mutetikka et al. (2000) that practices that heavily draw on farmers' leisure time may inhibit their adoption of on-farm technologies.

The age of farmer assumed a quadratic function which implies that farmers' rate of adoption is low at both the younger and older ages. It was found to be significant at the 5% level. At the younger age, farmers may not be able to adopt, especially for capital intensive technologies because of the fact that they might not have adequate resources to do so. At an older age, farmers' volume of economic activities reduced hence they may be unable to pay for technologies. Besides, older farmers have accumulated years of experience in farming through experimentation and observations and may find it difficult to leave such experiences for the new technologies.

The level of education of farmers was found to have a positive relationship with the probability of adoption and significant at 5% level. The implication of this is that farmers who are educated are more likely to adopt agricultural modernization technologies than those who are not educated. This is consistent with the literature that education creates a favourable mental attitude for the acceptance of new practices, especially of information-intensive and management-intensive practices (Waller et al., 1998; Caswell et al., 2001).

The gender of farmer was found to be positively related to the adoption of agricultural modernization by farmers. This was found to be significant at 1% level. This means that male farmers are more likely to adopt modernization of their farming activities than their female counterparts. The reason for this is that men are the people who make production decisions in the study area. This finding is inconsistent with that of Doss and Morris (2001) who in their study on factors influencing improved maize technology adoption in Ghana as well as Overfield and Fleming (2001) studying coffee production in Papua New Guinea show insignificant effects of gender on adoption.

Access to information on SADA and its activities were found to have a positive relationship with the probability of adoption of agricultural modernization. This was found to be significant at the 1% level. This means that SADA must make information on its agricultural modernization initiative available to farmers to be able to make their decisions as to whether to adopt the initiative or not. This is because access to information is said to reduce the uncertainty about a technology's performance hence may change individuals' assessment from purely subjective to objective over time (Caswell et al., 2001). Access to extension services was also found to be positively related to the adoption of agricultural modernisation but insignificant.

CONCLUSION AND RECOMMENDATIONS

The factors that influence the adoption of new technologies such as the agricultural modernization initiative of SADA are broadly categorised into economic factors, social factors and institutional factors. The economic factors include farm size, cost of technology or modernization, expected benefits from adoption of the technology, and off-farm activities. All the economic factors will have significant influences on farmers' decision to adopt the agricultural modernization initiative of SADA. The social factors that influence probability of adoption of new technologies by farmers include age of farmers, level of education and gender. All these social factors were found to significantly influence the decisions of farmers to adopt agricultural modernization by farmers. Access to information was the only institutional factor which will have significant influence on farmers' probability of adopting SADA's agricultural modernisation technologies. Based on the findings, the following recommendations are made:

1. Farm size positively influences the probability of adoption of agricultural modernisation by farmers. It is therefore recommended that SADA should be mindful of the fact that farmers are heterogeneous in terms of farm sizes and hence should come out with specific technologies that meet the needs of small, medium and large size farms.
2. Cost of modernisation or technology has the potential to inhibit adoption of the SADA agricultural modernisation initiative. As such, SADA should aim at delivering low cost technologies to the target farmers.
3. Farmers expect that the agricultural modernisation initiative by SADA will bring them higher benefits than their current farming practices. It is recommended that SADA should take advantage of this good will from farmers by ensuring that the expectations of farmers are met.
4. Age of farmers has quadratic relationship with the adoption of agricultural modernisation technologies. It is recommended that SADA should target the economically active age group in its attempts to modernise agriculture in its operational areas.
5. Access to information and extension services play critical influences on farmers' decision to adopt new technologies. It is recommended that SADA should make information on its operations readily available to its clientele group. It should also ensure that extension services on new technologies are integral part of the SADA agricultural modernisation initiative.

REFERENCES

Abara IOC, Singh S (1993). Ethics and Biases in Technology Adoption: The Small Farm Argument. Tech. Forecast. Social Change 43:289-300.

Adesiina AA, Baidu-Forson J (1995). Farmers' Perceptions and Adoption of New Agricultural Technology: Evidence from Analysis in Burkina Faso and Guinea, West Africa. J. Agric. Econ. 13:1-9.

Alston JM, Norton GW, Pardey PG (1995). Science under Scarcity: Principles and Practice of Agricultural Research Evaluation and Priority Setting. Ithaca: Cornell University Press.

Baidu-Forson J (1999). Factors Influencing Adoption of Land-enhancing Technology in the Sahel: Lessons From a Case Study in Niger. J. Agric. Econ. 20:231-239.

Boahene K, Snijders TAB, Folmer H (1999). An Integrated Socio-Economic Analysis of Innovation Adoption: The case of Hybrid Cocoa in Ghana. J. Policy Model. 21(2):167-184.

Caswell M, Fuglie K, Ingram C, Jans S, Kascak C (2001). Adoption of Agricultural production practices: Lessons learned from the US. Department of agriculture area studies project.Washington DC. US Department of Agriculture, Resource Economics Division, Economic Research service, Agriculture Economic Report. P. 792.

Daku L (2002). Assessing farm-level and aggregate economic impacts of olive integrated pestmanagement programs in Albania. PhD. Dissertation. Virginia Polytechnic Institute and State University.

Doss CR, Morris ML (2001). How Does Gender Affect the Adoption of Agricultural Innovation? The Case of Improved Maize Technologies in Ghana. J. Agric. Econ. 25:27-39.

Ehler LE, Bottrell DG (2000). The illusion of Integrated Pest Management. Issues in Science and Technology. Bell and Howell Information and Learning Company. pp. 61-64.

El-Osta HS, Morehart MJ (1999). Technology Adoption Decisions in Dairy Production and the Role of Herd Expansion. Agric. Res. Econ. Rev. 28:84-95.

Feder G, Slade R (1984).The acquisition of information and the adoption of new Technology. Am. J. Agric. Econ. 66:312-320.

Feder G, Just ER, Zilberman D (1985). Adoption of Agricultural Innovations in Developing Countries: A Survey. Econ. Dev. Cult. Change 33:255-298.

Fernandez-Cornejo J (1998). Environmental and Economic Consequences of Technology Adoption: IPM in Viticulture. Agric. Econ. 18:145-155.

Fernandez-Cornejo J (1996). The Microeconomic Impact of IPM Adoption: Theory and Application. Agric. Res. Econ. Rev. 25:149-160.

Gabre-Madhin EZ, Haggblade S (2001). Success in African Agriculture: Results of an Expert Survey. International Food Policy Research Institute. Washington DC.

Green DAG, Ng'ong'ola DH (1993). Factors Affecting Fertilizer Adoption in Less Developed Countries: An Application of Multivariate Logistic Analysis in Malawi. J. Agric. Econ. 44(1):99-109.

Greene WH (2008). Econometric Analysis, 6th Edition, Upper Saddle River, New Jersey, Prentice-Hall, New York University.

Griliches Z (1957). Hybrid Corn: An Exploration in the Economics of Technological Change. Econometrica 25:501-522.

Harper JK, Rister ME, Mjelde JW, Drees BM, Way MO (1990). Factors influencing the adoption of insect management technology. Am. J. Agric. Econ. 72(4):997-1005.

Hill L Kau P (1973). Application of multivariate probit to a threshold model ofgrain dryer purchasing decisions. Am. J. Agric. Econ. 55:19-27.

IFPRI (2001). Applying science to sub-Saharan Africa's food needs. Accessed from http://www.ifpri.org/pubs/books/ufa/ufa_ch26.pdf on May 29, 2011.

ISSER (2010). The State of the Ghanaian Economy in 2009, University of Ghana, Legon.

Kasenge V (1998). Socio-economic factors influencing the level of Soil Management Practices on Fragile Land. In Proceedings of the 16th Conference of Soil ScienceSociety of East Africa (Eds.: Shayo-Ngowi, A.J. G. Ley and F.B.R Rwehumbiza), 13th-19th, Tanga, Tanzania. pp.102-112.

Khanna M (2001). Sequential Adoption of Site-Specific Technologies and its Implications forNitrogen Productivity: A Double Selectivity Model. Am. J. Agric. Econ. 83(1):35-51.

Klotz C, Saha A, Butler LJ (1995). The Role of Information in Technology Adoption: The Case of rbST in the California Dairy Industry. Rev. Agric. Econ. 17:287-298.

Lionberger HF (1960). Adoption of New Ideas and Practices. Iowa State: University Press.

Lowenberg-DeBoer J (2000). Comments on Site-Specific Crop Management: Adoption Patterns and Incentives. Rev. Agric. Econ. 22(1):245-247.

McGuirk AM, Preston WP and Jones GM (1992). Introducing Foods Produced using Biotechnology: The Case of Bovine Somatotropin. Southern J. Agric. Econ. pp. 209-223.

McNamara KT, Wetzstein ME, Douce GK (1991). Factors Affecting

Peanut ProducerAdoption of Integrated Pest Management. Rev. Agric. Econ. 13:129-139.

Mugisa-Mutetikka M, Opio AF, Ugen MA, Tukamuhabwa P, Kayiwa BS, Niringiye C, Kikoba E (2000). Logistic Regression Analysis of Adoption of New Bean Varieties in Uganda (Unpublished).

Nkonya E, Schroeder T, Norman D (1997). Factors Affecting Adoption of Improved Maize Seed and Fertilizer in Northern Tanzania. J. Agric. Econ. 48(1):1-12.

Overfield D, Fleming E (2001). A Note on the Influence of Gender Relations on the Technical Efficiency of Smallholder Coffee Production in Papua New Guinea. J. Agric. Econ. 2001:153-156.

Pindyck SR, Rubinfeld LD (1998). Econometric Models and Economic Forecasts, 4th Edition. New York: McGraw-Hill.

Rogers EM (1995). Diffusion of Innovations,4th Edition. New York: The Free Press.

Shakya PB, Flinn JC (1985). Adoption of Modern Varieties and Fertilizer Use on Rice inthe Eastern Tarai of Nepal. J. Agric. Econ. 36(3):409-419.

Waller BE, Hoy CW, Henderson JL, Stinner B and Welty C (1998). Matching Innovations with Potential Users: A Case Study of Potato IPM practices. Agric. Ecosyst. Environ. 70:203-215.

Wayo AS (2002). Agricultural Growth and Competitiveness under Policy Reforms in Ghana. ISSER Technical publication. P. 61.

Yaron D, Dinar Aand Voet H (1992). Innovations on Family farms: The Nazareth Region in Israel. Am. J. Agric. Econ. 74:361-370.

11

Extension communication channels' usage and preference by farmers in Benue State, Nigeria

Oto Jacob Okwu* and Shimayohol Daudu

Department of Agricultural Extension and Communication, University of Agriculture, Makurdi, Benue State, Nigeria.

The study assessed farmers' use and preference of extension communication channels in Benue State, Nigeria. A sample of 316 farmers was selected through a stratified random sampling technique and interview schedule was used in eliciting information from them. Descriptive and inferential statistics were used for data analyses. Interpersonal communication channels were generally found to be more available, accessible and used by the farmers than the mass media to obtain information on improved farm technologies. Relatives/friends/neighbours constituted the most regularly available, accessible and used interpersonal channels although extension agents and television were mentioned by the farmers as the most preferred interpersonal channel and mass medium, respectively. Chi-square analysis shows that there is significant relationship between frequency of communication channels use by farmers and their educational level, sex, farm size and farming experience. It is recommended that rural radio booster stations and community rural television stations are established to feature special programmes targeted at rural farmers. To ensure regular availability and accessibility of extension agents (the farmers most preferred channel), efforts should be made to employ more extension agents.

Key words: Farmers, extension, communication channels, usage, preference.

INTRODUCTION

Background statement

In the post-second World War years, development cooperation was based on transfer of technologies to developing countries. According to the model of social labour division in western cultures, technical knowledge was generated by science and industry, transferred by extension services and utilized by farmers (Roth, 2001). The introduction of western technology to non-western farmers was intended to increase production capacity and improve the market position of agriculture.

According to Israel and Wilson (2006), developing an understanding of extension sources and channels used by clients to obtain information is a pre-requisite for efficient educational programming because messages that go unheard or unseen cannot lead to change.

Though early extension efforts were based on direct communication with clients, changes in society and technology have resulted in programmes using diverse array of communication channels to reach clients, both directly and through surrogates.

Many clients, especially older people, continue to rely on more traditional channels for agricultural information while using newer technologies as a complement (Howell and Hebron, 2004; Vergot et al., 2005; Boz and Ozca, 2010).

Conceptual/theoretical frame work

The concept of 'use' of communication channels or media is of great importance because the knowledge of it will provide keys for understanding and predicting outcomes of communication process. Exposure to (or use of) various communication media or channels is a pre-condition for any effect of media content on people to occur. It seems right to state that the influence of any medium in a communication situation or on the message depends not merely on the type of media but also on how it is used, or the use to which it is put.

This study is based on some theoretical perspectives,

*Corresponding author. E-mail: oto079@yahoo.com.

including uses and gratification theory (Lin, 1999), rational choice theories which include the utility maximization framework and social exchange theory (Strebel et al., 2004 adoption-diffusion theory) (Rogers, 2003).

Uses and gratification theory perceives messages recipient as one, who selectively chooses, attends to, perceives and retains the media messages on the basis of his/her needs. The focus was thus shifted from media production and transmission functions to the media consumption functioning. The theory acknowledges that users are goal driven decision-makers who select information communication channels that meet their needs.

Social exchange theory suggests that extension clients are likely to use a particular source-channel combination when the social benefits are greater that the social costs. This theory also suggests that benefits are more likely to be realized when information is relevant to client's needs and when channels provide detailed individualized information.

Problem statement

Nigeria has elaborate agricultural research and extension systems. A wealth of research results therefore exist in the agricultural research system in Nigeria. However, these results are not fully made available to the end-users, the farmers. CTA (1998) has identified limited access to agricultural information as one of the most serious constraints to agricultural development in West Africa.

Benue State of Nigeria has abundant agricultural land estimated at 5.09 million hectares of which 3.8 million hectares is arable (Benue Agricultural and Rural Development Authority, 1999). The state is literally referred to as the "Food Basket" of the nation because of its endowed agricultural potential. But it is one of the poorest states in Nigeria (DFID, 1997). Some questions concerning the utilization and effectiveness of the various communication channels in agricultural extension services delivery there are therefore pertinent and need to be addressed empirically. Which communication channels are available and which ones are being used to disseminate farm innovation messages to the farmers? What is the relative usage of the channels by farmers? Which of the channels do the farmers prefer but constrained in use? Are the disseminated messages relevant to the agricultural information needs of the farmers? Attempts were made by Ilevbaoje (1998), Jeffery (2001) and Obinne et al. (2000) to address some of these questions. However, none of the studies was comprehensive enough to address all the questions raised. Moreover, there is need for generalizations about communication and the consequences for human activities to be continuously re-examined and re-tested against the realities of a changing social world. Thus Israel and Wilson (2006) have suggested that periodic studies be carried out to identify trends among audience segments as knowledge of clients' use of information channels and sources can have an impact on reaching them.

Objective of the study

The general objective of the study is to assess farmers' relative use and preference of communication channels in obtaining information on improved agricultural technologies in Benue State, Nigeria.

The specific objectives of the study are to:

1. Examine the socio-economic characteristics of the farmers in the study area;
2. Find out the level of communication channels' availability and accessibility to farmers.
3. Assess the relative usage by farmers, of the available extension communication channels;
4. Determine communication channel preference of farmers;
5. Determine the relationship between farmers' personal and social characteristics and frequency of use of communication channels as sources of agricultural information.

The following null hypothesis was stated: There is no significant relationship between farmers' personal and social characteristics and their frequency of use of communication channels as sources of agricultural information.

For proper use of communication channels to disseminate information on agricultural innovations, the extension agents should observe which of the channels are used or preferred by farmers. The extent to which farmers use different channels need to be evaluated frequently. This study shall provide the necessary information on the appropriateness or otherwise of the use of communication channels in disseminating information on improved agricultural technologies to farmers in Benue State, Nigeria. The findings of the study shall also provide the necessary feedback mechanism to the media organizations, policy makers and the extension agents for necessary adjustments.

METHODOLOGY

The study area

The study area, Benue State, lies in the middle belt of Nigeria between longitude 6°35' E to 10°E of the Greenwich Meridian and latitude 6°30' to 8 10' N of the equator. With a population of4,218,244 (2006 National Population Census), the State occupies a land mass of 33,955 km^2. It is divided into three geo-political zones namely Zone A (Eastern Zone), Zone B (Northern Zone) and Zone C (Central Zone) and 23 local government areas (LGAs).

Most of the people in Benue State are farmers and the State is acclaimed "The Nation's Food Basket" because of its diverse rich agricultural produce which include yams, rice, beans, cassava, soy beans, benniseed , maize, sorghum, millet, tomatoes. Livestock such as poultry, goats, sheep, pigs and cattle are also reared.

Instrument for primary data collection

Cross sectional field survey was carried out to obtain required primary data for the study. An interview schedule was designed to elicit information from the farmers since most of them were not educated enough to be administered with questionnaire. Validity of the instrument was ascertained through Jury method (Kerlinger, 1973) where the entire instrument was subjected to the scrutiny of relevant experts. The test-retest method of affirming instrument reliability was employed.

Sample selection

The population for the study was made up of all the farmers in Benue State. A sample of 324 farm families made up of 36 each from nine extension blocks in Benue State, Nigeria was used for the study. A 4-stage stratified random sampling technique was used to select the farmer-respondents. This was in line with the agricultural and ecological division of the state into zones, blocks, cells and sub-cells. The three agricultural/ecological zones in the state formed the primary sampling strata and a 4-stage sampling was carried out in each of them. The drawing of 3 blocks from each of the 3 zones constituted the first stage while the selection of 2 cells from each of the chosen blocks was the second stage of the sampling. Selection of 2 sub-cells from each cell was the third stage and sampling of 9 respondents within each sub-cell constituted the last (4th) stage. The total number of farm families in each sub-cell was compiled and summed up to 2933 for the selected sub-cells and this constituted the sampling frame. From this frame, a random sample of 9 farmers was drawn from each of the selected sub-cells. This brought the total number of respondents interviewed to 324. A simple random sampling technique (lottery type) was used for the selection of sample units at each stage. Only data for 316 respondents were, however, utilized for analysis. The others were discarded for incomplete information and inconsistency.

Data collection procedure

Enumerators, specially trained for the work, were engaged for data collection. They were made up of some graduate assistants in the College of Agricultural Economics and Extension, University of Agriculture, Makurdi and extension staff of Benue Agricultural and Rural Development Authority (BNARDA).

Analytical techniques

Descriptive and inferential statistics were used in analyzing the data. Chi-square analysis was carried out to test the null hypothesis of no significant relationship between farmers' personal and social characteristics and their frequency of communication channels usage. SPSS was used in the analysis of the data and testing of the hypothesis.

RESULTS AND DISCUSSION

Personal and social characteristics of respondents

Frequency distribution of the respondents' personal and social characteristics is contained in Table 1. Age distribution of farmer-respondent showed that the middle-aged group of 40 to 59 years has the highest frequency of 183 respondents constituting 57.83% of the total number of respondents. In other words, majority of them were between 40 to 59 years'. The young age bracket (farmers of 39 years and below) constituted only 22.75% which is an indication of declining involvement of youth in agriculture.

A substantial proportion of the farmers (30.65%) had no formal education. Those with primary and secondary education constituted the highest percentage (56.56%) of the respondents. Only a small fraction of the respondents (12.66%) had post-secondary education.

The distribution of household size among the farmers showed that majority of them (56.25%) had between 1 and 10 people per household. The medium class had between 11 to 20 people per household and constituted 32.23% of the respondents. Those with 21 people and above constituted 11.38% of the total respondents.

Most of the respondents were males as they constituted 97.64% of the total number of respondents. This was due to the fact that heads of the households were the ones selected for the study and were in most cases, males except in the case of windows. Most of the respondents (97.33%) were married. Only 2.21% were single while only one (0.32%) was a widow. Majority of the farmers (61.94%) had farm size of 4 ha and less while 33.50% of them (respondents) had 5 to 10 ha. A small number of the farmers (4.42%) had 11 ha and above.

The farmers' personal and social characteristic as revealed by this study were similar to those found by Tologbonse et al. (2006) in their study carried out in the middle belt of Nigeria.

Communication channel availability/accessibility to farmers

Interpersonal channels were generally more available and accessible to the respondents as illustrated by Table 2. Within the IPCs, friends/neighbours/relations were the most regularly available/accessible interpersonal channels of communication (45.89%), followed by extension agents (38.61%). About 31% of the respondents mentioned contact farmers as regularly available and accessible to them as source of agricultural information and 22% indicated opinion leaders as regularly available and accessible to them as sources of agricultural information.

Radio was the most available/accessible mass medium for obtaining agricultural information as 81.01% of the respondents indicated its availability/accessibility. Only 28.80% of the respondents, however, indicated its regular availability/accessibility while 52.21% indicated its availability/accessibility only sometime. Television was the least regularly available/accessible source of agricultural information with only about 1% of the respondents

Table 1. Percent distribution of respondents' personal and social characteristics

Respondents characteristics	Frequency		Percentage
Age (years)			
≥ 39 (young)	72		22.75
40-59 (middle)	183		57.83
60 and above (old)	61		19.27
Educational level			
No formal education	97		30.65
Primary/secondary	179		56.56
Post-secondary	40		12.66
Household size			
≥ 10 (small)	178		56.25
11-20 (medium)	102		32.23
21 and above (large)	36		11.38
Mean		11	
Farm size (ha)			
≥ 4 (small)	196		61.94
5-10 (medium)	106		33.50
11 and above (large)	14		4.42
Mean		4.51	
Sex			
Male	309		97.64
Female	7		2.21
Marital status			
Single	7		2.21
Married	308		97.33
Widowed	1		0.32

N = 316.

mentioning it as regularly available and accessible. On average, about 62% of the respondents indicated that mass media were never available and accessible to them as sources of agricultural information.

Frequency of communication channels' usage as sources of agricultural information

The interpersonal channels have higher relative usage by farmers than mass media. Table 3 gives the details. Friends/relations/neighbours (41.14%), constituted the most regularly used interpersonal channels followed by extension agents (37.02%) and then the contact farmers (22.15%).

Generally, majority (66.77%) of the respondents indicated use of radio in obtaining agricultural information but only 10.44% of them indicated its regular use while 56.33% indicated that they used it only some times. Newspaper, television and film show have very low frequency of usage as 87.02, 81.01 and 76% of the respondents had respectively indicated their non-use at all as sources of agricultural information. The low or non-usage of mass media could be due to low socio-economic status of the farmers and lack of facilities such as electricity to operate them. The low or non-usage of print media such as newspapers, extension bulletins/ newsletters can be attributed to low literacy level of the rural farmers. These findings corroborated those of Yahaya (2002) and Tologbonse et al. (2006) which reveal that television, extension publications (bulletins, news-letters, posters and hand bills) were not considered as important sources of agricultural information among the farmers in Nigeria while friends/neighbours/relations, extension agents and contact farmers were considered important in terms of availability and usage.

Table 2. Percent distribution of channel availability/accessibility.

Communication channel	Availability/accessibility					
	Regularly		Sometimes		Never	
	Frequency	Percentage	Frequency	Percentage	Frequency	Percentage
Mass media						
Radio Benue	91	28.80	165	52.21	60	18.99
Television	3	0.95	73	23.10	240	75.95
Newspapers	8	2.53	78	24.68	230	72.78
Ext. bulletin/news letter	8	2.53	95	30.06	213	67.41
Film show	4	1.26	81	25.63	231	73.10
Posters/Handbills	5	1.58	114	36.07	197	62.34
Interpersonal channel						
Extension agents	122	38.61	153	48.42	35	11.07
Contact farmers	98	31.01	152	48.10	85	26.90
Opinion leaders	70	22.15	133	42.09	75	23.73
Friends/ neighbours	145	45.89	155	49.05	5	1.58

N = 316.

Table 3. Communication channels' usage frequency in accessing agricultural information.

Communication channel	Channel usage					
	Regularly		Sometimes		Never	
	Frequency	Percentage	Frequency	Percentage	Frequency	Percentage
Mass media						
Radio Benue	33	10.44	178	56.33	105	33.23
NTA Makurdi	3	0.95	57	18.04	256	81.01
Nigerian Voice	2	0.63	39	12.34	275	87.02
Ext. bulletin/news letter	7	2.21	76	24.05	233	73.73
Film show	2	0.63	73	23.10	241	76.26
Posters/handbills	4	1.26	97	30.70	215	68.04
Interpersonal channel						
Extension agents	117	37.02	157	49.68	42	13.29
Contact farmers	70	22.15	155	49.05	91	28.80
Opinion leaders	50	18.67	132	41.77	125	39.56
Friends/ neighbours	130	41.14	157	49.68	29	9.17

N = 316.

Communication channels preference by the respondents

Table 4 shows the channel preference by the respondents. The respondents were asked to indicate the most preferred channel should all listed channels be made available and accessible to them. Extension agents were mentioned by majority of them (54.75%) as the most preferred channel. Radio occupied the second position in the ranking of the most preferred channels with 17.41% of the respondents mentioning it as their most preferred channel. Friends/neighbours/relations and television jointly occupied the third position in the ranking order. Contact farmers came next (5th position) with 6.33% of the respondents indicating it as their most preferred channel. Newspapers occupied the last position in the ranking of the most preferred channels. In fact, no respondent mentioned it as the most preferred channel.

The high preference rating for extension agents may be attributed to the interpersonal interaction and immediate feedback enjoyed by the farmers. The less preference shown for newspapers, posters/handbills and extension

Table 4. Percent distribution of channel preference.

Most preferred channel	Frequency	Percentage	Ranking
Mass media			
Radio Benue	55	17.41	2
Television	26	8.23	3
Newspapers	0	0.00	10
Extension bulletin/news letter	8	2.53	6
Film shows	4	1.26	7
Posters/hand bills	2	0.63	8
Interpersonal channel			
Extension agents	173	54.75	1
Contact farmers	20	6.33	5
Opinion leaders	2	0.63	8
Friends/neighbours	26	8.23	3
Total	316	100.00	

Table 5. Chi-square analysis of farmers' characteristics and their level of use of communication channels in obtaining agricultural information.

Variable	X^2	CC	P- Value	Remark
Age	63.87	0.410	0.106	NS
Education	122.70	0.529	0.000	S
Household	194.60	0.618	0.000	S
Farm size	370.00	0.734	0.000	S
Income	64.980	0.413	0.581	NS
Farming experience	70.103	0.426	0.039	S
Sex	40.82	0.338	0.001	S
Marital status	21.945	0.255	0.945	NS

$P<0.05$; S: Significant; NS: Not significant.

bulletin/news letter may be probably due to the high illiteracy level among the respondents.

Relationship between farmers' personal/ social characteristics and frequency of communication channels' usage

The null hypothesis of no significant relationship between farmers' personal and social characteristic and their frequency of use of communication channels as sources of agricultural information was subjected to chi-square test. The results as presented in Table 5 show that farmers' level of education, household size, farm size, farming experience and gender were significantly related to at 5% level of probability. These results, are to some extent, in agreement with those of Yahaya (2002) which showed significant relationship between farmers' education level, sex, farm size and frequency of use of traditional as well as conventional media. The results also support the findings of Boz and Ozcatalbas (2010) which show that farmers' educational level and farm size have significant affect on their use of modern information sources. Similarly, Howell and Hebron (2004) found farm size, educational attainment, gender and household size influence both patterns of information use and methods of delivery.

The results from this study are consistent with social structure literature (Tucker and Napier, 2002) and diffusion research (Rogers, 2003) in showing that factors such as farm size, complexity, gender, etc make a difference in determining the extent to which a client uses information channels or sources. Extension communicators and educators who design extension programmes that account for those contextual effects can increase coverage of the targeted audience and subsequently, the impact of their programmes.

CONCLUSION AND RECOMMENDATIONS

Interpersonal channels were generally found to be more available, accessible and used by the farmers than the mass media to obtain information on improved farm technologies in Benue State, Nigeria. Relatives/friends/ neighbours constituted the most regularly available, accessible and used interpersonal channels. Extension agents were not readily available and used by the farmers although it was mentioned as the most preferred communication channel.

The study shows limited use of mass media by the farmers in obtaining farm information. Only Radio Benue featured regular agricultural programmes and it was the most often used mass medium by the farmers. Television and newspapers were hardly accessible to and used by the farmers. Many farmers indicated television as the most preferred channel if it were available and accessible to them. Most of the client attributes or characteristics had significant relationship with the frequency of use of the communication channels.

In order to solve the problem of low availability and accessibility or complete unavailability and inaccessibility of mass media as channels of agricultural information to rural farmers, the available mass media in Benue State (the study area) including: radio, television, newspapers, should devise ways of reaching out to the rural areas instead of concentrating their operations or attention in urban centers. Radio Benue took a good step in this direction by opening some booster stations in Zones A and C of the state. The booster stations, have, unfortunately, gone moribund. They should be resuscitated. Community rural television stations should be established to feature special programmes targeted at rural farmers.

Public television viewing and radio listening centers should be established to facilitate patronage of agricultural programmes featured by these mass media. Benue State Government and Federal University of Agriculture, Makurdi, through their extension agencies, can establish and manage such centers. To further enhance the patronage and comprehension of the programmes, viewers/listeners groups should be formed to discuss the topics of the programmes immediately after viewing or listening to them.

To ensure regular availability/accessibility of extension agents (farmers' most preferred channel), efforts should be made to fill the vacancies created by retracement; resignation, termination or death of extension agents.

There are presently many empty cells (that is, cells not manned by extension agents) in the study area.

REFERENCES

Benue Agricultural Rural Development Authority (1999). Report on adoption rates of proven technologies in Benue State, 35p.

Boz I, Ozcatalbas O (2010). Determining information sources used by crop producers: A case study of Gazian province in Turkey. Afr. J. Agric Res., 5(10): 980-987

CTA (1996). The role of information for rural development in ACP countries: review and perspectives. Proceedings of an international seminar, Montpellier, France, 12 – 16 June, 1995. Aniane: Louma Production, pp. 117 - 132.

DFID (1997). Improved farmer participation in research and extension in Benue State. West and North Africa Department, ODA, DFID, 16p.

Howell JL, Hebron GB (2004). Agricultural landowners lack of preference for internet extension. J. Ext., 42(6). retrieved April 1, 2011 from http://www.joe.org/joe/2004december/a7.php

Ilevboje IE (1998). Determinants of utilization of selected information sources by farmers in Benue and Plateau states of Nigeria: J. Agric. Rural Dev., 6(1): 1 -7

Israel GD, Wilson KM (2006). Sources and channels of information used by educational program clients. J. Appl. Comm., 90 (4): 55-78

Jeffery M (2001). "Mass media monitoring and evaluation". Cooperative Extension centre/Improved Farmer Participation in Research and Extension, Benue, University of Agriculture, Makurdi.

Kerlinger NF (1973). Foundations of behavioural research. New York: Holt Rinehart and Winston Inc. pp. 616 – 617.

Lin CA (1999). Uses and gratification. In stone G, Singletary M, Richmond V (eds) Clarifying communication theories: a hands-on approach (pp 199-208). Ames: Iowa State University press

Obinne CP, Ndubilo EO, Ozowa VN (2000) . Mass media and agricultural innovations diffusion in Benue State, Nigeria. J. Soc. Pol. Stud., 1: 54 – 62.

Rogers EM (2003). Diffusion of innovations (5th ed.) New York: free press.

Roth G (2001). The position of farmers' local knowledge within agricultural extension, research and development cooperation. Indigenous Knowl. Dev. Monit., 9(3): 10 – 12.

Strebel J, Erdem T, Swait J (2004). Consumer search in high technology markets: exploring the use of traditional information channels. J. Cons. Psychol., 14(1/2): 96-104

Tologbonse EB, Mesini O, Tsado JH (2006). Farmers perception of sources of information in relation to adoption of improved technology by farmers in inland valley swamps of middle-belt zone of Nigeria. J. Agric. Ext., 9:63-73

Tucker M, Napier TL (2002). Preferred sources and channels of soil and water conservation information among farmers in three Midwestern U.S. watersheds. Agric, Ecosyst. Environ., 92: 291-313.

Vergot III P, Israel G, Mayo DE (2005). Sources and channels of information used by beef cattle producers in twelve counties of the northwest Florida extension district. J. Ext., 43(2). Retrieved April 1 2011 from http://www.joe.org/joe/2005april/rb6.php

Yahaya MK (2002). Gender and communication variables in agricultural information dissemination in two agro-ecological zones of Nigeria. Research Monograph. Ibadan:Corporate Graphics Ltd.

Impact of Olam organisation extension programme on socio-economic life of cocoa farmers in Ondo State, Nigeria

Akinnagbe O. M* and Ajayi A. R

Department of Agricultural Extension, University of Nigeria, Nsukka, Nigeria.

The study was undertaken to assess the impact of Olam organization extension programmes on socio-economic life of cocoa farmers in Ondo state, Nigeria. Data were collected from 60 respondents made up of 30 Olam farmers and 30 non-Olam farmers through the use of structured interview schedule. Multistage random sampling technique was used in selecting respondents. Data were analyzed using mean statistics, standard deviation, t-test, chi-square, performance index and propensity score matching technique. The findings showed that the net income per capital of Olam farmers was significantly different from that of non-Olam farmers with a gain of ₦16, 466.47. Also, the project made some appreciable socio-economic impact on cocoa farmers in the following areas: ease of selling cocoa products; knowledge on cocoa production and marketing; increase in number of farmers with knapsack sprayers and other basic assets. Based on the findings of this study, Olam organization extension programmes have the potential to boost cocoa production in Ondo State. It is therefore recommended that Olam organization should involve more farmers in her programme to compliment the effort of government in cocoa resuscitation programme in Nigeria.

Key words: Impact, programme, Olam, cocoa farmers, Ondo State.

INTRODUCTION

The relevance of cocoa to most developing economies cannot be overemphasized as cocoa is produced by more than fifty developing countries across Asia, Africa and Latin America; all of which are in tropical or semi-tropical areas (Ogunleye and Oladeji, 2007). In Nigeria, cocoa has been the main agricultural stake of Nigeria economy until early 1970's when the crude oil was discovered in the country in commercial quantity. However cocoa has remained a valuable crop and a major foreign exchange earner among agricultural commodity exports of the country. The contribution of the cocoa sub-sector to Nigeria's total agricultural export earnings averaged 70.6% between 1971 and 1975, 89.8% between 1976 and 1980, 84.6% between 1985 and 1987, 76.8% between 1986 and 1990, and 53.3% between 1992 and 1996 (Agboola, 2005). Although these figures indicated a declining trend, a lot of factors have been attributed to the decline in cocoa production. These include: over-dependence on oil exports, the problem of depleted soil fertility, poor control of pests and diseases, use of poor planting materials, poor maintenances of cocoa farms, lack of credit facility to support production practices, defective methods of harvesting and poor handling of post harvest processes and inefficient agricultural extension services.

Recent years have seen a renewed commitment by a number of West African governments and their development partners to promoting a healthy cocoa sector based on the recognition of the significant role cocoa can play in the fight against poverty and in promoting rural economic growth. The Federal Government of Nigeria has embarked on various resuscitation strategies which now appear to be yielding

*Corresponding author. E-mail: wolexakins@yahoo.com.

desired fruits. According to Tunde (2007), the inability of the government to cope with the development of agriculture in the country alone, has led to her deliberate efforts of encouraging private individuals and organizations to participate maximally in agricultural development process. Among these are the Olam Nigeria Limited (ONL), Saro Agro Allied Limited and Sustainable Tree Crops Programme (STCP) - Nigeria. However, this study focused on Olam Nigeria Ltd (ONL).

Olam, a private, commercial and multi-product company was established in 1989 by the Kewalram Chanrai Group (KCG), with its headquarters in Singapore and principal office in Lagos, Its regional offices are found in Kano and Akure. It provides what could be called specialized extension services and uses specific agricultural products (such as cocoa, coffee, cashew, sheanuts, sesame, rice and teak wood) as industrial materials. Its cocoa business began in Nigeria in 1992 and has participated in all aspects of cocoa business like production, marketing and processing, in the cocoa-producing states such as Ondo, Osun, Ekiti, Edo, Ogun, Cross River, Taraba and Adamawa (Olam, 2007). Available records revealed that ONL had implemented model farming concept under the nucleus estate initiatives; an out grower's programme, by reviving old farm settlements and model farms in all cocoa growing states. Besides, it has launched training and awareness programmes, covering about 6,000 farmers and capacity buildings across the cocoa producing states. The farmers' plots are taken as demonstration plots on which the practical trainings are done in demonstration farms being managed in Ondo, Osun and Ekiti states (Akinnagbe, 2008).

From the foregoing and after years of existence of the programme, the activities of this organization needed to be evaluated to confirm the acclaimed performances and impact on socio-economic life of the cocoa farmers. According to Horton et al. (1993), evaluation is the act of judging, appraising or determining the value or quality of a programme, whether it is on-going or completed. It is also a process of providing reliable, valid, relevant and useful information to decision makers about the operation and effects of social programme. Impact evaluation assesses changes in the well-being of individuals, households, communities or firms. It also deals with the effects of the intervention programmes outputs on the target beneficiaries. An evaluation model called targeting outcomes of programme (TOP) presented by Bennett and Rockwell (1995), as an integrated approach to planning and evaluation was used in this study. TOP uses a single model to target outcomes, track the extent they have achieved and evaluate programme performance towards achieving them. It measured impacts by collecting information on people reaction, changes in knowledge, attitude, behaviours and ultimately the social, economic and environmental condition, which are targets of the social impact approach used in this study.

Purpose of the study

The overall purpose of this study was to assess the impact of Olam Organisation extension programme on socio-economic life of cocoa farmers in Ondo State. Specifically, the study was designed to:

1) Describe the socio-economic characteristics of the respondents;
2) Assess the performance of the Olam organization with respect to farm rehabilitation and distribution of inputs; and
3) Determine the impact of the programme on socio-economic life of the cocoa farmers.

Hypotheses

(1) There is no significant difference between the income of Olam farmers and non-Olam farmers.
(2) There is no significant difference between the farm input and basic assets possessed by Olam farmers and non-Olam farmers.

THEORETICAL FRAMEWORK

Impact studies aim to measure not only the reactions of the beneficiaries and the outputs generated by them, but also the proportion of any discernible change attributable to the project. In any project, and throughout the project cycle, there is need not only for routine collection of data through monitoring or continuous assessment, but also for evaluation and assessment of impact. The nature and purpose of a study determines the type of analysis that can be employed. Impact evaluation of this nature will rely on the construction of a counterfactual – an attempt to estimate what a given outcome would have been for the beneficiaries of a programme if the programme had not been implemented; impact evaluations thus address causality and allow result to be attributed to specific interventions (Rawlings and Schady, 2002). Normally, if the comparison group is correctly identified, the difference between the treatment and comparison groups isolates the effect of the intervention (Chase, 2002).

While calculation of the rates means, frequency distribution and percentages may be adequate for some exploratory studies, more detailed and higher level analysis will be required for case studies and sample surveys especially those that deal with quantitative data (Eboh, 1998) like impact analysis. For analyzing dependence, multiple regressions are used. For analyzing relationships whose dependent variables assume a discrete or dichotomous value, qualitative choice models are used. In such relationships, the probability of an event occurring is a function of a set of non-stochastic explanatory variables and a vector of unknown parameter. Following Amemiya (1981), the general form of the univariate dichotomous choice model can be expressed as:

$$P_i = P_i (Y_i = 1) = G (X_i \Phi) (I = 1,2 ..,n) \ldots \quad (1)$$

The Equation 1 states that the probability of an outcome, P_i (Y_i = 1) is a function of the vector of explanatory variables X_i and unknown parameter vector Φ. Because the functional form of G is unknown, practical applications of the model are not feasible (Amemiya, 1981). Then, an explicit functional specification of G is necessary. Three functional relationships often specified are the linear probability, probit and logit models. The dichotomous dependent variable model that will be used in the study is logit (the standard normal distribution function).

The model is specified in the general form thus:

$$\text{Log } P/1\text{-}P = \text{Log } O_i = a_i + \beta_i X_i + \beta_2 X_2 + \beta_3 X_3 + \ldots + \beta_n X_n \quad (2)$$

Where Log P/1-p = Log of odds of participating in the intervention programme

a = constant

β = coefficients and

X_n = socioeconomic characteristics of the respondents.

In most impact estimation, the problem is the issue of taking care of observables and unobservable factors that influence impact. Ordinarily, an OLS regression of the outcome indicator (for example per capita) on participation and other explanatory variables used in targeting participants can be determine by the effect of participation. However, the unobserved variables that influences participation conditional on some observables may also be influencing expenditure per capita, the outcome indicator, thereby making it difficult to isolate the effect of benefiting from the programme. In order words, the error term in participation equation may be correlated with the error term in expenditure/capita equation for example in

$$Pi + a + eZi + vi \ \ldots \quad (1) \text{ and}$$

$$E/C = a + cPi + cxi + ei \ \ldots. (2),$$

vi may be correlated with ei.

Some methods, which can deal with these problems, have been identified. Hence, in rigorous estimation of impact, the methods normally used are randomization, propensity score matching, instrumental variables and double differences (Ravallion, 2001; Rawlings and Schady, 2002). Randomization assigns "treatment" through some sort of balloting or lottery allowing researcher to construct treatment and control group. In order words, the selection into treatment and comparison groups is random in some well-defined set of people (Ravallion, 2001). Randomization is immensely appealing because if the sample is large enough, this method controls for all differences, observable and unobservable, between the treatment and control group; simple differences in charges in outcomes, can then be credibly interpreted as the impact of the treatment on the treatment on the treated (Rawlings and Schady, 2002).

Propensity score matching in its simplest form involves predicting the probability of treatment on the basis of observed covariates for both the treatment and the control groups' samples (Rawlings and Schady, 2002). In propensity score matching, one picks an ideal comparison group from a larger survey then matches the comparison group to the treatment group on the basis of set of observed characteristics on the predicted probability of participation given observed characteristics (propensity score). The observed characteristics are those used in selecting individuals but not affected by programme participation. Propensity score matching is a better method of dealing with differences in observables. However, a few tests that have been done suggest that with good data, propensity score matching can greatly reduce the overall bias and outperforms regression-based methods (Ravallion, 2001).

Instrumental variables (IV) are variables that matter to participation but not to the outcome given participation (Ravallion, 2001). Instrumental variable are correlated with the probability of treatment but correlated with unobserved determinant of outcomes. IVs estimates are predicted entirely on the validity of the instrument and unobserved determinant of treatment effects can result in serious biases. The instrumental variables are first used to predict programme participation, and then one sees how the outcome indicator varies with the predicted value conditional on other characteristics.

In double difference one compares treatment and comparison group (first differences) before and after a programme (second difference) (Ravallion, 2001). The method can be descriptive as a basis for assessing impact if there is no control group and only the treatment group is compared before and after interaction. Exploratory factor analysis will also be used for grouping constraining factors for informed policy recommendation. The step in propensity score matching for the programme included:

1) Pooling of two groups of individuals (that is, the treatment and comparison group as already indicated). After the pooling, a logit model of participation as a function of some socio-economic variables was estimated. The variables selected were those that were used in selecting participants and those that were not affected by participation. The socio-economic variables included were age, household size, number of years of schooling, gender. The equation was put thus:

$$Pi = \log \frac{Pi}{1-Pi} = \log Oi = a_i + b_i age + b_2 housesize + b_3 school + b_4 gender$$

2) From the logit regression, a predicted value of the probability of participation was created, these are propensity scores. Each individual had a propensity score. Non-participants with low propensity scores that were outside the range for participants were excluded.
3) For each individual participant, a non-participant that had the closest propensity score, as measured by the absolute difference in scores, referred to as nearest neighbour was obtained. For more precise estimate, the nearest five neighbours were used.
4) The mean value of the outcome indicators (per capital expenditures, income share spent on food, and income) for the nearest five neighbour was calculated. The differences between the mean and actual value for the participants (beneficiaries) was the estimate of the gain due to the programme.
5) The mean of individual gains was calculated to obtain the average overall gain.

METHODOLOGY

Study area

The study was carried out in Ondo State. The state lies between latitudes 5°45[1] and 7°52[1] north of the equator and longitudes 4°20[1] and 6°5[1] east of Greenwich Meridian. The State is bounded in the north by Ekiti and Kogi States; in the east by Edo State; in the west by Osun and Ogun States; and in the south by the Atlantic Ocean. Its land area is about 14,793 square kilometers (http://www.ondostategovernment.com). Ondo State has a population of 3,460,877 persons made up of 1,745,057 male and 1,715,820 female (NPC, 2007). The climate of the area is highly favourable for the agrarian activities of her teeming population who grow crops such as cocoa, kola nut, palm tree, rubber and arable crops like maize, yam and cassava. The annual rainfall is between 1000 and 1500 mm with a high daily temperature of about 30°C and relatively high humidity.

Population and sampling procedure

All cocoa farmers in the Ondo State constituted the population for the study. Multistage sampling technique was employed in the selection of the cocoa farmers for the study. Firstly, the farmers in the state were grouped into 2; Olam farmers and non-Olam farmers. Olam farmers are those that participated in Olam Nigeria limited extension programme, while the other cocoa farmers in the state constituted the the second group that were not used by Olam, hence were regarded as non-Olam farmers. The second stage involved the selection of the community that produced cocoa and at the same time were used by Olam as a contact community. From the list of 70 communities used by Olam organization in Ondo State, 10 cocoa producing communities (Aponmu, Irese, Laagba, Bagbe, Oboto, Igbo-Ore, Igodo, Arogbo, Ago paanu and Agogo) were randomly selected. The third stage involved the selection of the cocoa farmers. A list of Olam's model farmers was obtained from the management of Olam organization. From the list, 3 farmers each were selected through simple random sampling technique, from the 10 selected communities, making a total of 30 Olam farmers (OFs). The identification of OFs was done with the help of Olam staff.

Also from the same 10 communities, 3 Non-Olam farmers (NOFs) were randomly selected, making a total of 30 NOFs, this category served as a pipeline comparison (control) group to estimate the impact of the programme on the participant and non-participant farmers. Thus, the total sample size for this study was 60 cocoa farmers in the state.

Instrument for data collection

Data for the study were collected from the respondents through the use of structured interview schedule. The interview schedule contains relevant questions based on each of the objectives of the study. Data on Olam extension programme were collected from the organization zonal office in Akure. The information needed was retrieved from the organization's publication and report. Since majority of the farmers were not used to keeping records of events, their responses were purely on memory-recall. However, Casley and Lusy (1982) and Ladele (1991) maintained that any memory-recall data, collected as satisfactory as possible, are valid for use in social research. Face validity was carried out to ensure that the instrument collects the data it is meant to collect. This was done by lecturers in the Department of Agricultural Extension, University of Nigeria, Nsukka before field administration.

Measurement of variables and data analysis

To assess the performance of the organization's programme, a performance index model, used by Ajieh and Ajayi, 2007 was used.

The model is given as: $I_A = \frac{Q_A}{Q_o} \times 100$

Where: I_A = Availability index

Q_A = Quantity actually available

Q_o = Optimum quantity at crucial period

When $I_A \geq 0.50$ or $\geq 50.0\%$ shows that the performance is on the average or high.

To determine the impact of Olam organization programme on socio-economic life of the farmers, 'before' and 'after' comparison, recommended by Gittinger (1982) as one of the approaches used to isolate the changes arising from an intervention, was used to compare farmers well being before inception of the Olam and afterwards. He has recommended the 'with' and 'without' comparison as the more appropriate tool for this purpose because of its ability to take care of the changes inherent in a project and those identified by Maitima et al. (2007) as arising out of external factors that are not related to the project. To determine the impact of Olam organization on income of cocoa farmers, propensity scores matching was used to determine the differences in their net income.

Data analysis

Data were analyzed using frequency, percentage, mean, t-test and Chi-square statistics. Statistical Package for the Social Science (version 11) and STATA were the software packages used for the analysis.

Table 1. Distribution of respondents according to personal and socio-economic characteristics.

Variable	Olam farmers		Non Olam farmers		T-test value
	Mean $\bar{x}$	Standard deviation	Mean $\bar{x}$	Standard deviation	
Age (years)	58.03	12.57	57.03	12.41	0.31
No of years spent in school	6.63	4.64	8.60	3.79	-1.79
Household size (number)	5.83	3.01	5.30	1.91	0.81
Age of cocoa farm (years)	36.10	13.56	32.83	11.70	0.99
Farming experience (years)	36.57	10.49	30.37	11.28	2.20*
Total farm size in 2007 (hectare)	7.67	3.57	6.93	11.10	0.34
Cocoa output in 2007 (bags)	8.97	190.96	7.20	3.59	1.54
Estimated income from the sale of cocoa product in 2007 (naira)	219556.00	172246.08	101632.00	59174.19	3.54*

* Significant (P≤ 0.05).

RESULTS AND DISCUSSION

Socio-economic characteristics of the respondents

The result of descriptive statistics of the socio-economic variables used in the analysis of propensity score matching technique is presented in Table 1. Some of the variables were included in logistic regression of the estimation of the propensity score. The average age of the respondents for OFs and NOFs were 58 and 57 years, respectively. The results further show that there was no significant different (t = 0.31; p≤0.05) between their ages. The average number of years spent by OFs and NOFs in school were 6 and 9 years, respectively. This shows that there was no significant different (t = -1.79; p≤0.05) between the number of years spent in school for both OFs and NOFs. The average household sizes were 6 and 5 persons and the mean ages of cocoa plantation were 36 years and 32 years, respectively. Also, there was no significant different (t = 0.81 and 0.99; p≤0.05) between household size and age of cocoa farms of both OFs and NOFs, respectively. The average farming experiences were 36 and 30 years showing a significant different (t = 2.20; p≤0.05). The total farm size in 2007 for both OFs and NOFs was put at 7.67 and 6.93 ha respectively, showing no significant difference (t = 0.34; p≤0.05), while the total cocoa output was about 9 bags for OFs and 7 bags for NOFs. There was also no significant difference (t = 1.54; p≤0.05) between the total farm size of OFs and NOFs. The average income of the cocoa farmers were ₦219556.00 and ₦101632.00 for OFs and NOFs respectively. There was a significant difference in the estimated mean income from the sale of cocoa products for both OFs and NOFs. The implication of these findings is that, the project made an appreciable impact on annual income of the OFs.

Performance of the ONLs as regards rehabilitation of cocoa farms and distribution of inputs to cocoa farmers

A performance index model was used in assessing the performance of the ONL as regards rehabilitation of farms and distribution of inputs to farmers in Ondo State. Like every organization, Olam set a target of activities in 2007 for rehabilitation of farms and distribution of inputs. Available record shows that, Olam proposed to achieve the following activities in 2007. These include; regular visit to cocoa farmers for dissemination of innovation on cocoa, regular training programmes on pre-planting, planting and post planting operations of cocoa, rehabilitation of cocoa farms and distribution of cocoa hybrid seedlings.

At the end of 2007, results in Table 2 show the target level of the organization, its achievement level and performance index. At the end of 2007, Olam revive 70 old farm settlements and model farms out of the proposed 70 old farms. The index score was 100.0%. This implies that, Olam performance as regards reviving of old farm settlements, model farms was high. Seventy (70) demonstration plots were targeted and achieved with an index score of 100.0%. This was also high. Olam proposed to train 1750 cocoa farmers. As at 2007, 1355 farmers were trained on improved agronomic practices (for example nursery preparation and maintenance, transplanting, weeding, chemical application, harvesting, fermentation and drying). Hence, the index score was 76.3%. This implies that the performance of the organization as regard training was high. About 70 farmers were given agrochemicals at a subsidized rate out of the proposed 80 farmers targeted. This represents an index score of 85.7%.

In 2007, 2050 hybrid cocoa seedlings were distributed free to farmers as against 2000 targeted. This represents an index score of 102%. This implies that the performance of ONLs with respect to provision of cocoa seedlings distribution was high. From the foregoing, the following extension activities/programmes of Olam to cocoa farmers were high. These include training programmes on pre-planting, planting and post planting operations of cocoa, rehabilitation of old cocoa farms and distribution of cocoa hybrid seedlings free to their target farmers. Hence, Olam organization extension programmes

Table 2. Performance indices of ONL as regards farm rehabilitation and distribution of inputs.

Activities	Target level (No)	Achievement level (No)	Performance index (%)
Reviving of old farm settlements and model farms (No / ha)	70	70	100.0*
Training of farmers on improved agronomic practices (for example nursery preparation and maintenance, transplanting, weeding, chemical application, harvesting, fermentation, drying etc)	1750	1355	76.3*
Demonstration plots (No)	70	60	100.0*
Distribution of agrochemicals at a subsidized rate	70	60	85.7*
Provision of hybrid seedlings to farmers (No)	2000	2050	102.0*

*Above average.

have the potential to boost cocoa production in Ondo State thereby contributing to reduction in the poverty of the state.

Impact of Olam organization extension programme on the net-income of the cocoa farmers

The impact of the organization's programme on net-income of the cocoa farmers was ascertained using propensity score matching technique (steps indicated in methodology). The logit result was significant at the Chi-square value of 9.67 at 0.05 level of probability (Table 3). Thus, at least one variable significantly influence participation in the programme. The analysis result in Table 3 shows that, farming experience was significant (2.57; $P \leq 0.010$) at 0.05 level of significance.

The detailed result in Table 4 shows the average overall gain in net income per capita (total income (Bags of cocoa produced x selling price) / household size} due to Olam's intervention after the nearest neighbour matches of propensity score obtained after the logit analysis. The result shows that, the average gain in net income per capital was ₦16, 466.47. The t-value of 2.153 for the test of difference between the net income per capita of the beneficiary (OFs) and non beneficiary (NOFs) with close propensity score from which the ATT was obtained, was significant at 0.05 level of probability. This shows that, the net income per capita of OFs was significantly different from that of NOFs with a gain of ₦16, 466.47. Thus, the programme of Olam had positively and significantly impacted on the Olam participation farmers in Ondo State.

Impact of the programme on ease of accessibility to modern farm inputs

Entries in Table 5 show that, majority (70.0 and 66.7%) of the Olam farmers (OFs) and Non Olam farmers (NOFs) do not have access to modern farm inputs (like herbicides and insecticides) in the years 2003. However, about 80% and 73% have easy access to these modern farm inputs. Further analysis show that there was no significant difference ($x^2 = 5.00$; $p \leq 0.05$) between the degree of accessibility to modern farm inputs by OFs before (2003) and after benefiting in Olam cocoa programme in 2007. Also, there was no significant difference ($x^2 = 2.13$; $p \leq 0.05$) between the degree of accessibility to modern farm inputs by NOFs between the year 2003 and 2007. The data further show that there was no significant difference ($x^2 = 0.59$; $p \leq 0.05$) between the degree of accessibility to farm inputs by the OFs and NOFs in 2007.

Impact of the programme on ease of selling cocoa products

Entries in Table 5 show that, 86.7 and 70.0% of the OFs and NOFs find it easy in selling their cocoa products in the years 2003, respectively. However, about 90.0 and 93.3% find it easy in selling their cocoa products in 2007, respectively. Further analysis show that there was a significant difference ($x^2 = 9.487$; $p \leq 0.05$) between ease of selling cocoa products by OFs before (2003) and after benefiting in Olam cocoa programme in 2007. There was no significant difference ($x^2 = 0.918$; $p \leq 0.05$) between the ease of selling cocoa products by NOFs between 2003 and 2007. There was a significant difference ($x^2 = 3.832$; $p \leq 0.05$) between ease of selling cocoa products by OFs and NOFs in 2007. The observed change in ease of selling cocoa products is an indication of positive impact of the Olam farmers on the cocoa farmers.

Table 3. Logistic regression of the estimation of the propensity score.

Variable	Coefficient	Standard error	z	P>/z/	(95% Coefficient interval)	
Age	-0.0526131	0.0380617	-1.38	0.167	-0.1272127	0.0219866
Farming experience	0.1126725	0.0437844	2.57	0.010	0.0268566	0.1984884
Source labour	-0.0267601	0.8432545	-0.03	0.975	-1.679509	1.625988
Source of fund	0.7992571	0.8523841	0.94	0.348	-0.8713851	2.469899
Family size	-1.041005	1.604543	-0.65	0.516	-4.185851	2.103841

Log likelihood = -36.754763; Number of obs = 60; LR chi2 (4) = 9.67; Prob > chi2 = 0.0464; Pseudo R2 = 0.1162.

Table 4. Propensity score matching technique (PSMT) showing the overall gain in income per capital of OFs .

No of treatment	Number of control	ATT (Net income per capital)	Standard error	t-value
30	21	16466.476	7649.694	2.153*

*Significant (P ≤ 0.05); the numbers of treated and control refers to actual nearest neighbour matches; ATT refers to average treatment effect on the treated using nearest neighbour matches.

Impact of the programme on the knowledge of cocoa production and marketing

Entries in Table 5 show that, 50.0 and 46.7% of the OFs and NOFs have fair knowledge on cocoa production and marketing in the years 2003, respectively. However, about 73 and 53% have adequate knowledge on cocoa production and marketing in 2007, respectively. Further analysis show that there was a significant difference (x^2 = 20.50; p≤0.05) between the knowledge level by OFs in 2003 and after benefiting in Olam cocoa programme in 2007. Also, there was a significant difference (x^2 = 12.61; p≤0.05) between the knowledge level by NOFs between the year 2003 and 2007. The result of the findings further reveal that there was a significant difference (x^2 = 13.14; p≤0.05) between the knowledge level of the OFs and NOFs in 2007. A greater proportion of OFs had adequate knowledge about cocoa production and marketing. It is possible to conclude that the Olam programme had improved the knowledge of OFs through regular training.

Impact of the programme on perceived standard of living

Data in Table 5 show that, 63.4 and 73.3% of the OFs and NOFs rated their standard of living as good as others in the years 2003, respectively. However, about 37 and 43% of OFs and NOFs rated their standard of living as better than others in 2007, respectively. Further analysis show that there was a significant difference (x^2 = 24.09; p≤0.05) between the rating of standard of living by OFs before (2003) and after becoming a project farmers in 2007. Also, there was a significant difference (x^2 = 18.938; p≤0.05) between the rating of standard of living by NOFs between the year 2003 and 2007. The result of the findings further reveal that, there was no significant difference (x^2 = 20.411; p≤0.05) between the standard of living of the OFs and NOFs in 2007. It could be deduced from these findings that the OFs had a positive change in the perception of their standard of living after becoming project farmers. It is concluded that the standard of living of the OFs and NOFs due to the existence of Olam programme are not the same.

Impact of the programme on basic assets of the cocoa farmers

Data in Table 6 show that, 60.0% of OFs and NOFs had knapsack sprayer in year 2003, respectively. On the other hand, about 80 and 67% of OFs and NOFs had knapsack sprayer, respectively in 2007. This implies that, the proportion of OFs and NOFs with knapsack sprayers increased over the year. Further analysis show that there was a significant difference (x^2 = 11.250; p≤0.05) between the number of OFs with knapsack sprayer before (2003) and after becoming a project farmers in 2007. There was no significant difference (x^2 = 2.500; p≤0.05) between the number of NOFs with knapsack sprayer between the year 2003 and 2007. Meanwhile, there was significant difference (x^2 = 3.750; p≤0.05) between OFs and NOFs with knapsack sprayer in 2007. The implication of these findings is that, more of OFs acquired knapsack sprayer after becoming a project farmer. This could be due to the importance of this tool for spraying cocoa farms for pest, diseases and weeds. The observed significances confirm the benefits of the project to the farmers.

Data in Table 6 show that, 60.0 and 63.3% of the OFs and NOFs had fan in year 2003, respectively. About 83 and 63% of OFs and NOFs had fan in 2007, respectively. This shows that the number of cocoa farmers that purchased fan increase in 2007. Further analysis show

Table 5. Chi-square analysis of the impact of the Olam extension programme on socio-economic life of cocoa farmers.

Socio-economic variable	Olam farmer (OFs) n=30			Non-Olam farmers (NOFs) n = 30			OFs	NOFs	X^2-value
	2003	2007	X^2-value	2003	2007	X^2 value	2007	2007	
Degree of accessibility to modern farm inputs									
Not access at all	70.0	13.3	5.000	66.7	26.7	2.131	13.3	26.7	0.597
Easy accessibility	30.0	80.0		33.3	73.3		80.0	73.3	
Very easy accessibility	-	6.7		-	-		6.7	-	
Ease of selling cocoa products									
Very easy	86.7	90.0	9.487*	70.0	93.3	0.918	90.0	93.3	3.832*
Difficult	10.0	10.0		30.0	6.7		10.0	6.7	
Very difficult	3.3	-		-	-		-	-	
Knowledge on cocoa production and marketing									
Poor knowledge	16.7	10.0	20.509*	30.0	30.0	22.65*	10.0	30.0	13.146*
Fair knowledge	50.0	16.7		46.7	16.7		16.7	16.7	
Adequate knowledge	33.3	73.3		23.3	53.3		73.3	53.3	
Rating of standard of living as compared with others in the community									
Worse than others	-	-		3.3	-		-	-	
As good as others	63.4	33.3	24.098*	73.3	50.0	18.938*	33.3	50.0	20.411*
Better than others	23.3	56.7		10.0	36.7		56.7	36.7	
Don't know	13.3	10.0		13.3	13.3		10.0	13.3	

*significant.

that there was a significant difference (x^2 = 9.00; p≤0.05) between the number of fan owned by OFs before and after becoming a project farmers in 2007. Also, there was a significant difference (x^2 = 6.11; p≤0.05) between the number of fan owned by NOFs between the year 2003 and 2007. The result of the findings further reveal that there was no significant difference (x^2 = 0.02; p≤0.05) between the number of fan owned by OFs and NOFs in 2007. It is concluded that the number of OFs and NOFs that purchased fan are the same.

Data in Table 6 show that, majority (73.3%) of the OFs and NOFs had radio in year 2003, respectively. About 80 and 77% of OFs and NOFs had radio in 2007, respectively. This implies that, the proportion of farmers with radio increase over the year for both OFs and NOFs respectively.

Further analysis show that there was a significant difference (x^2 = 20.625; p≤0.05) between the number of OFs with radio before and after becoming a project farmers in 2007. On the other hand, there was a significant difference (x^2 = 25.109; p≤0.05) between the number of NOFs with radio between the year 2003 and 2007. The possible reason for increase in the number of people with radio among the NOFs could be as a result of significant roles radio played in disseminating information to the farmers. Significant difference (x^2 = 0.23; p≤0.05) existed between the number of OFs and NOFs with radio in 2007. Hence, as a result of the influence of the programme, a greater proportion of OFs had radio.

Data in Table 6 show that, 53.3% of OFs and 36.7% of the NOFs had television in the year 2003. On the other hand, about 56.7 and 58.6% of

Table 6. Chi-square analysis of the impact of the Olam extension programme on basic assets of cocoa farmers.

Basic assets of cocoa farmers	Olam farmer (OFs) n=30			Non-Olam farmers (NOFs) n = 30			OFs	NOFs	X^2-value
	2003	2007	X^2-value	2003	2007	X^2 value	2007	2007	
Knapsack sprayers									
Yes	60.0	80.0	11.250*	60.0	66.7	2.500	80.0	66.7	3.750*
No	40.0	20.0		40.0	33.3		20.0	33.3	
Fan									
Yes	60.0	83.3	9.000*	56.7	63.3	6.111*	83.3	63.3	0.029
No	40.0	16.7		43.3	36.7		16.7	36.7	
Radio									
Yes	73.3	80.0	20.625*	73.3	76.3	25.109*	80.0	76.3	15.093*
No	26.7	20.0		26.7	23.3		20.0	23.3	
Television									
Yes	53.3	56.7	13.274*	36.7	58.6	2.876	56.7	58.6	1.094
No	46.7	43.3		63.3	41.4		43.3	41.4	
Refrigerator									
Yes	30.0	43.3	16.813*	6.7	13.8	2.368	43.3	13.8	0.738
No	70.0	56.7		93.3	86.2		56.7	86.2	
Bicycle									
Yes	53.3	46.7	11.059*	50.0	36.7	1.292	46.7	36.7	0.741
No	46.7	53.3		50.0	63.3		53.3	63.3	
Motorcycle									
Yes	36.7	65.5	5.058*	20.0	43.3	4.887*	65.5	43.3	0.012
No	63.3	34.5		80.0	56.7		34.5	56.7	
Car									
Yes	23.3	30.0	7.462*	3.3	16.7	5.172*	30.0	16.7	0.286
No	76.7	70.0		96.7	83.3		70.0	83.3	
Wall clock									
Yes	86.7	96.7	6.724*	90.0	96.7	9.310	96.7	96.7	30.000*
No	13.3	3.3		10.0	3.3		3.3	3.3	

*significant.

OFs and NOFs had television, respectively in 2007. This implies that, the proportion of OFs with television increased more than the proportion of NOFs with television. Further analysis show that there was a significant difference (x^2 = 13.27; p≤0.05) between the number of OFs with television before and after becoming a project farmers in 2007. On the other hand, there was no significant difference (x^2 = 2.87; p≤0.05) between the number of NOFs with television between the year 2003 and 2007. Meanwhile, there was no significant difference (x^2 = 1.09; p≤0.05) between the number of OFs and NOFs with television in 2007.

Data in Table 6 show that, 30.0% of OFs and 6.7% of the NOFs had refrigerator in the year 2003. On the other hand, about 43.3 and 13.8% of OFs and NOFs had refrigerator, respectively in 2007. This implies that, the proportion of OFs with refrigerator increased more than that of NOFs. Further analysis show that there was a significant difference (x^2 = 16.81; p≤0.05) between the number of OFs with refrigerator before and after becoming a project farmers in 2007. On the other hand, there was no significant difference (x^2 = 1.29; p≤0.05) between the number of NOFs with refrigerator between the year 2003 and 2007. Meanwhile, there was no significant difference (x^2 = 0.74; p≤0.05) between the number of OFs and NOFs with refrigerator in 2007. Hence, the numbers of people that purchase refrigerator during the programme intervention are the same.

Data in Table 6 show that, 53.3% of OFs and 50.0% of the NOFs had bicycle in year 2003. On the other hand, about 46.7 and 36.7% of OFs and NOFs had bicycle, respectively in 2007. This implies that, the proportion of OFs and NOFs with bicycle decreased over the year. Further analysis show that there was a significant difference (x^2 = 11.05; p≤0.05) between the number of OFs with bicycle before and after becoming a project farmers in 2007. On the other hand, there was no significant difference (x^2 = 1.29; p≤0.05) between the number of NOFs with bicycle between the year 2003 and 2007. Meanwhile, there was no significant difference (x^2 = 0.74; p≤0.05) between the number of OFs and NOFs with bicycle in 2007.

Data in Table 6 show that, 36.7% of OFs and 20.0% of the NOFs had motorcycle in year 2003. On the other hand, about 65.5 and 43.3% of OFs and NOFs had motorcycle, respectively in 2007. This implies that, the proportion of OFs and NOFs with motorcycle increased over the year. Further analysis show that there was a significant difference (x^2 = 5.05; p≤0.05) between the number of OFs with motorcycle before and after becoming a project farmers in 2007. On the other hand, there was a significant difference (x^2 = 4.88; p≤0.05) between the number of NOFs with motorcycle between the year 2003 and 2007. Meanwhile, there was no significant difference (x^2 = 0.01; p≤0.05) between the number of OFs and NOFs with motorcycle in 2007.

Data in Table 6 show that, 3.3 of OFs and 3.3% of the NOFs had car in year 2003. On the other hand, about 3% of OFs and NOFs had car, respectively in 2007. This implies that, the proportions of OFs and NOFs with car are the same over the year. Further analysis show that there was a significant difference (x^2 = 30.00; p≤0.05) between the number of OFs with car before and after becoming a project farmers in 2007. On the other hand, there was a significant difference (x^2 = 30.0; p≤0.05) between the number of NOFs with car between the year 2003 and 2007. Meanwhile, there was no significant difference (x^2 = 0.03; p≤0.05) between the number of OFs and NOFs with car in 2007.

Data in Table 6 show that, 86.7% of OFs and 90.0% of the NOFs had wall clock in year 2003. On the other hand, about 97% of OFs and NOFs had wall clock, respectively in 2007. This implies that, the proportion of OFs and NOFs with wall clock increased over the year. Further analysis show that there was a significant difference (x^2 = 6.72; p≤0.05) between the number of OFs with wall clock before and after becoming a project farmers in 2007. On the other hand, there was a significant difference (x^2 = 9.31; p≤0.05) between the number of NOFs with wall clock between the year 2003 and 2007. Meanwhile, there was a significant difference (x^2 = 30.0; p≤0.05) between the number of OFs and NOFs with wall clock in 2007.

CONCLUSION AND RECOMMENDATIONS

The findings of the study have shown that, there was a significant difference in the estimated mean income from the sale of cocoa products for both OFs and NOFs at the end of 2007. Also, the performance of ONL in terms of provision of farm inputs to OFs was encouraging; the highest score was recorded in distribution of agrochemicals at a subsidized rate and provision of hybrid seedlings to farmers. The result of the impact of Olam organization on socio-economic life of cocoa farmers showed that, the net income per capital of OFs was significantly different from that of NOFs. Also, significance differences existed between OFs and NOFs in the following: ease of selling cocoa products; knowledge on cocoa production and marketing; rating of standard of living; number of farmers with knapsack sprayers and radio. Thus, the Olam programme had positively and significantly impacted on the socio-economic life of the participating farmers (OFs) in Ondo State.

Based on the findings of this study, Olam organization extension programmes have the potential to boost cocoa production in Ondo State. It is therefore recommended that Olam organization should involve more farmers in her programme to compliment the effort of government in cocoa resuscitation programme.

REFERENCES

Agboola AF (2005). Socio-economic assessment of farmers' usage of indigenous and non-indigenous pest control technologies in cacao agro-ecosystems of Ondo and Osun States, Nigeria. Ph.D Thesis. Department of Agricultural Extension and Rural Sociology, Obafemi Awolowo University, Ile-Ife, Nigeria, pp. 1-228.

Ajieh PC, Ajayi AR (2007). A preliminary Assessment of the Rapid Food Production Project (RFPP) in Delta State. Nig. J. Sci. Agric. Food Technol. Environ., 7(1): 30-37.

Akinnagbe OM (2008). The Impact of Cocoa Resuscitation Extension Activities of the Olam Organisation on Socio-economic Life of the Farmers in Ondo State, Nigeria. Pre-Ph.D Seminar, Department of Agricultural Extension, University of Nigeria, Nsukka, pp. 1-63.

Amemiya T (1981). Qualitative Response Models: A Survey. J. Econ. Lit., 19: 1483-1536.

Bennett C, Rockwell K (1995). Targeting Outcomes of programmes (TOP): An Integrated approach to Planning and Evaluation. University of Nebraska, pp. 18-23.

Casley DJ, Lusy DA (1982). Monitoring and Evaluation of Agricultural and Rural Development Projects. A World bank Publication. Balthimore and London: The Joins Hopkins University press, p. 145.

Chase RS (2002). Supporting communities in transition; The impact of the Armenian social investment fund. World Bank Econ. Rev., 16(2): 219-240.

Eboh EC (1998). Social and Economic Research: Principles and Methods. Lagos: Academic Publications and Development Resource Ltd.

Gittinger JP (1982). Economic Analysis of Agricultural Projects (2nd Ed.). Baltimore, Maryland 218 U.S.A: The Johns Hopkins University Press, p. 505.

Horton D, Ballantyne P, Perterson W, Uribe B, Gapasin D, Sherikda K (1993). Monitoring and Evaluating Agricultural Research: A Source Book. The Hague, Netherland: CAB International, pp. 5-204.

Ladele AA (1991). A socio-economic analysis of the Impact of Agricultural Cooperative Organisations on Rural Development in Nigeria. A Ph.D research Thesis, University of Ibadan, Nigeria, pp. 1-265.

Maitima JM, Rodriguez LC, Kshatriya M, Mugatha S (2007). Guidelines for assessing environmental and socio-economic impacts of tsetse and trypanosomiasis interventions. International Livestock Research Institute (ILRI). Nairobi, Kenya, p. 37.

National Population Commission (NPC) (2007). Population Figure. Federal Republic of Nigeria, Abuja. http://www.npc.gov.

Ogunleye KY, Oladeji JO (2007). Choice of cocoa farmers in Ila local government area of Osun State. Middle-East J. Sci. Res., 2(1): 14-20.

Olam (2007) Bulletin of Olam Nigeria Limited, Akure, Nigeria. pp. 1-10.

Ravallion M (2001). The Mystery of the Vanishing benefits: An Introduction to Impact Evaluation. World Bank Econ. Rev., 15(1): 115-140.

Rawlings L, Schady NR (2002). Impact evaluation of social funds: An Introduction. World Bank Econ. Rev., 16(2): 213-217.

Tunde O (2007). Reducing Nigeria's dependence on oil through cocoa production and consumption. Nigerian Tribune, March, 8 2007, p. 32.

Vegetable field workers provide insight for improving farm labor retention at the US-Mexican border

Kurt D. Nolte*[1] and Jorge M. Fonseca[2]

[1]Yuma County Cooperative Extension University of Arizona 2200 W. 28th St., Suite 102
Yuma, Arizona 85364 USA.
[2]Department of Plant Sciences, Yuma Ag Center, Yuma, Arizona, USA

A 60-question assessment was administered to 305 field workers during 2007 - 2009 to identify factors that could aid US southwest vegetable producers with retaining current field labor. The population surveyed included 38% females and 62% males with over 50% crossing the US/Mexican border daily. Age distribution of the survey participants was 0.9% (< 18 years old), 14.9% (18 - 24 years old), 10.3% (25 - 34 years old), 20.6% (35 - 44 years old), 29.0% (45 - 54 years old), 21.5% (55 - 64 years old) and 2.8% (> 65 years old). Contrary to the conventional belief that salary is the only true mechanism to retain field labor, the study revealed several factors that may also be important. Among workers who do not recommend field labor, over 30% indicated the need for additional work breaks, and up to 23% sought better health care benefits. There was also a marked increase in field work dissatisfaction with increasing age among both male and female workers, a reflection of the physical nature of field work. Overall, this study provides a direction for agricultural labor leaders to initiate efforts aimed at retaining field labor in the southwest, USA.

Keywords: Mexico, immigration reform, Arizona, benefits, job shortage, field labor, USA agriculture, vegetable production.

INTRODUCTION

Due to its warm winter climatic conditions, Yuma, located in the southwest portion of Arizona, USA, is essentially the only agricultural region in the nation where the field production of vegetables is possible during the period between November and April. The region represents the bulk of the United States' winter leafy greens, providing food, income, and revenue for a state that ranks second nationally in the production of fresh produce (US Department of Agriculture, 2006). To meet the demands during the peak periods of production, a daily labor pool of over 50,000 workers is required to either work directly in field production or within other agricultural related activities (US Department of Labor, 2005). During the 1990's, vegetable production revenue in the region increased by over 120%, partially due to utilization of more modern technologies, greater per capita consumption and increased population growth (US Department of Agriculture, 2006). During the last few years, however, industry growth peaked, due in part to a lack of a stable supply of field labor. In fact, the present labor status in the desert southwest has sparked a federal and national debate concerning the implementation and feasibility of a national guest worker program (Rodriguez, 2007; Ruiz, 2006; Schiff, 2004). Although the level of potential economic impact from a reduced field labor supply is uncertain, the possibility of a significant reduction in the future is estimated to be high (Gonzalez, 2006).

As a result, there is a clear identified need for an in-depth understanding of the immigrant worker's culture, their satisfaction at work and the ensuing impact on worker continuity/retention. The US farm workforce has increasingly become an immigrant workforce (Gabbard et al., 2002) which warrants a different human resource management approach to assure maximum efficiency. In the case of Arizona, a border state in the southwest region of the USA, the situation may even be more complex as thousands of people are legally allowed to work within

*Corresponding author. E-mail: knolte@ag.arizona.edu.

the USA commute daily to reach their jobs, especially in agriculture (Gonzalez, 2006). In this light, the present effort will focus on the potential mechanisms to retain current and recruit future workers in crop field jobs in Arizona were investigated. The approach examines the opinion and attitude of the current workers and determines the feasible options growers and labor contractors have to create an environment that encourages labor continuity. The objective of this study aims to identify factors that could be useful for future decisions regarding guest worker legislation, and immigration reform. Specifically, four key objectives of the assessment effort were aimed at clarifying: 1) The background of the workers (origin, gender, age, immigration status, etc); 2) The level of job satisfaction (e.g. whether they recommend the job); 3) General factors which could improve their work environments (health benefits, salary, commuting time, etc); 4) Broad-based suggestions and recommendations to improve job satisfaction.

METHODOLOGY

The study site

Labor assessments were conducted within the agricultural areas of Yuma County (Latitude, 32º 44' – 32º 41'N; Longitude, 114º 58' – 113º 44'W), situated in southwest Arizona (USA). The region encompasses approximately 14,200 square kilometers of desert with 93,000 hectares of arable land clustered within valley regions. The location is known for having warm winters, a dry climate and ample surface water, resulting in a thriving agricultural business that emphasizes the production of winter vegetables. The study site is immediately adjacent to the Colorado River and the international boundary with Mexico on its West and Southern boundaries, respectively, and includes the small cities of Somerton, Gadsden and the City of San Luis, the Arizona border city with Mexico.

Study logistics and preliminary overview

An assessment tool specifically targeted for field workers and their daily lifestyles within the desert southwest was developed. To do so, typical agricultural practices encountered in vegetable crop operations, and particularly those focused on activities that involved field labor were examined. And, the US/Mexican border town of San Luis was frequently visited to observe the highly energetic activity that takes place before sunrise, and highlighted by hundreds of people processed by US Customs and Border Protection prior to US entry. Also appraised were fields in production within the survey area to determine the ideal location for assessment. As a result of the pre-assessment period, it was determined to personally interview workers directly in fields, located by contacting area labor contractors. Conducting personal, one-on-one interviews in the field gave two main advantages: (1) it allowed sufficient time to complete the survey, and (2) it provided for a relatively more private survey environment. Generally, crew supervisors allowed 15 min of time per worker, which was sufficient to respond to all survey questions. During the process of conducting the field labor interviews, approximately 50% of the workers were removing weeds at the moment of the survey, 30% were harvesting 5% in multiple activities (such as packing or driving field equipment) and 15% were having lunch (but working that day in any of those activities indicated above).

Design of the survey tool

In developing a strategy for assessment, the aim was to develop a specific tool that would allow for a thorough understanding of field labor attitude towards their job with the intent of gaining insight for labor recruitment and retention. A questionnaire was initially developed with input from different scientists, growers and contractors and piloted with workers sub-contracted to work at the Yuma Agricultural Center in Yuma, Arizona (USA). According to preliminary results (12 interviews) and further discussion with several industry representatives, the survey tool was modified which resulted in a 60-question survey which was administered face-to-face in Spanish to fieldworkers (n = 305) during the leafy vegetable season, from December 2007 to March 2009. The sample size was estimated to account for over 0.7% of the entire field labor population located in the region. The survey consisted of questions concerning different aspects of field work ranging from worker demographics, age, income, job duties, housing, job satisfaction, working conditions and work schedule. The survey was administered by a technician who also noted additional comments, as needed, from respondents during the interviews.

Statistical analysis (frequency distribution, regression) with SAS software (SAS Institute, Cary, NC, USA) was used when considered appropriate. However, the outcomes of this work are reported as descriptive results and meaningful metrics since statistically significant differences among responses may not be useful for decision makers. For example, a significant difference between the importance of work schedule and health insurance would have little value when both are nevertheless mentioned by a large percent of the respondents, suggesting that both factors are important and need to be addressed.

RESULTS AND DISCUSSION

Validity and reliability of measurements

Embedded within the assessment tool was a set of questions which collectively, measured different facets of each of the 4 main objectives of this study. Alternate form analysis was used to statistically evaluate the correlation of the related survey data and was reflected in Cronbach's alpha correlations. For all objectives evaluated, the Cronbach's alpha (Cronbach, 1951) ranged from 0.82 to 0.87, suggesting that the survey conducted was done with high reliability. A concurrent validity assessment was used that compared the results from this study to similar studies conducted in the neighboring states of California (Martin, 2001), New Mexico (Wood and Skaggs, 2005) and in the US (Levine, 2009).

Worker demographics and job satisfaction

One of the advantages that this effort provided was drawing from the attitudes and views obtained from those directly involved with field work. Given the large amount of workers surveyed randomly across the area and the fact that a great majority of the local labor contractors were included (a total of 13 companies), it is thought that the data reveals the reality of the entire field labor population within the southwest Arizona, an area representing

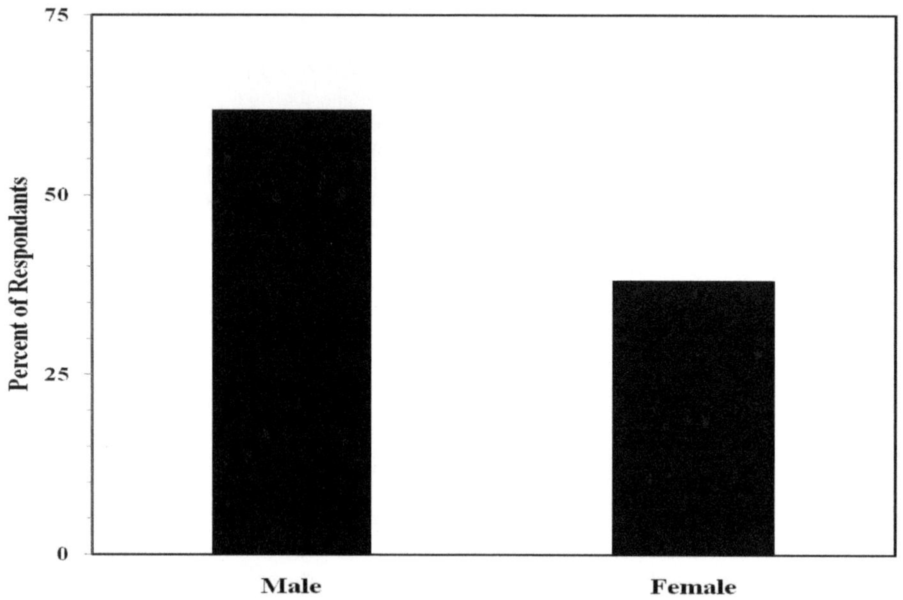

Figure 1. Overall relative age percentages of male and female field workers in southwest Arizona, USA.

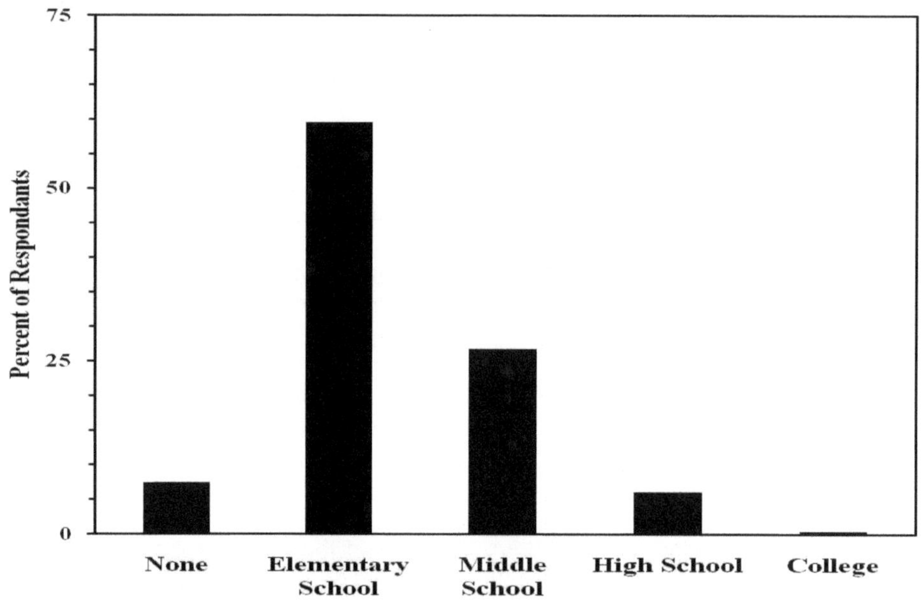

Figure 2. Highest degree completed by interviewed workers in southwest, USA

over 80% of Arizona's overall leafy green production. The survey population consisted of 38% females and 62% males (Figure 1), and nearly 70% had 4 -7 years of education (Figure 2). In addition, it was found that 90% of the farm worker population either live in the US/Mexican border town of San Luis, Arizona or maintain permanent residence in Mexico and commute across the US/Mexican border daily. This later group accounts for nearly half of the field crews working in vegetable fields within the region assessed (Figure 3).

More than half of the respondents, irrespective of gender, were over the age of 45 (Figure 4). This appears unique for this region, as it has been documented that the US field labor workforce is composed more of young, single males (Mines et al., 2001). It is expected that many questions in any survey to workers are a function of the age of the population. In this regard, the assessment showed that over 50% of current female field workers over the age of 25 years and males over the age of 45 years, would not recommend field work to others (Figure 5).

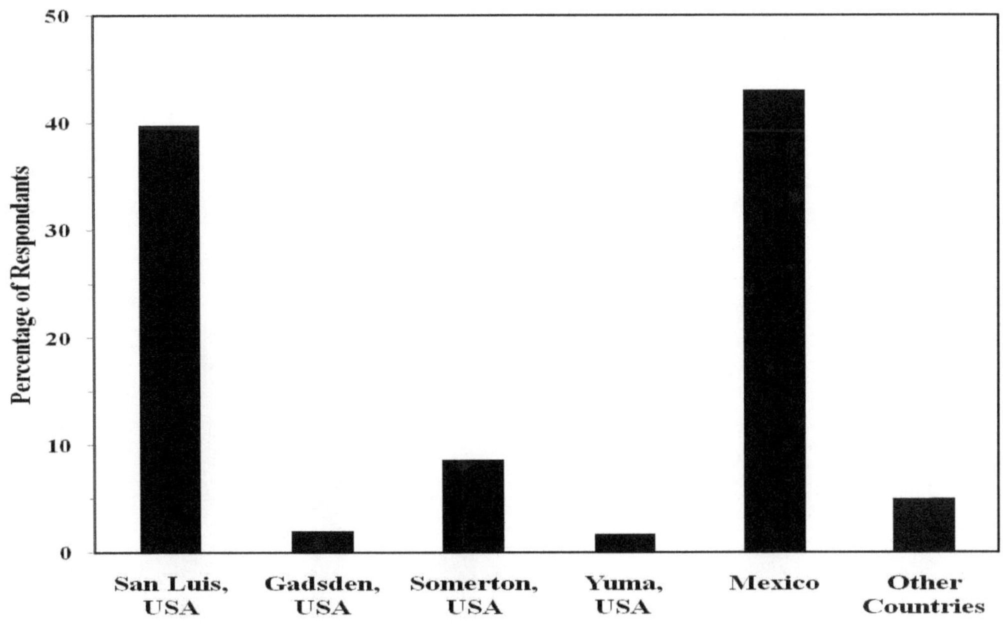

Figure 3. Relative residency status of vegetable field workers in Arizona, USA.

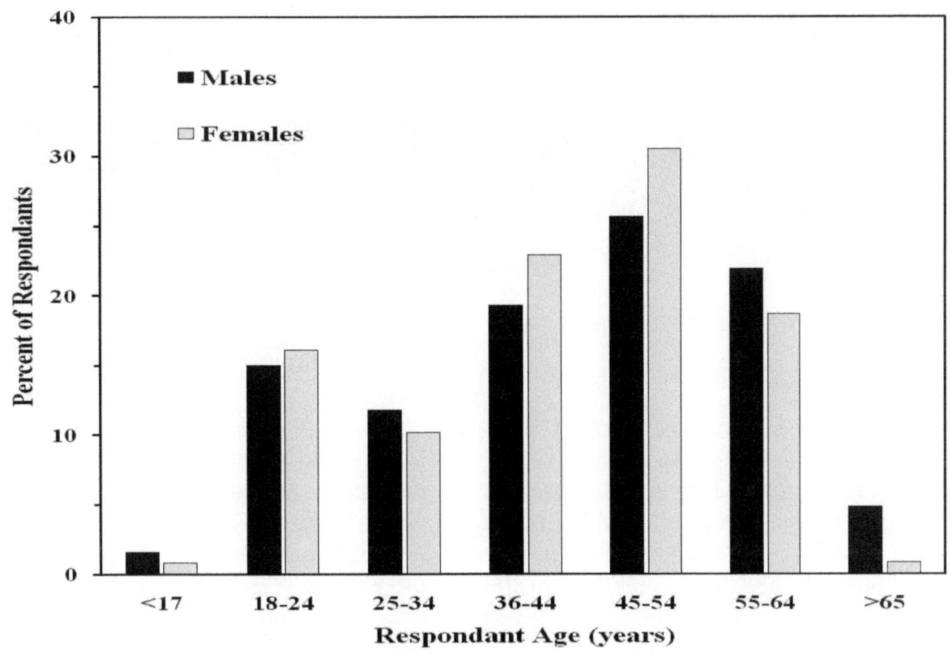

Figure 4. Age breakdown of southwest USA fieldworkers

To this end, there is a marked increase in field work dissatisfaction with increasing age (Figure 5) among both male and female workers, which could be considered concerning, since job recruitment for field labor can increase through employee referrals (Shinnar et al., 2005). Elucidating the factors that affect either recommending or not recommending field work seems paramount (Bitsch and Hogberg, 2005, Herzberg et al., 1957) to better understand and support the ongoing labor shortage in the southwestern USA. The results showed that older workers are more likely to discourage others from choosing field work as a source of income (Figure 5). This attitude was attributed to a combination of factors including workers age, the physical nature of the job, a need for more frequent work breaks and, more importantly, a lack of adequate health care benefits for a group that could value

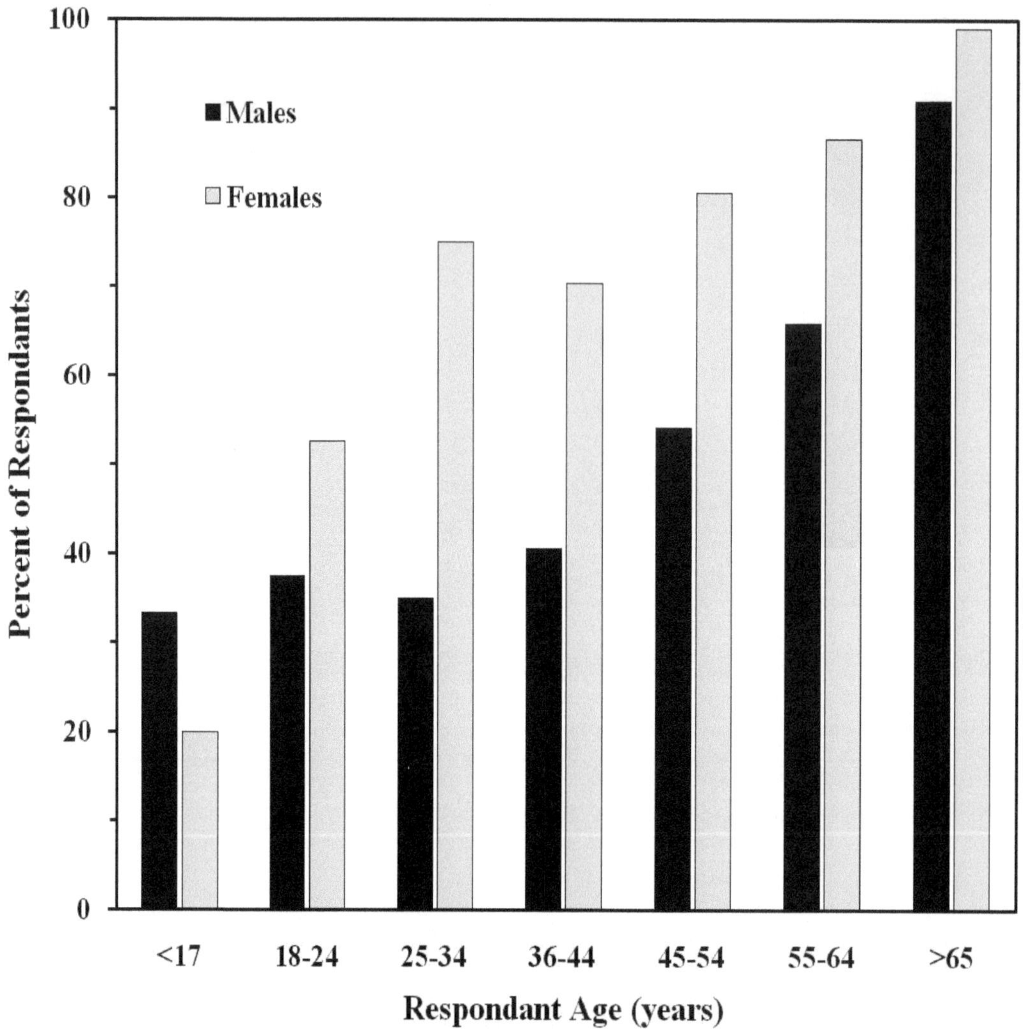

Figure 5. Male and female workers who do not recommend field work, relative to age.

it the most (Findeis et al., 2005). It seems clear that in the study, as found in other professions (Bachmann, 2000; Delong, 2004; Fullerton et al., 2001; Rix, 2004), crop field labor is represented by an aging workforce which could be invigorated and possibly maintained with greater contributions from a younger labor base. Aside from the growing risk for short term field labor shortages due to insufficient young workers to replace those that are aging, there is also an emerging risk that older labor may discontinue working in fields sooner due to the physical nature of the work and/or associated health-related issues. Although young workers (<25 years old) tend to have a higher injury rate than older workers worldwide (Salminen, 2004), older people suffer more from chronic disease and are more susceptible to fatigue (Crandall et al., 1997). Thus, continuing to build an industry which relies on the dependence of an aging workforce could result in additional labor challenges in the future despite progress industry wide in productivity, plant genetics and technology.

Current economic research on the well-being of US farm workers has largely focused on wages, income, monetary bonuses, paid leave, and other immediate worker effects (Taylor, 1992). However, Findeis et al. (2005) suggested that the long-term effects of field labor on worker well being, such as greater access to adult education, job advancement and comprehensive health care, warrants further investigation. Yet, enhancing the long-term well being of workers may prove difficult as evidence from industry suggests that firms are reluctant to invest in training workers (Chiswick, 1986), especially if they are unauthorized or if there is a high employee turnover (Taylor, 1992). In the next five years, the fact that the highest numbers of workers who anticipate working in the southwestern region of the USA are those who travel into Mexico (Figure 6) provides a warning that an even greater proportion of field work will increasingly rely on people who cross the border on a daily basis. Thus, consequences of slower border crossings, inefficient processing of immigrant documentation, debate over guest

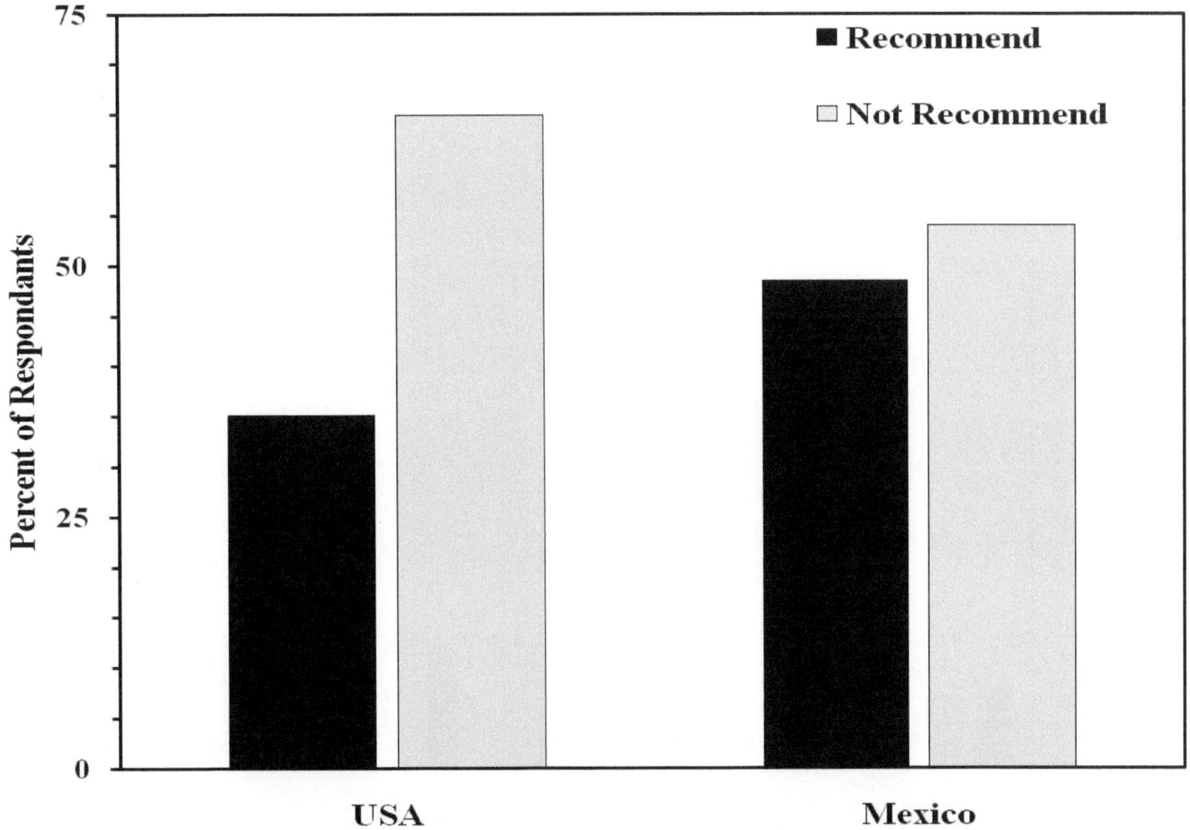

Figure 6. Recommending or not recommending field work as function of where a field worker lives (that is, USA or Mexico).

worker legislation and more stringent regulation of daily commuters into the USA may have a direct effect in the commercialization of winter vegetable crops.

RECOMMENDATIONS FOR IMPROVEMENT

In countries like Sweden, it is known that agricultural wages are defined by not only job performance, but also the duration that workers have been at the job and that these factors have dramatic effects on long-term earning potential, job satisfaction and length of service (Aslund and Rooth, 2007). In this study, however, it was learned that of the major elements which affect worker satisfaction, a field workers wages was not necessarily the most important factor. It is clear, however, that workers who commute daily from Mexico are more likely to recommend field level jobs than US workers (Figure 6), as wages can be 5 - 7 times higher than what employees receive in the Northern region of Mexico. Thus, worker satisfaction in southwest USA may differ greatly from other agricultural regions, and specifically from those areas where production is done in greenhouses (Aslund and Rooth, 2007). In the southwestern region of USA, and particularly within the winter vegetable industry, it is common for workers to have seasonal employment and in most cases it could be the only job found in the area. Thus, due to the nature of the workers themselves, their skill level (Bellenger et al., 2008), the close proximity to Mexico, and the relative scarcity of yearly employment, field workers do not appear to prioritize wage adjustments despite lengthy work histories, experience and legal immigration documentation (Ise and Perloff, 1995).

While wages did not seem to be the highest priority for improving job satisfaction among those surveyed, more than half of those assessed, regardless of gender, suggested that longer and more frequent work break periods (Figure 7) would improve job fulfillment. Other suggested improvements included greater work schedule flexibility, fewer hours and better benefits. In fact, it was found that workers who do not recommend their job to others seem to have relatively fewer breaks (data not shown). It has been suggested that food consumption during work, a very important part of physical working environments, is often neglected (Grunert et al., 2007). A partial explanation why field workers request more frequent work breaks intervals could be related to the need for more quality time, and consistent energy sources during the course of a workday, in addition to accomplishing personal needs, like access to drinking

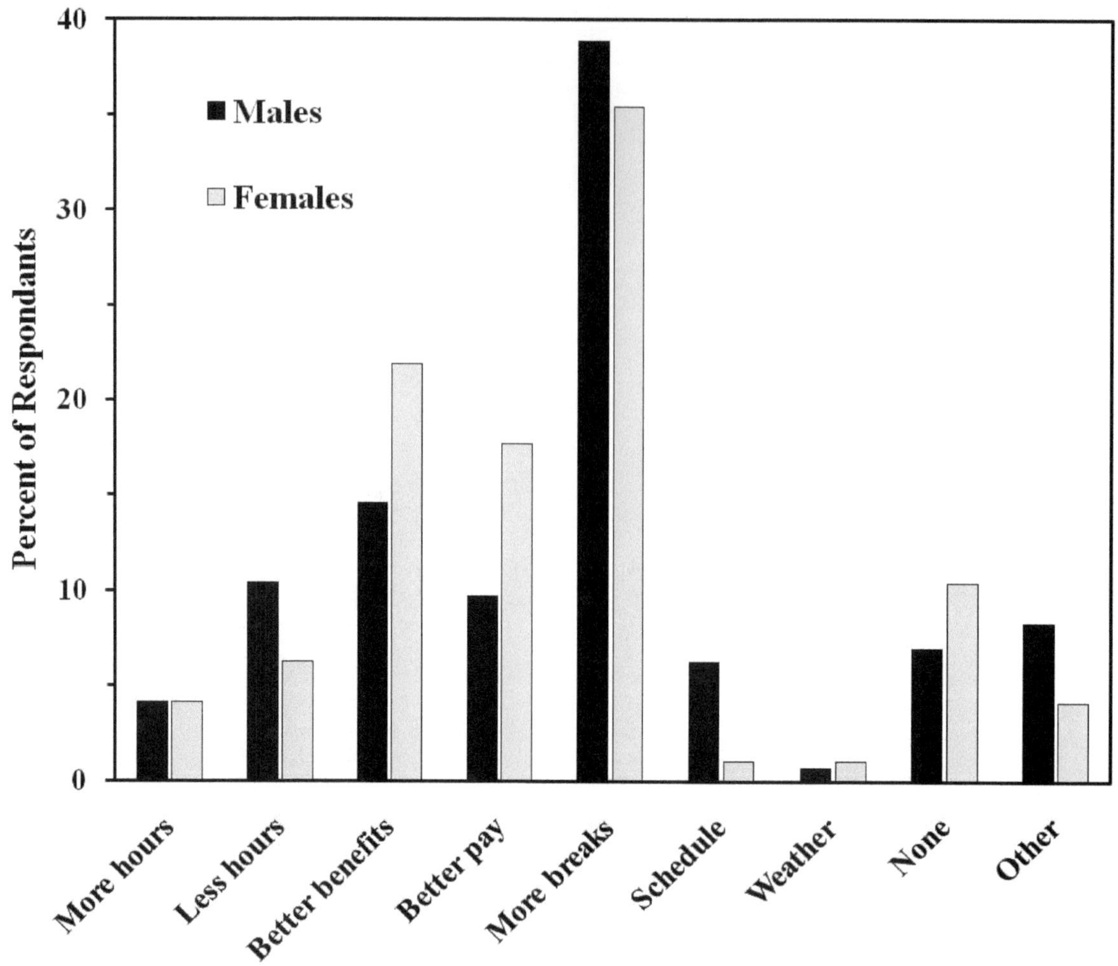

Figure 7. Suggestions by surveyed field workers to make work more appealing as a function of gender.

water, restroom facilities and a short rest. The similar pattern of responses among males and females is somewhat unique to the southwest USA as opinions, attitudes and job duties often differ between men and women in agriculture (Kritzinger and Jan Vorster, 1996). In desert vegetable production, however, some job responsibilities are not gender specific (hand thinning, cultivation and some aspects of harvest) which could promote similar levels of enthusiasm and interest towards job satisfaction. With the relatively large number of female field workers in the area surveyed, the similar wage distribution among tasks and gender, and that, in many cases, men working side-by-side with women, it is not surprising to find similar opinions towards job improvement. Health care appears to be an essential benefit that most field workers request from employers since, in Arizona, USA, field crews do not typically receive this form of assistance from employers.

The issue is not new within agriculture and it has been a concern in the USA for decades. Assessments conducted in 1978 and 1989 showed that field workers remain in serious need of health care or access to financial assistance programs (Slesinger and Ofstead, 1993). Although many field workers would feel more satisfied at work with additional or longer break periods, more noticeable was the observations associated with health insurance, as almost twice as many respondents who do not have health insurance did not recommend their jobs (Figure 8). It has been reported that employers could use work-related health insurance or health care provisions to ensure worker productivity or attract immigrant workers. However, permanent farm workers, and non-international commuters, are more likely to receive employer-provided benefits and public health assistance (Findeis et al., 2005). Clearly, this could be a consideration as it has been shown that, although a great majority of farmers have health insurance coverage (Jensen, 1983), insurance to field workers is something not well established in Arizona, and perhaps agriculture in general. In the US during the late 1990s, a period of great success, the likelihood of receiving employer-provided benefits by farm workers declined across all forms of assistance (Findeis et al., 2005). Interestingly, by the end of this study, it was determined that two labor

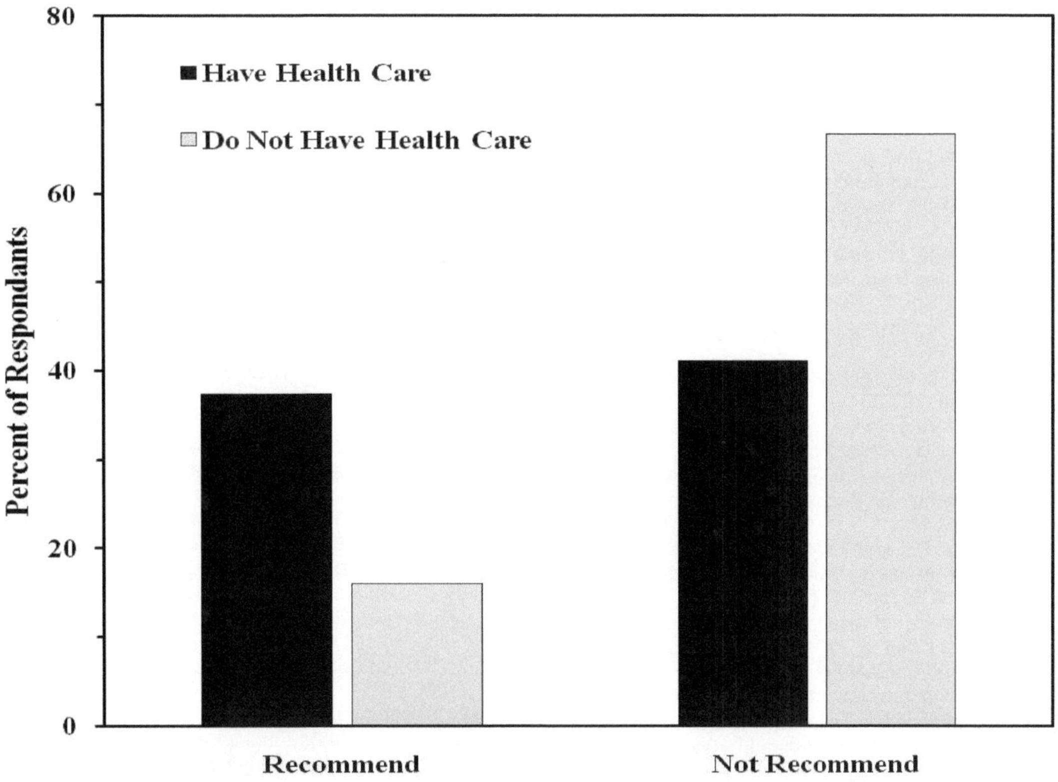

Figure 8. Field worker views on either recommending or not recommending field work as a function of having or not having health insurance.

contractors, in addition to one that had already identified early in the study, decided to provide health care benefits to their workers as a competitive means for field worker recruitment efforts. In taking this action, it seems clear that current market competitiveness could include measures which were never considered as vital to the industry. Thus, in addition to improving working relationships with workers, greater outreach needs to be done with labor contractors and supervisors to improve work environment, for both social and economic reasons.

Conclusion

This study revealed promising options and insight for improved field worker retention in the desert southwest. Some aspects that survey participants suggested as critical would implicate additional costs for production operations, however, some options (e.g. more rest periods) may not be as costly as an across the board increase in wages or greater worker related benefits. Moreover, this work needs to be placed in the proper context, since several suggestions may only be applicable in production areas that are in immediate proximity of an international border. Further, it was observed that several significant points in human resource management seem to be ignored by employers in this region. Significant job orientation and training, for example, known for its importance in providing long-term work efficiency (Breaugh and Starker, 2000) is an aspect that employers do not seem to embrace in southwestern Arizona, and is a probable reflection of history or because a top priority is to find enough people to do the work needed. Overall, this work emphasizes the potential for agricultural labor leaders to initiate efforts aimed at retaining an aging field labor workforce in the desert southwest. As workers appear to have little long term commitment under current working conditions, they consider including health-care benefits, greater frequency of work breaks, and a more convenient work schedule as paramount if they are to continue to provide a stable workforce in the region.

ACKNOWLEDGMENTS

This work was funded by The Western Center for Risk Management Education. The authors thank Bianca Rios, Jonathan Dinsmore and Ramiro Galvez for their assistance in elaborating and conducting this study.

REFERENCES

Aslund O, Rooth DO (2007). Do when and where matter? Initial labour market conditions and immigrant earnings. Econ. J. 117: 422-448.

Bachmann K (2000). Work-Life Balance: Are Employers Listening? Ottawa. Conference Board of Canada.
Bellenger M, Fields D, Tilt K, Hite D (2008). Producer preferences for migrant labor and the wage, hours, and gross sales effects in Alabama's horticulture industry. Hort. Technol. 18: 301-307.
Bitsch V, Hogberg M (2005). Exploring horticultural employee's attitudes toward their jobs: a qualitative analysis based on Herzberg's theory of job satisfaction. J. Ag. Appl. Econ. 37: 659-671.
Breaugh JA, Starke M (2000). Research on employee recruitment: So many studies, so many remaining questions. J. Mgmt. 26: 405-434.
Chiswick BR (1986). Illegal Aliens: A preliminary report on an employee-employer. Amer. Econ. Rev. 76: 253-262.
Crandall CS, Fullerton L, Olson L, Sklar DP, Zumwalt R (1997). Farm-related injury mortality in New Mexico. Accident Anal. Prevention, 29: 257-261.
Cronbach LJ (1951). Coefficient alpha and the internal structure of tests. Psychometrika, 16: 297-334.
DeLong DW (2004). Lost Knowledge: Confronting the Threat of an Aging Workforce. Oxford University Press, Inc.
Findeis J, Snyder A, Jayaraman A (2005). The well-being of U.S. farm workers: Employee benefits, public assistance and long-term effects. Rev. Ag. Econ. 27: 361-368.
Fullerton HN, Toossi M (2001). Labor Force Projections to 2010: Steady Growth and Changing Composition. Monthly Labor Rev. 124: 21-38.
Gabbard S, Fermandez-Mott A, Carrol D (2002). Examining farm worker images. In: The dynamics of hired farm labour: Costraints and community responses. Findeis J, Vandeman A, Larson J, Runyan J eds. pp. 15-24, UK: CAB International, 2002.
Gonzalez D (2006). Arizona growers race against the clock to find winter harvest workers. USA Today. Issue of November 20. http://www.usatoday.com/news/nation/2006-11-20-winter-harvest_x.htm
Grunert KG, Moira D, Raats MM, Nielsen NA, Lumbers M (2007). A measure of satisfaction with food-related life. Appetite, 49: 486-493.
Herzberg F, Mausner B, Peterson RD, Capwell DF (1957). Job attitudes: Review of research and opinion. Pittsburgh: Psychological Service of Pittsburgh.
Ise S, Perloff JM (1995). Legal status and earnings of agricultural workers. Legal status and earnings of agricultural workers. Am. J. Ag. Econ. 77: 375-386.
Jensen HH (1983). Farm people's health insurance coverage. Rural development research report-United States Dept. Agric. Econ. Res. Ser., p. 39.
Kritzinger A, Vorster J (1996). Women farm workers in South African deciduous fruit farms: -Gender relations and the structuring of work. J. Rural Stud. 12: 339-351.
Levine L (2009). Farm Labor Shortages and Immigration Policy. Congressional Res. Service Rep. Number RL30395.
Martin P (2001). Farm Labor in California: Then and Now. The Center for Comparative Immigration Studies Report Number 37, University of California.
Mines R, Mullenax N, Saca L (2001). The binational farmworker health survey: An in-depth study of agricultural worker health in Mexico and the United States. Davis, CA: California Institute of Rural Studies.
Rix SE (2004). Aging and Work: A View From the United States. Washington, D.C: AARP.
Rodriguez CM (2007). Guest Workers and Integration: Toward a Theory of What Immigrnats and Americans Owe One Another (July 28, 2008). New York University School of Law. New York University Public and Legal Theory Working Papers. p. 77.
Ruiz O (2006). Migration and Borders. Latin Am. Perspect. 33: 46-55.
Salminen S (2004). Have young workers more injuries than older ones? An international literature review. J. Saf. Res. 35: 513-521.
Schiff M (2004). When migrants overstay their legan welcome: a proposed solution to the guest-worker program. Institute for the Study of Labor (IZA), Discussion Paper, No. 1401.
Shinnar RS, Young CA, Meana M (2005). The motivations for and outcomes of employee referrals. J. Bus. Psy. 19: 271-283.
Slesinger DP, Ofstead C (1993). Economic and health needs of Wisconsin migrant farm workers. J. Rural Health, 9: 138-148.
Taylor JE (1992). Earning and mobility of legal and illegal immigrant workers in agriculture. Am. J. Ag. Econ. 74: 889-896.
US Department of Agriculture (2006), National Agricultural Statistics Service, Agricultural Statistics Board, (04/17/ 1986). Published Estimates Database Files, hdl 1902.5/601816 National Archives and Records Administration. Accessed July 20, 2008 and September 25, 2008.
US Department of Labor, Office of the Assistant Secretary for Policy. March, (2005). Findings From the National Agricultural Workers Survey: A Demographic Employment Profile of United States Farmworkers (Research Report No. 9). Available at: http://www.doleta.gov/agworker/report9/toc.cfm. Accessed July 15, 2008.
Wood JA, Skaggs R (2005). An Analysis of Farm Labor Contracting in New Mexico. New Mexico Chile Task Force, Report Number 18. New Mexico State University.

14

Effective extension methods for increased food production in Kakamega District

Adijah M. Ali-Olubandwa[1]*, N. J. Kathuri[2], Timothy E. O. Wesonga[3]

[1]Department of Applied Community Studies and Extension, Egerton University, Kenya.
[2]Department of Agricultural Education and Extension, Egerton University, Kenya.
[3]East Africa Community, Arusha, Tanzania.

Despite the importance of agriculture to Kenya's economy, the Kenya government's budgetary allocation to the sector has declined over the years. This resulted in the sector being resource constrained with a high farmer to staff ratio, hence the need for extension methods that can effectively reach more clients at the lowest cost. This paper reviews effective extension methods which have the least constraints and most cost effective that can be used by extension staff in Kakamega district. Kakamega district experiences low food production, despite the fact that it is one of Kenya's food baskets. One hundred extension staffs were systematically sampled from the department of livestock and the department of agriculture from the Ministry of Agriculture and Livestock Development. Data was analysed using descriptive statistics, cross tabulations, correlations and Chi-square. The research revealed that group demonstrations were the most cost effective extension methods with the least constraints. The study therefore recommended that group demonstrations and a combination of group demonstration and individual farmer follow-up be used to pass extension methods in Kakamega district.

Key words: Extension methods, food production, Kenya, Kakamega district, extension service.

INTRODUCTION

Kenya's economic and social development challenges include unemployment, poverty and food insecurity (Ministry of Agriculture (MOA), 2005). Food insecurity is a condition in which a population does not have access to sufficient, safe and nutritious food over a given period to meet dietary needs and preferences for an active life. The world food summit defines food insecurity as a state of food security that leads to lack of physical and economic access to safe and nutritious food for an active and healthy life (Food and Agriculture Organisation (FAO), 2000). FAO's most recent estimates put the number of hungry people in the world at 923 million in 2007, an increase of more than 80 million since the 1990 to 1992 base period (FAO, 2008).

Kenya like most of the countries in Intergovernmental Authority on Development (IGAD) region is one of the most foods insecure in the world (United States Agency for International Development (USAID), 2009a).

According to the USAID (2009b), in East Africa, food insecurity results from among other factors, below average harvests, limited storage capacity, lack of post harvest services, high food prices, conflict and insecurity. The Government of Kenya (2010) has further attributed food insecurity in Kenya to limited storage capacity, lack of post harvest services and poor access to input market (GOK, 2010). Kakamega district, which is one of Kenya's food baskets, also experiences chronic food shortages despite favourable weather conditions.

One of the government's strategy to address the imbalances caused by unemployment, poverty and food insecurity, has been to promote economic growth through public sector programme and measures in support of small-holder and large scale farming (Ministry of Agriculture (MOA), 2005). Extension service is one such programme that could bring about improvement in agricultural production and subsequently food security.

The extension service is charged with the responsibility of ensuring sustainable food production that is sufficient for domestic use and for export. Thus the objectives of extension service include: Ensuring sustainable

*Corresponding author. E-mail: olubandwaadijah@yahoo.com

agricultural development; Introduction of high quality, appropriate agricultural technology; maintaining the ecological balance in the natural environment; establishment of export oriented agro-processing industries; Sustaining and balancing production, consumption and income, and reduction of rural poverty (Republic of Kenya, 2005).

The Kenya extension service, Kakamega district inclusive, is severely resource constrained. Extension staffs have limited operating funds. Funds for transport, vehicle maintenance and fuel, field days, telephone communication and basic stationary are inadequate. Extension staff's pay and morale is low, yet at times, officers spend their own money on fuel to go to the field or their own materials for field days (Kodhek, 2005; Nyoro and Muiruri, 2001). This may be as a result of the low budgetary allocation to the agricultural sector (USAID 2010; KEPCO and CGD, 2010). Though the sector contributes significantly to the gross domestic product (GDP), the Kenya Governments' budgetary allocation to the sector has declined over the years, from 13% in 1983 to 5.6% in 2010 (KEPCO and CGD, 2010).

In Kakamega district, like all parts of Kenya, farmer to extension staff ratio continues to remain high particularly during the structural adjustment programme due to reduction of the number of extension staff through natural attrition and a freeze on new hiring (Nyoro and Muiruri, 2001; Kodhek, 2005). This makes the service ineffective in reaching many small scale and large scale. This study, therefore, identified extension methods that are effective in reaching many clients, had least constraints and most cost effective to be used by extension staff in Kakamega district.

Purpose of the study

The purpose of the study was to determine effective extension methods used in Kakamega district that have least constraints and are cost effective.

Objectives of the study

The study aimed at determining:

1. Frequency at which extension staff used various extension methods.
2. The extension methods with the least constraints.
3. The extension methods with the most constraints.
4. Cost effective extension methods.

Definitions of terms

1. Constraints: Limitations such as lack of funds, low pay and morale, lack of materials and equipment, low staff to farmer ratio among others.
2. Cost effective extension methods: These are methods that can be used to teach a large number of clients on new or improved practices, using minimum resources.
3. Farm and home visits: Involves meeting and teaching individual or farm worker on the farm or home
4. Field days: A trade show to which all stakeholders in agricultural sector are invited to display their products and pass information to.
5. Group demonstrations: This is made up of both result demonstration and method demonstration. In result demonstration, the farmer is shown the end product of two practices, that is, a new or recommended practice and an old practice for them to compare. In method demonstration, the group is shown how to carry out a practice step by step.
6. Group demonstration and farmer follow up: The extension staffs follows up the clients to their farms or homes after a group demonstration to establish whether they have adopted the innovation and what challenges they are facing in adopting the innovation.
7. Group meeting: A formally arranged gathering between extension staff and the clients.
8. Seminars: Comprise a small group of trainees engaged in specialised study under the leadership of an expert in this case an extension staff.
9. Workshop: An educational seminar emphasizing interaction and exchange of information among a usually small number of participants.

METHODOLOGY

Research design

The research design used was ex-post facto based on one-time cross-sectional survey, with the data collected at one point in time. The design was selected because the independent variables cannot be manipulated by the researcher. In any case, manifestations of these variables have already occurred. The ex-post facto design uses naturally occurring treatments or subjects having a self-selected level of the independent variable (Kathuri and Pals, 1993; Kerlinger, 1976) which is the condition that existed for this study. The researcher also dealt with human subjects in the study, which was free to choose what they would or would not participate in. The independent variable of the study was the personal characteristics of the extension staff, while the dependent variables were the extension methods used and the frequency at which the extension staff used the extension methods.

Study population

The study was carried out in Kakamega district, which has a high potential for food production, however, the district experiences chronic food shortages. The target population which was also the accessible population was the extension staff both in the department of livestock and agriculture in the ministry of agriculture and livestock development.

Table 1. Association between the extension staff's personal characteristics and the extension methods used.

		Level of deployment	Area of deployment	Gender	Age	Designation	Department	Years of service
Extension method used	X^2	146.713*	113.806	334.856*	22.674	98.199	279.047*	30.03
	DF	104	104	260	26	78	208	26

Sampling procedure and sample size

One hundred extension staffs were proportionately sampled from the department of livestock and agriculture in the ministry of agriculture and livestock development. The staffs were then systematically sampled.

Data collection and analysis

Questionnaires containing closed and open ended questions were administered to the extension staff with a request that they fill and return to the researchers. Data was analysed using descriptive statistics, cross tabulation, correlations and Chi-square with the help of Statistical Package for Social Sciences.

RESULTS AND DISCUSSION

Profile of extension staff

Eighty eight extension staff filled and returned the questionnaires, of the extension staff studied, 77.9% were men while 22% were women. Thirty percent were deployed at the district level, 37.3% at the divisional level, 10.5% at location level, 18.6% at the sub location level and 3.6% at training centre. Among the extension staff interviewed, the highest percentage (39%) had been trained in general agriculture, 21% in animal production and 10% in animal health. Few extension staff (1%) had been trained in the other fields including agricultural extension and education.

More than half (53%) of the extension staff had also worked for over 11 years. This indicates that the extension staff interviewed had served for long in the service thus they could give informed judgement of the field situation.

Forty one percent of the extension staff interviewed had a certificate in agriculture related courses, 25% said they held a diploma, 23% held a bachelors degree, 1% held a masters degree while 2% had been trained on the job. Due to the long years of service indicated by most extension agents and the fact that most of them were certificate holders, there was need for refresher courses and upgrading training for extension staff in order to increase their capacity to understand, utilize and disseminate new research findings.

The researchers further sought to find out the association between the extension staffs' personal characteristics and the extension methods used, using Chi- square. The results revealed that the extension method used was not significantly related to the level of deployment, gender, age, nor the department in which the extension staff was deployed (Table 1).

However, the method of extension used was dependent on the years of service, area of deployment of extension staff and the designation of the staff as shown on Table 1.

Frequency at which extension staff used various extension teaching methods

The extension staffs were asked to state whether they used the various extension teaching methods and the frequency at which they used these methods. Chi-square test was used to determine the association between the extension staffs' personal characteristics and the frequency of using various extension methods (Table 2).

The results revealed that there was no statistically significant association between the extension staffs' personal characteristics and the frequency at which the officers used farm and home visits, group demonstration and group demonstration, and individual farmer follow-up. However, there was a statistically significant relationship between the extension staffs level of deployment and the frequency at which they used field days, workshops, seminars and tours.

Sixty three percent (63%) of the extension staff at the district level frequently used field days. This was because more officers from the district often dealt with large numbers of farmers, hence had to use a method that could reach more. In addition, field days allowed for other stake holders such as manufacturing, institutions of higher learning and research, to reach the same at the same time. On the other hand, sixty nine percent (69%) of the extension staff at the divisional level rarely or occasionally used the field days to reach the farmers. Similarly more staff from the district level (41%) used workshops either often or very often as compared to the 7% of the extension staff at the divisional level who used the method either often or very often. This is because the extension staff at the district level are the subject matter specialists in the various departments, and therefore are charged with the responsibilities of designing messages that will be passed to the extension staff at the divisional, location and sub location levels, to be able to pass this messages effectively to these extension staff, the extension staff at the district level need to use workshops.

Table 2. Association between extension staffs' characteristics and the frequency at which they used various extension methods.

Frequency of using extension method		Extension staffs' personal characteristics						
		Years of service	Level of deployment	Area of deployment	Gender	Age	Designation	Department
Field days	x^2	16.995	26.202	50.524	5.731	12.550	44.071	4.796
	DF	16	16	36	4	12	32	4
Group demonstration	x^2	11.719	19.029	58.952	8.192	17.201	25.802	4.735
	DF	16	16	40	4	12	32	4
Group meetings	x^2	16.626	10.550	71.977*	1.243	11.268	33.295	9.998
	DF	16	16	40	4	12	32	4
Group demonstrations and farmer follow-up	x^2	7.040	11.733	48.979	9.260	12.219	32.386	6.960
	DF	16	16	40	4	12	32	4
Field days	x^2	23.334	31.014*	93.124*	3.163	14.216	41.730	4.414
	DF	16	16	40	4	12	32	4
Workshops	x^2	9.457	33.406*	70.430*	3.671	11.679	64.582*	11.203*
	DF	16	16	40	4	12	32	4
Seminars	x^2	15.795	52.373*	46.701	5.738	14.550	34.108	2.588
	DF	16	16	40	4	12	32	4
Tours	x^2	12.518	52.213*	43.918	5.622	17.952	43.906*	4.550
	DF	12	16	40	3	9	24	3

*Significant association.

It is interesting to note that only11% of the extension staff at the district level used seminars. Owing to the time that the extension staff have been in service (81% of the extension staff at all levels were found to have been in service for more than seven years), the staff at the district level should organise regular seminars to up grade their knowledge and that of staff at divisional, locational and sub locational levels. All the extension staff at the district, divisional and locational levels never, rarely or occasionally used tours. On the other hand, 50% of the extension staff at the farmer training centres used tours very often. There was also a statistically significant relationship between the area in which the extension staffs were deployed and the frequency at which they used group meetings, field days and workshops. More staff in horticulture (80%) and soil and water conservation (60%) used group meetings either frequently or very frequently. While few extension staff (24%) in crop production, 29% in home economics, 25% in animal production and 20% in animal health used group meetings either frequently or very frequently. Likewise, more extension staff in horticulture (100%) preferred field days. In addition, 83% of the extension staff in marketing used field days either often or very often. However, most of the extension staff in the other departments never, rarely, or occasionally used field days.

The results also revealed that there was a statistically significant relationship between the extension staff's department and the frequency at which they used workshops (Table 2). A high percentage (75%) of extension staff form horticulture preferred to use workshops as an extension method

Extension methods with the least constraints

To be able to understand why the extension staff preferred certain extension methods, the study sought to determine the extension methods with the least constraints (Table 3). Five percent of the respondents did not indicate their first choice of the method with the least constraints, 20% did not indicate their second choice and 23% did not indicate their third choice. The extension method considered to be the first choice method having the least constraints by the extension staff was farm and home visits (30%). This was followed by group

Table 3. Extension method with the least constraints.

Extension method with the least constraint	Order of preference			
	First choice (%)	Second choice (%)	Third choice (%)	Total
Farm and home visits	30	7	8	45
Group demonstrations	25	19	15	59
Group meetings	16	16	13	45
Group demonstration and individual follow-up	14	5	6	25
Field days	2	17	13	32
Workshops	6	8	6	20
Seminars	2	7	9	18
Tours	0	1	7	8
No response	5	20	23	48

Table 4. Extension methods with most constraints.

Extension method	Order of preference		
	First choice (%)	Second choice (%)	Third choice (%)
Farm and home visit	35.9	5.4	5.4
Group demonstration	9.8	3.3	7.6
Group meetings	5.9	14.	5.4
Group demonstration and individual farmer follow-up	5.9	9.8	8.7
Field days	15.2	6.5	9.8
Workshops	2.2	12.	5.4
Seminars	17.4	18.5	13
Tours	1.4	12	14.1
Barazas	1.4	0.0	1.2
No response	3.3	18.5	29.3

demonstrations (25%), group meetings (16%); and group demonstration and individual farmer follow-up (14%) as seen in Table 3. It is worth noting that though most officers chose farm and home visits as first choice method with least constraints, few officers chose it as a second and third choice. On the other hand, more officers chose group demonstrations and group meetings as the first, second and third choice. This implies that officers would find it easier to use group approach. This is confirmed by the information on Table 4 whereby few extension staff chose group demonstrations, group meetings, and group demonstration and individual follow-up as the first, second and third choice of method with most constraints. A correlation was drawn between the extension staff's personal characteristics and the ranking of the extension teaching methods with the least constraints. The results showed that there was a statistically significant relationship ($r = 0.227$; < 0.031) between extension staff's department and the first choice extension method with least constraints. Similarly there was a statistically significant relationship ($r = 0.301$; < 0.004), ($r = 0.227$; < 0.008) and ($r = 0.210$; < 0.045) between area of deployment and second choice extension methods with least constraints and extension staff's designation and second choice extension method with least constraints respectively.

More respondents from department of livestock (23.5%) as compared to the 7% from the department of agriculture considered group demonstrations and individual follow-up as first choice extension teaching methods with the least constraints. Similarly, most of the extension staff who had served for various years considered farm and home visits, group demonstrations and group meetings as the first choice extension methods with the least constraints. However, more extension staff (33.3 and 25%) who had served for less than two years and 3 to 6 years, respectively, considered group demonstrations and individual follow-up as the first choice extension teaching methods with the least constraints.

Cross tabulations revealed that most of the extension staff at all designations chose group demonstrations and group meetings as second choice teaching method with the least constraints. Similarly, most of the extension staff in different areas of deployment chose group demonstrations and group meetings as second choice teaching

Table 5. Cost effective extension methods.

Extension method	Order of preference		
	First choice	Second choice	Third choice
Farm and home visit	7.6	4.3	16.3
Group demonstration	29.3	10.9	8.7
Group meetings	8.7	8.7	10.9
Group demonstration and individual farmer follow-up	10.9	6.5	8.7
Field days	12.0	26.1	6.5
Workshops	8.7	12.0	9.8
Seminars	4.3	15.2	10.9
Tours	14.1	3.3	12.0
Barazas	1.1	0.0	16.3
No response	3.3	13.0	15.8

method with the least constraints.

Extension methods with most constraints

The results revealed that a high percentage (42 and 37%) of extension staff at certificate and diploma level, respectively, chose home and farm visits as the first choice extension method with most constraints. This could be because the extension staffs at certificate and diploma levels were mainly the field staff that interacted with farmersat the farm level, and therefore were able to identify more constraints with this method.

When the extension staffs were asked to rank in order of preference the extension methods with the most constraints, about three percent of the respondents did not indicate their first choice method with the most constraints, 18.5% respondents did not indicate their second choice and 29.3% did not indicate their third choice method with the most constraints as seen in Table 4

It is worth noting that though the extension staff chose farm and home visits as extension method used very often, it ranks highest (35.9%) as the first method with most constraints. This could be because the extension staff are used to this method and hence have identified its constraints than any other method of extension. Alternatively, home and farm visits could be the best method of extension (method that encouraged more adoption of innovations) and therefore was used more though it had the most constraints.

Group meetings, seminars, workshops and tours were considered by a relatively high percentage of extension staff (14, 18.5, 12 and 12%) respectively as the second choice methods with most constraints (Table 4). Group meetings, workshops and seminars require that the clients be assembled at a common place and taught for a period of time. This entails planning for the venue, contacting the clients, buying teaching materials and probably providing some meals or snacks. This is expensive and time consuming.

Tours was also considered by relatively more officers as the second (12%) and third (14.1%) method with the most constraints. This explains why extension staffs do not use the method either often or very often. On the other hand, relatively few extension agents chose group demonstrations (first choice, 9.8%; second choice, 3.3% and third choice, 7.6%), group meetings (first choice 4.3%; second choice 14.1% and third choice, 5.4%), and group demonstrations and individual follow-up (first choice 4.3%; second choice, 9.8% and third choice, 8.7%), as the method with most constraints (Table 4).

Cost effective extension methods

The extension staffs were also asked to indicate in order of preference three extension methods that they considered to be most cost effective. Group demonstration was chosen by a relatively high percentage (29.3%) of extension staff as the first choice extension method that was cost effective as seen in Table 5. This could be because in group demonstrations, farmers are taught in groups hence the same demonstration materials are used as could have been used for an individual client. In addition, a client is taught skills in full, for example the group may be taught practically on all steps in maize planting among others; hence, adoption rates may be high. The group demonstration also has the advantage of multiplier effect (whereby if one client is convinced about an innovation he/she may convince the others to adopt). In addition, relatively more officers considered this method as their second and third choice of the method that was most cost effective.

Relatively high percentage of extension staff (26.1%) chose field days as the second choice extension method that was most cost effective while 16.3% chose tours as the third choice extension methods which were most cost effective.

Correlation analysis was run between extension staff personal characteristics and the extension methods considered to be cost effective. The results illustrated a statistically significant relationship between the first choice cost effective extension method and the extension staff designation (r = 0.308; < 0.003), professional qualification (r = 0.270; < 0.009) and extension staffs' years of service (r = 0.237; < 0.023).

It is important to note that few officers considered farm and home visits as first choice (7.4%), second choice (4.2%) or third choice (8.7%) most cost effective method. Farm and home visit was also considered by many officers as having most constraints (Table 4). This implies that it is expensive and cumbersome for the extension staff to use farm and home visits solely as a means of passing extension packages. Though field days (first choice, 12%; second choice, 26.1% and third choice 6.5%); and tours (first choice, 14.1%; second choice, 3.3% and third choice 12.0%) were considered by relatively more extension staff as the methods that were most cost effective, they were also considered by more extension staff as methods with the most constraints (Table 4) and by few officers as methods with least constraints (Table 3). This could be because tours and field days enable an extension staff to educate more at the same time, thus the same cost is used for several. However, these methods are cumbersome to organize, since they require a lot of capital and time.

Seminars and workshops were chosen as the second and third method that was the most cost effective by more extension staff. However, few extension staff considered them as methods with least constraints (Table 3). Seminars and workshops, like field days and tours, reach more at the same time hence cut down on the cost of passing innovations. But they are expensive and cumbersome to organize.

Conclusion

Extension staff preferred group meetings and group demonstrations, since most officers used these methods either often or very often. In addition, group meetings and group demonstrations were considered to be the methods with the least constraints. Group demonstration was also considered to be the method that was most cost effective. The farm and home visits were preferred by the extension staff though most extension staff considered it to be having the most constraints. Tours and field days were not preferred by the extension staff.

RECOMMENDATIONS

The researchers recommended that:

1. The extension service should be facilitated to use group demonstrations and group meetings to pass innovations. However, individual client follow-up should be encouraged to ensure that clients are able to practice what they were taught at the farms.
2. The Kenya government, donors and other stakeholders should provide enough funding to the extension service to enable them use group demonstrations and individual farmer follow-up to ensure increased adoption of innovations.

REFERENCES

FAO (2000). Food security: When people live with hunger and fear of starvation. The state of food insecurity in the World. Food and Agriculture Organisation of the United Nations.

FAO (2008). The state of food insecurity in the world. Vialle delle Terme di Caracalla (The baths of caracala), 00153 Rome Italy

Government of Kenya (2010). Agricultural sector development strategy 2010-2020. Nairobi, Kenya: Government Printers.

Kathuri NJ, Pals DA (1993). Introduction to educational research_Njoro, Kenya: Educational Media Centre, Egerton University

KEPCO AND CGD (2010). Resource allocation and utilisation in Kenya's Agricultural sector. Retrieved on 25th March 2011. http://www.usaid.gov/our_work/humanitarian_assistance/foodcrisis

Kerlinger FN (1976). Foundations of behavioural research (2nd ed.). New York: Holt, Reinhart and Winston.

Kodhek GA (2005). Contemporary issues determining the future of Kenyan agriculture: An agenda for policy and research. Retrieved on 25 October 2005. http://www.yahoo. Agenda_pol_resrch(1). htm

Ministry of Agriculture (2005). Strategic plan for the Ministry of Agriculture (2005-2009). Nairobi, Kenya: Government Printers

Nyoro JK, Muiruri HK (2001). Policies for agricultural sustainability in Kenya. A case study report for Policies that work for sustainable agriculture and regenerating rural economies. Tegemeo Research Institute, Nairobi, Kenya.

Republic of Kenya (2005). Review of the National agricultural extension policy (NAEP) and its implementation. Ministryb of Agriculture and Ministry of Livestock and Fisheries Development, Nairobi, Kenya

USAID (2009a). Kenya- Food nsecurity. http://www.usaid.gov/our_work/humanitarian_assistance/disaster_assistance.

USAID (2009b). Global food crisis. Retrieved on 9th December 2010. http://www.usaid.gov/our_work/humanitarian_assistance/foodcrisis

USAID (2010). Kenya- food insecurity. Retrieved on 25th March 2011. http://www.usaid.gov/our_work/humanitarian_assistance/foodcrisis.

15

Networks among agricultural stakeholders in the Southwestern Highlands of Uganda

Fungo B.[1]*, Clark L.[2], Tenywa M. M.[3], Tukahirwa J.[4], Kamugisha R.[4], Birachi E.[5], Wanjiku C.[5], Bizoza A. R.[6], Wimba B.[7], Pali P.[8], Adewale A.[8] and Olowole F.[8]

[1]Faculty of Forestry and Nature Conservation, Makerere University P. Box 7062 Kampala, Uganda.
[2]Institute of Development Studies, London NW6 4BT, UK.
[3]Makerere University Agricultural Research Institute, Kabanyolo (MUARIK), Uganda.
[4]African Highlands Initiative (AHI), Kabale Uganda.
[5]International Center for Tropical Agriculture (CIAT), Kigali, Rwanda.
[6]National University of Rwanda (NUR), Butare, Rwanda.
[7]INERA-Bukavu, Democratic Republic of Congo.
[8]Forum for Agricultural Research in Africa (FARA), Accra-Ghana.

The aim of this study was to explore the interactions that exist among agricultural stakeholders in the southwestern highlands of Uganda as a way of identifying opportunities and gaps for operation of Innovation Platforms (IPs) under the proof of concept of Integrated Agricultural Research for Development (IAR4D) research project. The specific objectives were to (i) characterize the agricultural stakeholders in the study sites (ii) determine the nature, diversity and relative importance of horizontal and vertical networks that exist among stakeholders in the Southwestern Highlands of Uganda. Data were collected from both stakeholder analysis and household interviews in Kabale and Kisoro Districts. Results show that extension staff, local governments and farmer groups accounted for approximately 75% of all categories of stakeholders in the area. Most of these organizations started after 10 to 15 years ago following the return of relative political stability in Uganda. Generally, stakeholder interactions in site with limited ARD intervention are more limited compared to their high-intervention counterparts. Sites with "good" market access have more institutions operating there but majority are isolated from each other. At household level, an individual household has networks with approximately two different organizations most of which are farmer groups or credit associations. The greatest proportion of horizontal networks that a household has is with fellow farmers. In order to make the value chain complete, establishment of IPs should pay special attention to including the private sector such as input and produce dealers. Facilitating IP actors to identify critical challenges and opportunities, and effectively articulate them will ensure cohesion. It is also critical to periodically monitor and evaluate stakeholders in terms of the quality of the networks to minimize conflict situations.

Key words: Networks, stakeholders, integrated agricultural research for development (IAR4D), Southwestern highlands, Uganda.

INTRODUCTION

Past approaches to Agricultural Research and Development (ARD) that followed the "technology generation-transfer-adoption" model and subsequent ones had a design flaw in focusing on the supply of new knowledge from research to farmers rather than providing a mechanism for nurturing the innovative capacity of multi-stakeholders to make markets work and address recurrent production and market risks in complex farming systems (Tenywa et al., 2011). Multi-stakeholder approaches to ARD are being encouraged because of the associated benefits such as improved social capital the value individuals derive from their connections, group processes, and knowledge utilization (Gulati et al., 2002; Raider and Krackhardt, 2002; Brass et al., 2004; Hall,

*Corresponding author. E-mail: bfungo1@yahoo.com

2006; Totterdell et al., 2008).

Hall (2006) noted that strengthening the linkages and interaction among ARD actors has been considered as key to improved efficiency and effectiveness of ARD efforts aimed at raising the level of economic performance of rural economy through increased productivity. Empirical studies of the consequences of networks have discovered associations between the number, structure, strength and content of network ties and a range of individual and organizational outcomes, including personal influence, job performance, innovation, career success, satisfaction and affect (Brass, 1984; Flap et al., 1998; Forret and Dougherty, 2004; Totterdell et al., 2004).

Integrated Agricultural Research for Development (IAR4D) is a new approach that is envisaged as a superior way to conducting ARD. The IAR4D approach involves five critical stakeholder categories (Research, Extension, Farmer, Policy maker and Private Sector) come together in and Innovation Platform (IP) to jointly identify and articulate the critical problems along a value chain, and be able to identify and implement potential solutions to overcome this problem. Ultimately, the income and livelihood of farmers will be improved. The Forum for Agricultural Research in Africa (FARA) coordinated sub Saharan Africa Challenge Program (SSA-CP) is conducting a study to test whether the IAR4D research approach is superior to conventional ARD initiatives. An Innovation Platform (IP) is the core of the Integrated Agricultural Research for Development (IAR4D) approach. The IP is a coalition, collaboration, partnership and alliance of ARD actors (that is, public and private scientists, extension workers, representatives of farmers, farmers' associations, private firms and non-governmental organizations, and government policy makers). These actors communicate, cooperate and interact to set priorities, develop concepts and plans to promote agricultural productivity and profitability (FARA, 2009). This is because increasing agricultural productivity supported by appropriate policies and markets without doing harm to the natural resource base can help improve the welfare of all members of society. It is envisaged that when the aforementioned actors work together as an IP, the benefits are greater than the aggregate of the individual actors working separately.

IPs can act as powerful tools for farmers to build social capital that helps them avert risks associated with agricultural investments. Networks among stakeholders help to ensure that the IP is cohesive and consequently, there is direct and continuous interaction, communication and knowledge sharing among the IP actors, quick and continuous feedback from end users (farmers) at all stages of research for development, and timely integration of new knowledge into the innovation process using experiential learning, monitoring and evaluation and the continual feedback. The number, diversity and strength of the networks among actors are critical factors determining the sustainability of the IP.

The sub Saharan Africa Challenge Program (SSA-CP), coordinated by Forum for Agricultural Research in Africa (FARA) is conducting a study to test whether the IAR4D research approach is superior to conventional ARD initiatives. To be able to foster effective partnerships among IP members ahead of the IAR4D study, there is urgent need to understand the nature of existing networks among various actors in the agricultural setting and the dimensions where improvement is required. This information is also useful for future impact studies to understand the contribution of IAR4D approach in improving interactions among stakeholders. In this study, we explored the interactions that exist among agricultural stakeholders in the southwestern highlands of Uganda where part of the study is being conducted. The objectives were to (i) characterize the agricultural stakeholders in the study sites (ii) determine the nature, diversity and relative importance of horizontal and vertical networks that exist among stakeholders in the Southwestern Highlands of Uganda. This study is part of the sub Saharan Africa Challenge Program (SSA-CP), which is testing the concept that the IAR4D approach is superior to conventional ARD approaches in delivering research outputs to farmers.

Study area

The study was conducted in four sub counties (Bufundi, Chahi, Nyakabande and Rubaya) in the Southwestern Highlands of Uganda, which is one of the three Pilot Learning Sites (PLS) of the SSA-CP. The Lake Kivu Pilot Learning Site (LK PLS) is located on the border region between Uganda, Rwanda and the DR Congo (Figure 1). It is a highland agro-ecological zone (AEZ) characterized by steep slopes, deep good volcanic soils and ample rainfall that offers good potential for agriculture. The area receives mean annual rainfall ranging between 900 to 2,200 mm and has medium to long growing periods (180 to 270 days) (Komutunga and Musiitwa, 2001). The principal food crops grown include sorghum, millet, Irish and sweet potatoes, peas, maize, beans, and bananas. Some cash crops such as tea and coffee are also grown in addition to other tree species for fruit and forestry products. Despite the apparent rich natural resource endowment, the Lake Kivu (LK) Region is considered one of the poorest and most densely populated areas of Africa, with densities ranging from 400 to 700 persons/km^2 (Pender et al., 2004). This has led to over-exploitation of the natural resources. Over 90% of the population derives their livelihood from agriculture and other natural resource based enterprises on less than 0.6 ha per household. Nearly 60% of the land area is intensively cultivated and poverty in the region is directly linked to the low and deteriorating productivity and profitability of these enterprises. The region has also

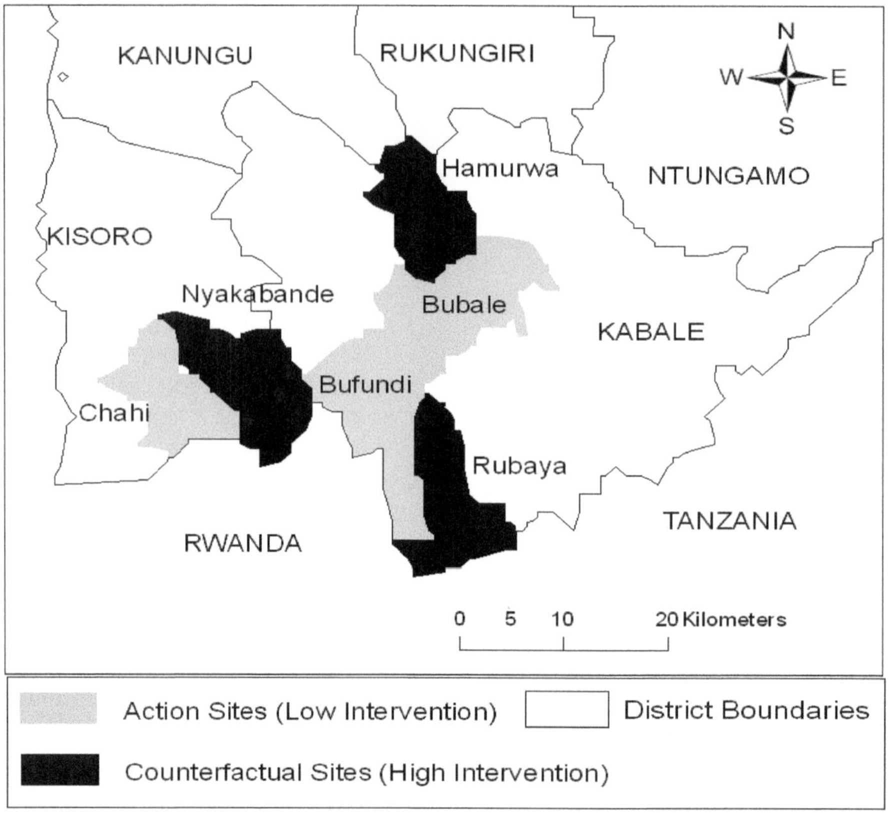

Figure 1. Study sites in southwestern Uganda.

experienced recurrent volatility of conflicts with sporadic conflict still continuing in some parts of DR Congo. The principal challenge in LK is thus to contribute to improved food and nutrition security, increased household incomes and improved quality of the (Pretty et al., 2006) natural resource base by applying IAR4D to develop, test and promote technological, institutional and policy innovations based on integrated watershed management (IWM) concepts.

Methods

Stakeholder analysis was conducted in the four sub counties in May 2008. The selection of the sub counties was based on previous level of intervention by development projects and access to market. The detailed methodology of the site selection process can be found in Farrow et al. (2009). Eight sub counties were selected but for this paper, only four were use because of the uniformity of the biophysical and socio-economic environment. The selected sub counties were grouped in pairs, one representing the low (action sites) and high (counterfactual sites) interventions by development projects.

In each of the sites, provisional lists of relevant institution stakeholders were developed during consultative meetings with local contact persons and opinion leaders. The identified individuals and institutions were invited for a stakeholder meeting that took place at the sub county headquarters of each of the sub counties. During the meeting, members were introduced to the SSA-CP, and its relevance to agricultural and natural resource research was highlighted. Participants were exposed to the assessment questionnaire that was developed to map the stakeholders and their activities. In a plenary session, participants were then guided on how to complete the questionnaire. In situations where a participant was unable to read and/or write, facilitators interviewed the respondents and completed the tool on their behalf. In order to understand the nature of benefits that arise from the existing interactions (social capital), households were asked to provide information about the role, type, frequency and perception of strength of the interaction in the last twelve months. The questions focused on interactions for exchange of agricultural information.

The household study followed a stratified random sampling design to compare the baseline situation of households and communities under the following three categories of communities; (i) IAR4D, (ii) conventional and (iii) no intervention of ARD. The first-level strata consisted of sub counties while second-level consisted of villages. These strata represent the extent to which agricultural research and development projects have intervened in the study area. The details of the site selection and basis for this categorization can be found in Farrow et al. (2009). Briefly, south western Uganda was divided into "good" and "poor" access. Each of these was further divided into two treatments referred to as Intervention (IP) and control (Counterfactual) with low and high level of intervention by research and development organizations, respectively. In each of the categories, 10 villages were selected randomly from a list of existing villages. Using a sampling frame comprising of names of household heads in the village as provided by local council leaders, 10 households were selected using random numbers for inclusion in the household interviews. Therefore, a total of 600 households were interviewed.

Table 1. Stakeholder representation by category and location in south-western Uganda.

Stakeholder category	Action		Counterfactual		Total	(%)
	Bufundi	Chahi	Nyakabande	Rubaya		
Farmer groups/Association	10	17	15	8	50	39.4
Local government Admin.	9	6	12	16	43	33.9
Extension	1	5	1	3	10	7.9
Faith-based organizations	1	3	1	4	9	7.1
Marketing	3	-	1	1	5	3.9
Financial and Credit	-	3	-	1	4	3.1
Research	-	1	-	2	3	2.4
International NGOs	1	-	2	-	3	2.4
Private sector (Input/produce dealers)	-	-	-	-	-	-
Grand total	25	35	32	35	127	100

Descriptive statistics were used to explore the characteristics of stakeholders' organizations and households. Network matrices showing ties between actors (group-to-group and group-to-individual) were generated. For each treatment network matrices centrality measures were computed using UCINET 6 for windows software (Borgatti et al., 2002).

RESULTS AND DISCUSSION

Nature of agricultural stakeholders

The categories of stakeholders analyzed during the study are shown in Table 1. Overall, Farmer groups had the greatest representation (39%), closely followed by Local government at 34%. Extension agents and faith-based organizations constituted 8 and 7%, respectively. The least represented category was the NGOs, followed by researchers. Credit organizations were represented by only 4% while the private sector was not represented at all.

The categorization of the stakeholders is not clear-cut, implying that a single stakeholder may fall in more than one category. For example, representatives from NGOs may also be doing extension work by the nature of the activities they engage in, extending services to farmers. For the purpose of this discussion, the principle service provided by the stakeholder was considered in the classification. In case of actors that provide multiple services, the major one is considered. Additionally, farmer organizations may grow and form a wide network across districts and even regions. For example, Kabale district Farmers Association and Kisoro District Farmers Association have a common umbrella association, the Uganda National Farmers Association (UNAFA) but also under Uganda National Farmers Federation (UNEFE). In the context of this study, however, farmer organizations means those that have agricultural and related activities as their major goal while CBOs may have a different goal such as health, persons with disabilities and orphans. International NGOs such as Africare also provide extension services to farmers such as training in soil conservation and post harvest handling. Overall, Bufundi, which is an action site, had the least number (17) of stakeholder categories compared to the other three sites (Chahi, Nyakabande, and Rubaya) that had almost similar numbers (24, 27 and 25, respectively). Extension agents, farmer associations, faith-based organizations, community-based organizations and individual farmers were distributed among all the four sites. However, there were no research organizations and private sector in the action sites (Bufundi and Chahi). As indicated earlier, the former organizations/ Institutions usually exist in many areas independent of their remote location.

This categorization of stakeholders may have several limitations. For example, the criteria used to select and invite the representatives may sometimes be centered on well-established and popular organizations/institutions. Also, the responses of the stakeholders about the organizations they belong to may have been strategic in the sense that many local people are used to projects that target specific groups. To be able to benefit from these project interventions, local communities associate themselves with the category of people that the project is likely to benefit directly. To ensure that the spectrum of stakeholders was as wide as possible the research team held consultative meetings with local leaders and other knowledgeable individual to indentify groups or individuals that would provide relevant information to initiate and make the IPs operational. A clear explanation of the intention and processes of the IPs to the participants of the stakeholder meetings helped to normalize their expectations.

The age of majority stakeholder representatives ranged between 26 and 65 years (Figure 2). This represents an age bracket where individuals are active and can hold responsible positions in the community. Below 25 years, individuals may be considered immature since they are just emerging adults with little experience. Above 65, a person may be considered old and unable to contribute

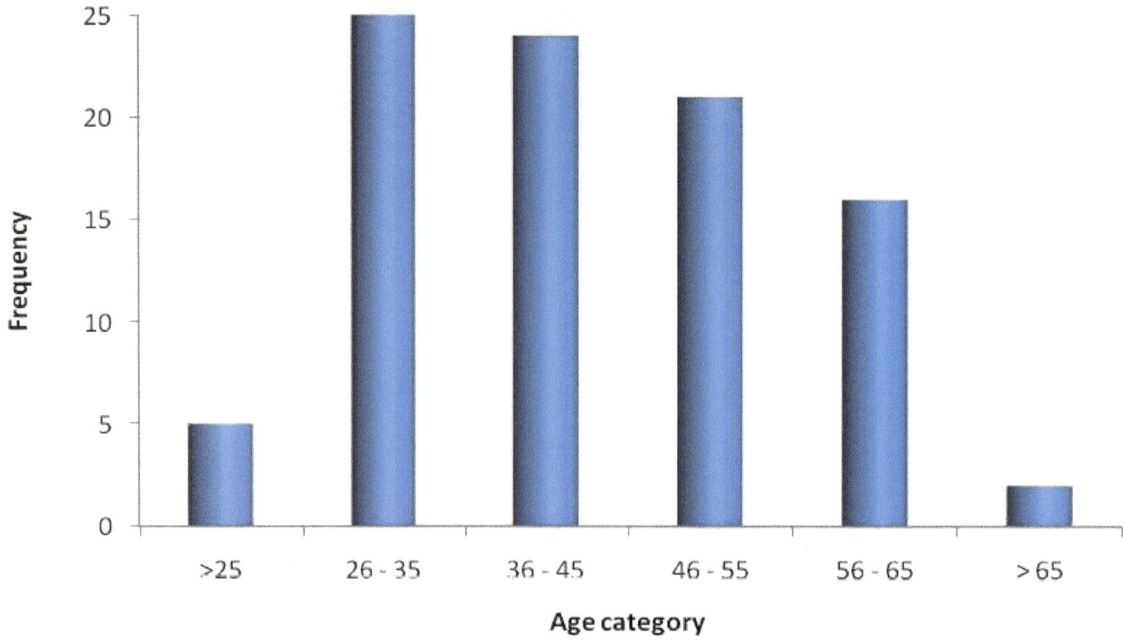

Figure 2. Age distribution of the age of respondents in study sites in Kabale and Kisoro under the Sub Saharan Africa challenge program.

much except in terms of experience. Additionally, there is a common practice for organizations to put much older people in positions of leadership but reserve the position of secretary, who is given responsibilities such as representation of the group, to a much younger and literate person.

The time partners have been in existence were classifies according to the political history of Uganda. According to the respondents, the organizations that existed before 1986 were mainly churches and local government (Table 2). Majority of CBOs, NGOs and credit organizations started between 1997 and 2006. This category or organizations represents institutions that are generally present in almost every community and are not much affected by political perturbations. Before 1986, the country was politically turbulent with a series of civil wars until the capture of power by the National Resistance army that has ruled since then. This shows that only a few organizations were working in this study area. The largest number of organizations (30%) started between 1986 and 1991 probably following the peace created after the end of the war. It would be expected that the number of organizations would continue increasing after 1991. Ironically, the years following 1991 experience the least number of emergent organizations. This could be related to the series of policy changes that were taking place during 1991-1996, including the making of the constitution in 1995.

The farmer groups that existed before 1986 were actually cooperative societies that were part of the country-wide arrangement that provide a joint voice for small-scale farmers at policy level. These arrangements were hierarchically well-organized but operationally wanting because of limited financial capability. Because of limited mobility of representatives as challenged by financial and in some cases topographic environments, representatives do not consult their subsidiaries but rather their personal views. Election of representatives was mainly through voting but also through other methods such as seconding of leaders of existing groups. Other important policies that were developed during this period include the Nation Agricultural Research Policy, National Environmental Action Plan and the National Policy for the Conservation and Management of Wetlands. Once dramatic policy changes are expected, the tendency for people to resent from making long-term plans that may be affected by the policy changes. In support this tenet, the years 1997-2001 through 2002 - 2006 experienced equally high emergence of new institutions at 24 and 25%, respectively.

Existing partnerships by site

Horizontal linkages

The interactions among institutions are shown in Figure 3. The stakeholders that were interviewed (Circles) indicated organizations they collaborate with (maroon squares) in the various aspect of agricultural value chain. The different colours show the category of stakeholder interviewed. Although the network suggests that there is

Table 2. Period of establishment of various stakeholder categories in south-western Uganda.

Stakeholder category	Before 1986	1986-1991	1992-1996	1997-2001	2002-2008	Grand total
Farmer groups/associations	2	8	6	8	23	47
Local government admin.	4	22	2	3	3	34
Extension	-	3	1	4	2	10
Faith-based organizations	2	-	-	3	1	6
Marketing	-	2	1	-	1	4
Financial and Credit	1	-	-	1	2	4
Research organizations	-	-	1	-	-	1
International NGOs	-	1	-	1	-	2
Grand total	9	36	11	20	32	108

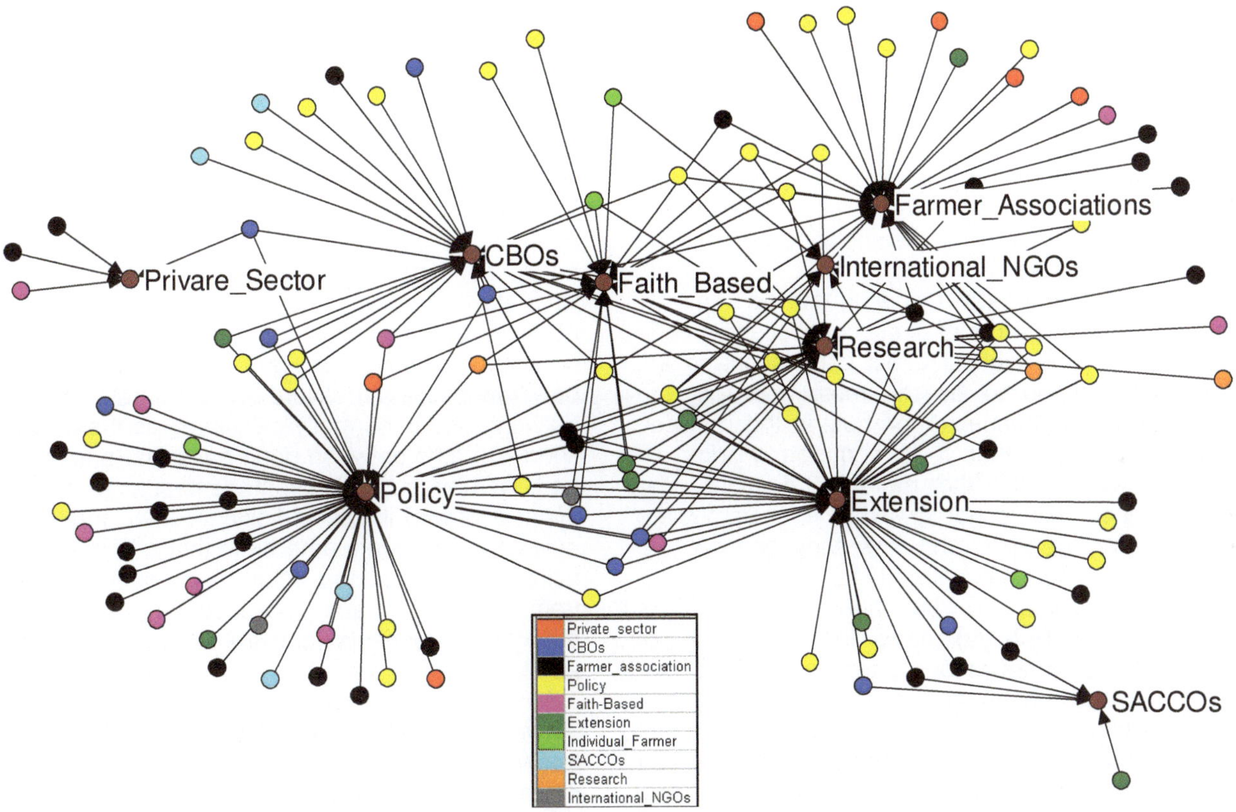

Figure 3. Interactions among agricultural stakeholder groups in southwestern Uganda (colour of the circle at arrow tail shows the organization that the respondent represents while the circle at the arrow head shows the type of organization that the respondent mentioned to be interacting with).

significant interaction among the organizations, it is important to note that some of the respondents interact as individuals rather than as institutions. For example, at the time these data were collected, NAADS had not started working in Chahi, but the respondents from Chahi mention having interacted with it. A closer look at the nature of interaction shows that they exchange not only agricultural information. Hence, it is logical to infer that these interactions have no definite boundaries. By introducing the Innovation Platforms, actors are better placed to identify sources of relevant information.

Additionally, the context of IAR4D is that the various partners discuss a common challenge/interest, rather than just interacting with a variety of partners. Data shows that approximately 36% of the respondents had participated in multi-stakeholder platforms. To better understand networks and their participants, an evaluation of the "location of actors in the network" using network

Table 3. Degree centrality measures for stakeholder interactions in Southwestern Uganda.

Statistic	Action		Counterfactual	
	Bufundi	Chahi	Nyakabande	Rubaya
Mean	1.9	2.0	5.0	4.7
Standard deviation	1.4	1.2	4.3	7.0
Sum	34.0	90.0	214.0	166.0
Variance	1.9	1.5	18.6	48.9
Minimum	0.0	1.0	1.0	0.0
Maximum	5.0	6.0	19.0	32.0
Centralization (%)	10.3	3.2	5.9	14.7
Heterogenuity (%)	8.5	3.1	4.1	9.7

metrics was conducted. Measuring the network location is finding the "centrality" of a node. These measures give us insight into the various roles and groupings in a network -- who are the connectors, mavens, leaders, bridges, isolates, where are the clusters and who is in them, who is in the core of the network, and who is on the periphery? We used the three most popular individual centrality measures to describe the social networks in the sites in Kabale and Kisoro.

Hanemann and Riddle (2010) define degree centrality as a measure of the number of direct connections a node has. A node with the highest direct connections in the network is considered to be the most active in the network. It is thus a 'connector' or 'hub' in the network. The authors observe that it is common wisdom in personal networks that "the more connections, the better." However, this is not always the case. Where those connections lead to and how they connect the otherwise unconnected is equally crucial.

Overall, institutions in the counterfactual sites (Nyakabande and Rubaya) have larger number of linkages with other organizations compared to the action sites (Bufundi and Chahi) as indicated by the degree centrality (Table 3). This was expected and in fact validates the site selection criteria that aimed at selecting contrasting sites in terms of actors and their interaction. As indicated in the network map, Chahi has relatively more organizations compared to Bufundi but the number of linkages for each organization does not exceed six. For a network to be considered fairly stable, the centralization should be 50% and above. All the sites generally have low level of network centralization that ranges from 3.2 to 14.7%. Low level of centralization implies that once one or a few actors leave, the network can very easily crumble.

Despite the generally low level of centralization of the four sites, Rubaya had the highest centralization of 14.7% followed by Bufundi, then Nyakabande and least was Chahi. In order to improve on the centralization of the networks, more of the relevant stakeholders should be identified and involved in the IPs. Suitable examples include transporters and credit organizations in Bufundi, who can contribute to farmer credit and overcome the transport challenge to help farmers reach the market more easily. The sole reliance on NAADS causes the network to be highly centralized rendering it prone to crumbling once such a partner exits the system.

Newman (2005) suggests that betweenness centrality is, in some sense, a measure of the influence a node has over the spread of information through the network. A node with high betweenness has great influence over what flows and what does not in the network. Such a node is 'between' two important constituencies of the network. According to Hennemann and Riddle (2010), betweenness is a measure of the centrality of a node in a network, and is normally calculated as the fraction of the shortest paths between node pairs that pass through the node of interest. The advantage of such an actor is that it plays a powerful role in linking certain members of the network. The unfortunate bit is that it is a single point of failure. Without it, the actors it connects will be cut off from information and knowledge in network or some network components (cluster) (Hennemann and Riddle, 2010)

Of all the four sites, Nyakabalde showed the highest betweeness centrality of 24 while the others had values ranging from 0 to 1.1 (Table 4). The netwok is disconnected, implying that the individual actors seem to operate on their own with no mediation role in the system probably because they do not benefit from the relationships or the cost of maintaining them is high in terms of time and resources. The implication for this study is that in as much as several organizations exist, they do not link up enough to be able to improve on social capital, group processes and knowledge utilization as shown in some studies (Brass et al., 2004; Gulati et al., 2002; Raider and Krackhardt, 2002). The brokerage role is not played by any of the organizations that exist in the study area. An innovation platform is expected to play a key role in linking several actors in order to ensure that information to one organization/individual is quickly and efficiently circulated to the other partner to realize the benefit of quick information feedback in agricultural research and development processes.

Table 4. Betweenness centrality of networks of agricultural stakeholders in Southwestern Uganda.

Statistic	Action		Counterfactual	
	Bufundi	Chahi	Nyakabande	Rubaya
Mean	0.3	0.0	23.9	1.1
Standard deviation	0.8	0.0	47.8	4.1
Sum	5.0	0.0	1027.0	39.0
Variance	0.6	0.0	2286.7	16.5
Minimum	0.0	0.0	0.0	0.0
Maximum	3.0	0.0	238.3	23.5
Centralization (%)	1.1	0.0	12.8	2.1

Table 5. Closeness centrality for action and counterfactual sites in South-western Uganda.

Statistic	Action		Counterfactual	
	Bufundi	Chahi	Nyakabande	Rubaya
Mean	6.1	2.3	3.8	3.2
Standard deviation	1.0	0.1	1.6	0.9
Sum	109.5	102.1	164.3	113.0
Variance	1.1	0.0	2.5	0.8
Minimum	5.6	2.3	2.3	2.9
Maximum	9.8	2.5	8.5	7.1

Closeness centrality defines the pattern of partners' direct and indirect ties that allows them to access all the stakeholders in the network more quickly compared to other members (Newman, 2005; Heinemann and Riddle, 2010). The partners have the shortest path to others and are in an excellent position to monitor the information flow in the network. Closeness centrality is inversely related with the ease of accessing a given partner in the network (Heinemann and Riddle, 2010). Again, Bufundi Sub County has a higher closeness centrality than the other sites (Table 5). This implies that it is more difficult for stakeholders in Bufundi to link up with each other in case they wish to discuss an issue. Ironically, Chahi has the best closeness centrality, probably because it is more accessible than even Rubaya. In terms of service provision, low closeness centrality helps partners to quickly provide or find information that they require to for a particular segment of the value chain.

Chahi, Nyakabande and Rubaya had almost equal number of stakeholder representatives but Bufundi had slightly less. Bufundi is the most disadvantaged in terms of access. Movement within the sub county and to the nearest township is more difficult in compared to other sites such as Nyakabande and Chahi. Although Rubaya is equally hilly, there have been several roads established under the phased-out Area-Based agricultural Modernization Program as well as Africare, which was not the case for Bufundi Sub County. There were 17% women among the stakeholder representatives and about half belonged to farmer groups.

Vertical linkages

Vertical interaction refers to the relationship between an individual, and institution at a lower level with institution or individual at a higher level. This relationship is important because it influences the flow of superior knowledge to lower levels and also information from lower to higher levels for inclusion in policy formulation. Results show that in the vertical dimension, farmers/households interact with various organizations working in the area (Figure 4). On average, an individual farmer interacts with five different organizations, most of which are Community-based or credit associations. National programs such as National Agricultural Advisory Services (NAADS) and Area-Based Agricultural Modernization Program (AAMP) have the highest interactions with households in Kabale and Kisoro. By their nature, these programs provide farmers with inputs such as seeds, fertilizers and farm implements. It is not unusual that farmers will obviously recognize these immediate benefits provided by these organizations. This could explain why these programs feature prominently in the networks mentioned. However, the sustainability of these hand-out schemes is questionable because farmers' capacity is sustained for as long as the program runs. To ensure sustainability, it is necessary to develop networks that contribute to stakeholders direct benefits such as access to credit, provision of transport for input/output, and frequent relevant information.

Experiences from similar programs suggest that the

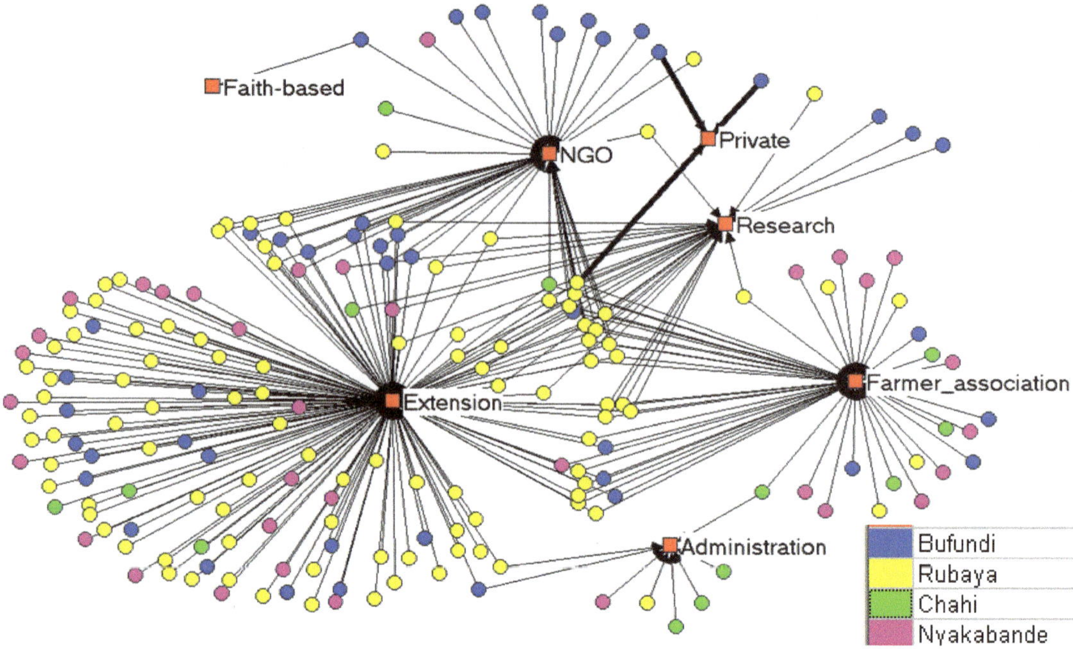

Figure 4. Vertical linkages between households (circles) and organizations (red squares) in the agricultural sector in Southwestern Uganda.

farmers farming systems reflected in the relatively high number of household that mentions it in their networks. At community level, this impact is evident from the focus group discussions where AHI was mentioned frequently for promoting *Acasia* ssp. under the land care the program.

Totterdell et al. (2008) found that managers and team leaders had greater propensity to connect with others than other employees in the organization, which implies that having this propensity may incline individuals to adopt or be adopted for certain roles within or among organizations. In particular, managers and team leaders were more inclined to form weak ties and bridging ties, and these are precisely the kinds of tie that have been associated with acquiring power and influence in organizations (Brass, 1984; Brass et al., 2004).

The organizations mentioned more frequently by households include NAADS, AAMP and Africare and account for approximately 80% of all the networks identified (Figure 5). The remaining percentage is shared among all the others with some having only one tie (those at the periphery of Figure 5). Generally, organizations that have a wider geographical scope of operation such as the aforementioned (NAADS, AMMP and AFRICARE) have relatively larger number of people they interact with compare to local ones. The more localized organizations, such as farmer groups, interact with only a few individuals and such individuals do not themselves interact with other organizations. However, AAMP and Africare have wound up their activities in the region but farmers still remember the activities undertaken, suggesting that they had significant impact on farmers' livelihoods.

By the nature of their operation, these dominant organizations a have specific focus in a given area. For example, NAADS organizes farmers to select a specific enterprise and focus on a particular point in the value chain such as "soil fertility improvement for potato production". In the same area, CARE may be promoting "improved varieties of beans". This implies that for each of these enterprises, information flow to and from such farmers is limited. Ideally, an IP in such an area should facilitate existing stakeholders to be aware of the on-going work so that they strive to solve the most critical problem first and take into account the entire value chain instead of solving isolated problems in various chains.

Once a particular chain if fully functional, the IP may switch to another enterprise. The challenge of such an arrange will always be how to facilitate dialogue among the various stakeholders to have common understanding of the critical problem. This can be overcome through continued training of IP actors in needs assessment and effective knowledge-sharing. More so, the quality of linkages needs to be periodically assessed to identity interests, roles and responsibilities in order to enhance the relevant ones and reduce or eliminate detrimental ones.

Benefits from stakeholder networks

The findings for all the stakeholder categories show

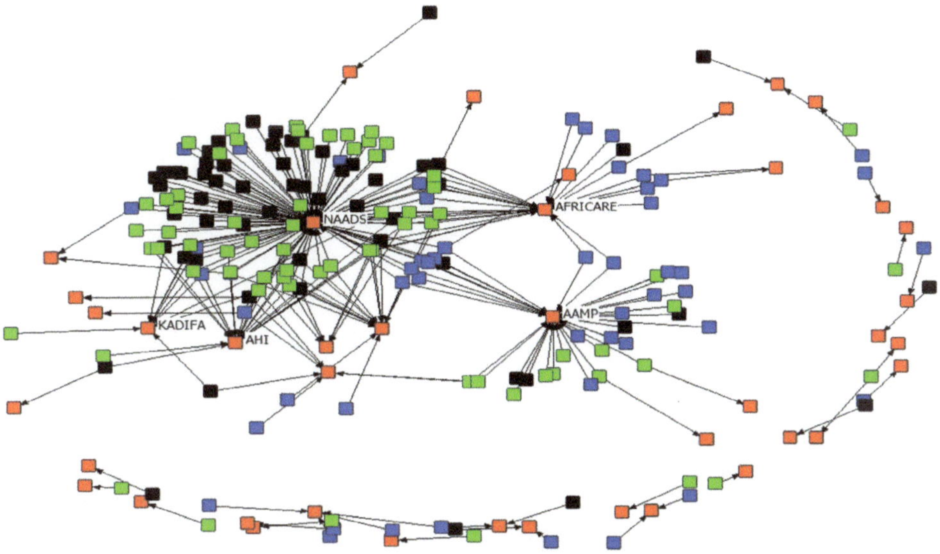

Figure 5. Most prominent agricultural stakeholders in southwestern Uganda. (Red = Organization mentioned; Green = Male headed-households and Blue = female-headed households).

Table 6. Reasons why farmers interact with organizations in Southwestern Uganda.

Stakeholder type	Purpose of interaction (N=570)				
	Information exchange	Commercial business transaction	Material exchange	Money exchange	Others
Administration	12	3	1	2	-
Extension	270	15	13	10	1
Faith-based	1	-	-	-	-
Farmer association	77	11	5	6	-
NGO	57	12	-	-	-
Private sector	1	6	-	-	-
Research	52	13	-	2	-

closure of the program is immediately followed by the recession in production. In Rubaya, for example, African Highlands Initiative (AHI) has made significant impact on that majority (83%) of the interactions involve mainly information exchange (Table 6). Other interactions involve materials exchange, money/credit exchange, commercial business transactions, and others which take 10, 3 and 2%, respectively. This situation is a positive one in terms of innovation platforms because information exchange is a critical ingredient of the knowledge economy, which we believe is the current global situation. The innovation platform is valuable in this case because it completes the gap in information loop of a given site. With respect to the first tenet stated in this paper, this is a positive situation and there is hope that the IAR4D approach will contribute to improved livelihoods that the conventional approach has failed to do. Overall, majority of the interactions (41%) take place on a daily basis while 25%, 16, 12 and 7% take place on a weekly, monthly and annual basis, respectively (Table 7).

General, the stakeholders perceive the interaction with their partners to be above moderate as majority indicated (Table 8). The results further show that the perceived strength of the interaction is independent of the type of partners interacting. As noted earlier, organizations interact significantly among themselves. Similarly, the nature of interaction that households have with fellow farmers comprises 80% to 90% (Figure 6). The other part of the interaction is with other actors such as community leaders, extension agents, researchers, traders and NGO staff. These results show that there is limited vertical interaction between the households and other stakeholders at a higher hierarchy in the network. This puts a limit on the extent that information can flow in the social network. This is the observed trend for all the four sites in this study.

The trend these days is that most organizations prefer to work with farmer groups rather that with individual farmers. Therefore, the high interaction between organizations working with farmers reflects this fact.

Table 7. Frequency of interaction between households and organizations promoting ARD interventions in Southwestern Uganda.

Stakeholder	Frequency of interaction					
	Daily	Weekly	Monthly	Every 6 months	Annually or less	>1 year
Administration	7	3	4	2	1	1
Extension	87	92	66	39	27	3
Faith-based	-	-	1	-	-	-
Farmer association	21	27	23	21	9	-
NGO	18	18	7	15	11	-
Private sector	-	7	-	-	-	-
Research	29	11	12	9	6	-

Table 8. Perception of households about the strength of interaction with organizations in Southwestern Uganda.

Stakeholder category	Perception of strength of interaction				
	Very weak	Weak	Moderate	Strong	Very strong
Administration	-	1	6	11	-
Extension	12	7	98	141	54
Faith-based	-	-	1	-	-
Farmer association	7	4	25	49	15
NGO	2	2	22	30	12
Private sector	-		1	3	3
Research	1	2	19	39	5

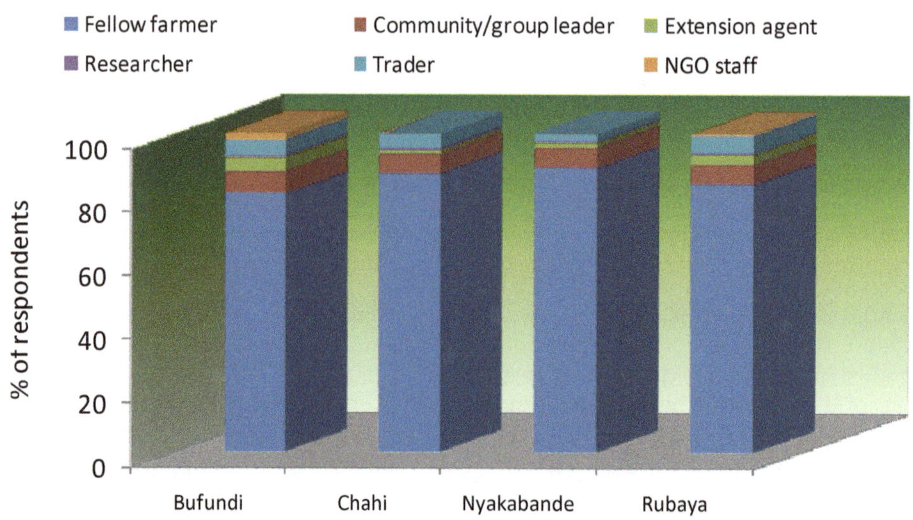

Figure 6. Role of farmers in household interaction in Kabale and Kisoro Uganda.

Latham (2000) suggests that vertical (hierarchical) relations such as those between individual households and people in authority (for example the workers of extension organizations), are the types of social relations relevant for understanding trust in authority. On the other hand, horizontal (equal, democratic) relations such as those between households or between organizations are relevant for understanding trust in civil society. He describes these as vertical and horizontal types of social capital. Recent empirical work (Knack and Keefer, 1997) challenges the significance of vertical and horizontal relations for social capital. It reinforces the need to understand empirically the impact of network hierarchies upon social capital. While the questions of whether social networks are characterized by vertical or horizontal relations – specifically how democratic a network is – and whether this is important for social capital remains a major theme in the social capital literature, this aspect of network relations also remains under-investigated. Those studies which have investigated this characteristic of networks typically ask about decision-making processes within networks or about the nature of rule enforcement as indicators of vertical and horizontal relations.

CONCLUSIONS AND RECOMMENDATIONS

Several types of organizations operate in southwestern Uganda but extension, local government and farmer groups, account for approximately 75%. Most of these organizations started after 1986 following the return of relative political stability in Uganda. Generally, action sites have fewer networks among organizations compared to the non-action sites. Even among actions sites, those with good market access have more institutions operating in the area. However, majority of the organizations operating in the action sites are isolated from each other. Overall, local government and national programs such as National Agricultural Advisory Services (NAADS) and Area-Based Agricultural Modernization Program (AAMP) have the highest number of networks, although they are mainly with farmer organizations and government extension agents. On average, an individual farmer about two networks with different organizations most of which are farmer groups or credit associations. The greatest proportion of networks that household in southwestern Uganda have are with fellow farmers. We recommend that the establishment of IPs should pay special attention to including the private sector such as input and produce dealers in order to make the value chain complete. For the IPs to make meaningful contribution to improving the vertical networks (among individual farmers and institutions), it is imperative to streamline communication mechanisms such as the use of mobile telephones, which most farmers have at the moment. A mechanism for coordinating the identification of new potentially relevant actors, monitoring and evaluating existing ones, as well as facilitating overall IP activities are inevitable central to improving useful networks among IP actors.

ACKNOWLEDGEMENTS

The study was funded by the European Union through the Forum for Agricultural Research in Africa (FARA). We are also grateful to International Center of Tropical agriculture (CIAT) for coordination of the Lake Kivu Pilot learning site where this study was conducted. Various project partners are greatly acknowledged for facilitating the identification and invitation of relevant stakeholders in the study area.

REFERENCES

Borgatti SP, Everett MG, Freeman LC (2002). Ucinet for Windows: Software for Social Network Analysis. Harvard, MA: Analytic Technologies.

Brass DJ (1984). Being in the right place: a structural analysis of individual influence in an organization. Admin. Sci. Q., 29: 518–539.

Brass DJ, Galaskiewicz J, Greve HR, Tsai W (2004). Taking stock of networks and organizations: A multilevel perspective. Acad. Manage. J., 47: 795–817.

Farrow A, Opondo C, Tenywa M, Rao KPC, Nkonya E, Njeru R, Lubanga L (2009). Selecting sites to prove the concept of integrated research for agricultural development: LKPL Site selection Annual Report (2008/2009).

Flap H, Bulder B, Volker B (1998). Intra-organizational networks and performance: A review. Comput. Math. Organ. Theory, 4: 109–147.

Forret ML, Dougherty TW (2004). Networking behaviors and career outcomes: Differences for men and women? J. Organ. Behav., 25: 419–437.

Gulati R, Dialdin D, Wang L (2002). Organizational networks. In: Baum, J.A.C. (Ed.), Companion to Organizations. Blackwell, Oxford, pp. 281–303.

Hall AJ, Yoganand B (2004). New Institutional Arrangements in Agricultural Research and Development in Africa: Concepts and case studies. In: A.J. Hall, B. Yoganand, R.V. Sulaiman, R.S. Raina, C.S. Prasad, G.C. Naik and N.G. Clark (eds.). Innovations in innovation: reflections on Partnership, Institutions and Learning, Crop Postharvest Research Programme/ICRISAT/national centre for Agricultural Economics and Policy Research, New Delhi and Andhra Pradesh, India, pp. 105-131.

Hanneman RA, Riddle M (2010). Introduction to Social Network Methods 1. Social Network Data. Department of Sociology at the University of California, Riverside. file:///C|/Documents%20and%20Settings/hanneman/My%20...ts/Network_Text/

Komutunga ET, Musiitwa F (2001). Climate. In: Mukiibi (Ed). *Agriculture* in Uganda – General information. NARO-Uganda, pp. 21-32.

Knack S, Keefer P (1997). Does social capital have an economic payoff? A cross-country investigation.' Q. J. Econ., pp. 1251-1288.

Krebs V (2008). Social Capital: The Key to Success for the 21st Century Organization. IHRIM J., 12(5): 38-42.

Latham M (2000). If only men were angels: Social capital and the Third Way.' In I. Winter (ed.) Social Capital and Public Policy in Australia, Australian Inst. Fam. Stud., Melbourne, pp. 192-224.

Newman ME (2005). A measure of betweenness centrality based on random walks. Social Netw., 27(1): 39-54.

Pender J, Nkonya E, Jagger P, Sserunkuuma D, Ssali H (2004). Strategies to increase agricultural productivity and reduce land degradation: Evidence from Uganda. Agric. Econ., 31(2/3): 181-195.

Pretty JN, Noble AD, Bossio D, Dixon J, Hine RE, Penning de Vries FWT, Morison JIL (2006). Resource-conserving agriculture increases yields in developing countries. Environ. Sci. Technol., 40(4): 1114 – 1119.

Raider H, Krackhardt DJ (2002). Intraorganizational networks. In: Baum, J.A.C. (Ed.), Companion to Organizations. Blackwell, Oxford, pp. 58-74.

Tenywa MM, Rao KPC, Tukahirwa JB, Buruchara R, Adekunle AA, Mugabe J, Wanjiku C, Mutabazi S, Fungo B, Kashaija NIM, Pali P, Mapatano S, Ngaboyisonga C, Farrow A, Njuki J, Abenakyo A (2011). Agricultural Innovation Platform As a Tool for Development Oriented Research: Lessons and Challenges in the Formation and Operationalization Learning Publics. J. Agric. Environ. Stud., 2(1): 117-146.

Totterdell P, Holman D, Hukin A (2008). Social networkers: Measuring and examining individual differences in propensity to connect with others. Social Netw., 30: 283–296.

Totterdell P, Wall T, Holman H, Epitropaki O (2004). Affect networks: A structural analysis of the relationship between work ties and job-related affect. J. Appl. Psychol., 89: 854–867.

Wendy S (2001). Measuring social capital; Towards a theoretically informed measurement framework for researching social capital in family and community life. Australian Institute of Family Studies. Research Paper No. 24, February.

16

Adoption behaviour of the farmers of Goalpara district of India towards aquaculture technology

Uttam Kumar Baruah, Jyotish Barman* and Benzamin Kaman

Krishi Vigyan Kendra Goalpara, National research Centre on Pig, Indian Council of Agricultural Research, Thekasu part - I, Damra Road, Dudhnoi - 783124, Goalpara, Assam, India.

A study was conducted during 2009 to 2010 to investigate the adoption behaviour of the fish growers towards the innovations in aquaculture technology in Goalpara district of Assam. The district produces 5,347 tonnes of fish per annum against its biophysical potential of 40,969.25 tonnes. Seven critical recommended practices of pond management were selected to study the extent of adoption and correlated with thirteen socio-economic and psychological variables of the growers. Statistical techniques and tools like frequency and percentage analysis, mean score, standard deviation (SD), Co-efficient of variance (CV) and simple correlation were used. The results revealed that the farmers of the district have not yet adopted the recommended package of practices and they lack entrepreneurial spirit in aquaculture operation. The study suggested the need of paradigm shift from the current perception on aquaculture as a poverty alleviation programme to a prestigious income generating enterprise.

Key words: Aquaculture, adoption behaviour, pond management, scaling up.

INTRODUCTION

The Indian Council of Agricultural Research (ICAR), Krishi Anusandhan Bhawan, New Delhi, India has established a Krishi Vigyan Kendra (farm science centre) in Goalpara district (250 5°′ N to 60°10′ N Latitude and 90° 00′ to 91° 15′ E Longitudes) of India in 2006 with the mandates of (i) conducting on farm testing for newly generated technologies, (ii) organising training for extension personnel, (iii) organising short and long term training for the farmers and rural youths with emphasis on 'learning by doing, and (iv) organizing frontline demonstrations proven technologies (www.icarzcu3.gov.in). The district has a population of 8, 22,306 with a density of 451 per sq. km (Anon, 2009). Almost all the people consume fish in their daily diet. The district produces 5,347 tonnes of fish from all resources against the biophysical potential of 40,969.25 tonnes per annum (Anon, 2009). The present demand for fish in the district is 9,045.37 tonnes per annum, considering the per capita requirement of fish at 11 kg/person/year recommended by the World Health Organisation (WHO) for the country. There are number of tested package of practices for fish farming (Assam Agricultural University, 1997), which are also suitable for the agro-ecological situations (Mandal et al., 1981) of the district. Basic information of the district and its agro-ecological characteristics are summarized in Tables 1 and 2, respectively. When there is substantial amount of research generated knowledge it becomes difficult to communicate more widely to the end users and scaling up becomes major concern. Therefore, strategies should primarily involve horizontal scaling up of suitable production technology. Success of scaling up shall invariably depend on adoption behaviour of the farmers. The district has 0.039 million ha water spread area, comprising 0.038 million ha (97.44%) of lentic and 0.001 million ha (2.56%) of lotic waters (Table 3). This amounts to 10.43% of state's fisheries resources and 0.53% of the Inland fisheries resources of the country ARDB (2011).

Most farm families have water bodies in their homesteads in the form of seasonal ponds, sumps,

*Corresponding author: E-mail: jyotish5@gmail.com, jyotish 77@rediffmail.com.

Table 1. District profile.

1	Geographical area (ha)	184262
2	Population	822306
	a) Male (No)	420707
	b) Female (No)	401599
3	Density (No/sq. km)	451
	Literacy (%)	58.56
4	Farm families (No)	106862
5	Caste wise family (No)	
	a) Schedule tribe	23219
	b) Schedule caste	5626
	c) Other backward classes	6971
	d) General	71046
6	Land holding wise family (No)	
	a) Large (> 2.00 ha) (No)	6987
	b) Small (1 - 2 ha) (No)	50479
	c) Marginal (<1.00 ha) (No)	35143
	d) Land less (<0.50 ha) (No)	10255
7	Net cultivated area (ha)	79560
8	Gross cropped Area (ha)	106089
9	Area shown more than once (ha)	27529
10	Cropping intensity	133.3 %
11	Forest coverage (ha)	38344
12	Land put on non agricultural use (ha)	21776
13	Barren and uncultivable land (ha)	32819
14	Grazing land (ha)	3576
15	Land under Misc. trees (ha)	5424
16	Cultivable waste land (ha)	1002
17	Fellow land other than current fellow (ha)	910
18	Chronically flood affected area (ha)	12136
19	Current fellow land (ha)	861
20	Sub-division (No)	2
21	Rural development block (No)	8
22	Gram Panchayat (No)	83
23	Revenue village (No)	861
24	Irrigation coverage (ha)	17142

Table 2. Agro-ecological conditions of Goalpara District, Assam.

1	Agro ecological zone	LBVZ
2	Soil type	A
3	Nitrogen content	M-H
4	Phosphate (P_2O_5)	L
5	Potash (K_2O)	M-L
6	Organic carbon (%)	1.5
7	pH	5.0-6.8
8	Texture	S-Sl
9	Rainfall (mm)	
	December-January	50
	March-May	500-600
	June-September	1000
	October-November	150
10	Sunshine (Av.daily hrs)	
	December -January	6-7
	March-May	6-7
	June-September	4-6
	October-November	6-7
11	Temp (Av. Max/min °C)	
	December.-January	29/10
	March-May	31/19
	June-September	31/25
	October-November	29/20

LBVZ= Lower Brahmaputra Valley Zone, A= alluvial, M = Medium, H = High, L = Low, S = Sandy, Sl = Sandy loam, Rl = Red loam, Lt = Lateritic, l = loam. (Source: NARP Document 39, 1981).

Table 3. Type and area of freshwater resources.

Type	Area (ha)
Ponds and tanks	973.30
Beel	1615.90
Low lying area	286.80
Paddy fish	34,973.50
River	1256.00
Integrated farming	15.30
Others	255.50
Total	39,377.30

roadside ditches, etc. (N RCP (Annual Report, 2006 to 2007).

The KVK during village survey observed apathy and pilferage (Talukdar et al., 1999) amongst the fish farmers. Individual farmers seem to exhibit unique adoption behaviour. The decision making process involve various decision stimuli *viz.,*level of physical capital, human capital, access to productive resources, risk attributes (Feder et al., 1985); agro-ecosystem and types of technology (Pingali et al., 2001), farming season (Moser and Barrett, 2003), as well as chance factors such as neighbours and village colleagues (Case, 1992; Munshi, 2004; Pomp and Burger, 1995; Zhang et al., 2002). Involvement of these factors in shaping the farmers' adoption-decision seems to vary. The present need for development of aquaculture in the district is the shift of paradigm from the current perception on aquaculture as a poverty alleviation programme to a prestigious income generating enterprise. The challenge faced by the KVK is: what approach should be appropriate to stimulate the

Table 4. Independent variables for the study on managerial problems.

S/n	Variable	Empirical measure	Methods
1	Age (X_1)	Chronological age rounded off to nearest year	Pareek and Trivedi (1964)
2	Education (X_2)	Socio-economic status scale-Rural	Pareek and Trivedi (1964)
3	Main occupation (X_3)	Socio-economic status scale-Rural	Pareek and Trivedi (1964)
4	Annual income (X_4)	Structured schedule	
5	Operational holding (X_5),	Structured schedule	
6	Localiteness-cosmopoliteness (X_6)	Localiteness-cosmopoliteness scale	Singh (1964)
7	Economic motivation (X_7)	Economic motivation scale	Singha (1991)
8	Decision making ability (X_8)	Decision making ability scale	Singha (1991)
9	Scientific orientation (X_9)	Scientific orientation scale	Singha (1991)
10	Interest (X_{10})	Structured schedule	
11	Information seeking behaviour (X_{11})	Structured schedule	
12	Knowledge on fish production technology (X_{12})	Structured schedule	
13	Attitude towards fish farming (X_{13})	Attitude scale developed for the study	

rural households of the district for judicious use of the existing water resources through aquaculture for sustainability of their livelihood and nutritional security. Therefore, the Krishi Vigyan Kendra (KVK) goalpara investigated the adoption behaviour of the fish farmers during 2009-2010 for preparing its work plan for the next five years.

MATERIALS AND METHODS

Both primary and secondary data were used in this study. Secondary data were collected from the published literature such as project reports, official documents etc. Primary data were collected through structured and validated questionnaire to elicit information from the respondents. The sampling procedure consisted of purposive selection of fifty (50) fish farmers from the fish growing areas of the district for investigating the extent of adoption of recommended package of practices for composite culture of carps (Assam Agricultural University, 1997). Seven critical recommended practices for pond aquaculture *viz.,* (i) desilting of pond bottom, (ii) preventing wild water from entering the pond, (iii) erosion control, (iv) pond liming (v) pond fertilization, (vi) stocking practice and (vii) feeding practice were selected. Test schedules were developed to study the extent of adoption. Weight age of the practices was decided by the judge's rating. Extent of adoption was measured as done most often (MO), often (O), seldom (S) and never (N) with assigned scores of 3, 2, 1 and 0, respectively. Final adoption scores were attained by multiplying the weight age of a practice with the corresponding extent of adoption scores.

A total of 13 socio-economic and psychological variables were selected to study the adoption behaviour of the farmers. These were - (i) age (X_1), (ii) education (X_2), (iii) main occupation (X_3), (iv) annual income (X_4), (v) operational holding (X_5), (vi) localiteness-cosmopoliteness (X_6), (vii) economic motivation (X_7), (viii) decision making ability (X_8), (ix) scientific orientation (X_9), (x) interest (X_{10}), (xi) information seeking behaviour (X_{11}), (xii) knowledge on fish production technology (X_{12}), and (xiii) attitude towards fish farming (X_{13}). Measurement scale used for independent variables are summarized in Table 4. Various descriptive and inferential statistical methods were employed to analyze the data following Panse and Sukhatme (1985). The main statistical techniques and tools used were – (1) Frequency and percentage analysis, (2) Mean score, (3) Standard deviation (SD), (4) Co-efficient of variance (CV) and (5) Simple correlation.

RESULTS

The study reveals that the farmers in the district operate aquaculture in an easy going manner and they lack the entrepreneurship spirit. Extent of adoption of the recommended practices is summarized in Table 5. The farmers of the district normally operate aquaculture in old ponds, where production is limited by anaerobic conditions. It is concluded from the study that 88% farmers do not remove silts from the pond bottom, which attribute to poor productivity. Ponds (80%) are generally well impounded and do not allow entry of wild water. However, 80% farmers are not cautious about erosion control measures. They do not take any measures to control erosion of the embankments. Only 5% farmers use sod cover to prevent erosion. Erosion caused by heavy rain and undermining of dykes is a major problem in the district and repairing becomes costly affair.

Fishes reared under composite culture of carps in Assam are *Catla catla* (Catla), *Cirrhinus mrigala* (Mrigal), *Labeo rohita* (Rohu), *Hypopthalmichthys molitrix* (Silver carp), *Ctenopharyngodon idella* (Grass carp) and *Cyprinus carpio* (Common carp). All these species need slightly alkaline water and pH ranging between 6.5 and 8.0. The soil and water of the district is acidic in reaction, which is the major limiting factor. The package of practices recommends application of agricultural lime at 2,100 kg per ha per annum in split doses. But the farmers have not adopted this practice. Majority of the fish farmers (74%) normally do not apply lime. Only 2% farmers regularly apply at recommended dose.

Growth of phytoplankton is essential for sustaining the primary productivity of the pond as the fish yield is the function of primary productivity (Sugunan and Sinha,

Table 5. Frequency and percentage distribution of respondents in different response categories against water quality management.
N = 50.

Recommended practices	Distribution of respondents				Mean	SD
	MO	O	S	N		
	(3)	(2)	(1)	(0)		
Regular desilting of pond	0 (0.00)	2 (4.00)	4 (8.00)	44(88.00)	0.160	0.46773
Prevent wild water	40 (80.00)	4 (8.00)	3 (6.00)	3(6.00)	2.620	0.85452
Erosion control	5 (10.00)	3 (6.00)	2 (4.00)	40(80.00)	0.460	0.99406
Liming at recommended dose	2 (4.00)	6 (12.00)	5 (10.00)	37(74.00)	0.460	0.86213
Fertilization at recommended	8 (16.00)	12 (24.00)	10 (20.00)	20(40.00)	1.160	1.13137
Stocking at recommended time with recommended species composition	0 (0.00)	0 (0.00)	0 (0.00)	50(100.00)	0.000	0.00000
Feeding with formulated feed	0 (0.00)	0 (0.00)	0 (0.00)	50(100.00)	0.000	0.00000

(Data in parentheses are percentage of frequencies).

2001). Application of raw cow dung (2000 kg/ha initially and 1000 kg/ha/month), urea (25 kg/ha/month) and single super phosphate (20 kg/ha/month) is recommended for maintaining the nitrate and phosphate content at optimum level. Pond fertilization is often done by 28 % and seldom by 72 %. None of the respondents fertilized their ponds regularly. Farmers stock their ponds arbitrarily and do not follow any norms in terms of species, size and even stocking period and time. Since, carps require water temperature above 28°C and optimum temperature prevails during the period from mid April to mid September, it is recommended that ponds should be stocked during April to May with stunted yearlings at the rate of 5000 numbers per ha. Recommended species composition includes silver carp 20%, catla 15%, rohu 15%, grass carp 10%, mrigal 20% and common carp 20%. None of the respondents follow this practice. Fish requires complete feed at 1% of body weight per day for maintenance (Paulraj, 1997). For maximum growth at a declining rate they need to be fed at 3 to 4% of body weight per day and for highest gross conversion efficiency they should be fed at 6 to 7% of body weight per day. The recommended practice for feeding is 3 to 5% of body weight per day with 1:1 mixture of rice bran and mustard oil cake or with formulated feed as per manufacturer's recommendation. No respondent follow this practice.

Majority (76%) of the respondents belonged to middle age category (29t o 58 years) followed by old (above 58 years) 22% and young (below 29 years) 2%. The majority (92%) of the respondents belonged to high category of educational status, that is above high school standard with technical training. Only 8% of the respondents belonged to medium education level that is between primary standard and high school standard. Only 18% of the respondents were fully engaged in fish farming. Others have taken fish farming as subsidiary occupation. While, 48% respondents had agriculture as major occupation, 30% had other business. Negligible section of respondents (4%) had government service as major occupation in addition to fish farming. Data on annual income reveals that 56% of the respondents had middle level of annual income (1326.29 GBP – 2652.45 GBP) followed by high category (24%) and low (20%) with annual income more than 2652.45 GBP and less than 1326.29 GBP, respectively. Amongst the respondents, 54% had medium operational holding (2.0 to 3.33 ha) followed by 38% low operational holdings (up to 2 ha) and 4% high (above 3.33 ha). All the respondents belong to medium level (52%) and low level (48%) of localiteness-cosmopoliteness. Economic motivation of the respondents showed that 90% were in medium level and 10% were in low level category. None of the respondents had high level decision-making ability. While majority (84%) respondents were in medium level, 16% were in low level.

Analysis of data on scientific orientation reveals mean score ($\bar{X}_9$) of 22.420, standard deviation (SD) of 1.907 and co-efficient of variation (CV) of 8.51%. The co-efficient of variation (8.51%) indicates that the respondents were quite homogeneous in respect of their scientific orientation. While 68% of the respondents exhibited a medium level of interest, 32% respondents exhibited high level of interest. None of the respondents exhibited high level of information seeking behaviour. The study reveals that 66% of respondents exhibited high level of

Table 6. Correlations amongst the independent variables.

	Pearson correlation				
Variable	**Age (X_1)**	**Education (X_2)**	**Main occupation (X_3)**	**Income (X_4)**	**Operational holding (X_5)**
Age (X_1)	1.0000	- 0.5275**	- 0.0577	0.3782**	0.3676**
Education (X_2)	- 0.5275**	1.0000	0.0761	- 0.0935	- 0.1454
Main occupation (X_3)	- 0.0577	0.0761	1.0000	0.3354*	- 0.1273
Annual income (X_4)	0.3782**	- 0.0935	0.3354*	1.0000	0.2284
Operational holding (X_5)	0.3676**	- 0.1454	- 0.1273	0.2284	1.0000
Localiteness-cosmopoliteness (X_6)	0.1051	- 0.0275	- 0.1695	- 0.0984	0.2105
Economic motivation (X_7)	0.3990**	- 0.0837	0.0838	0.6909**	0.2000
Decision-making ability (X_8)	0.3479*	- 0.1881	0.0058	0.6048**	0.1316
Scientific orientation (X_9)	- 0.0284	- 0.0125	- 0.0711	0.0506	0.0749
Interest (X_{10})	0.1718	- 0.1304	- 0.2314	0.1498	0.0976
Information seeking behaviour (X_{11})	0.0232	0.0994	0.1740	0.4573**	0.1193
Knowledge (X_{12})	0.0570	- 0.0888	0.1675	0.2107	0.1467
Attitude (X_{13})	- 0.1523	0.2829*	0.1598	- 0.0299	0.0204

** Correlation is significant at the 0.01 level (2-tailed); *Correlation is significant at the 0.05 level (2-tailed).

level of knowledge followed by 34% in the medium category. As many as 94% respondents had favourable attitude followed by 6% more favourable. Analysis of simple correlations amongst the independent variables were done and results are presented in Tables 6 and 7. Tables 6 reveals positively significant correlation between operational holding and age (r = 0.3676, significant at 1%), income and age (r = 0.3782, significant at 1%), economic motivation and age (r = 0.3990, significant at 1%), decision-making ability and age (r = 0.3479, significant at 5%), attitude and education (r = 0.2829, significant at 5%), income and main occupation (r = 0.3354, significant at 5%). Table 7 reveals positively significant correlations between economic motivation and income (r = 0.6909, significant at 1%), decision-making ability and income (r = 0.6018, significant at 1%), information seeking behaviour and income (r = 0.4573, significant at 1%), economic motivation and decision-making ability (r = 0.5769, significant at 1%), knowledge and information seeking behaviour (r = 0.3268, significant at 5%).

Negatively significant correlations were observed between interest and attitude towards fish farming {r = (-) 0.3393, significant at 5%}, age and education {r = (-) 0.5275, significant at 1%}. Individual independent characteristic and management practices and pond management practices adopted by the farmers had negative correlations with education (X_2) at 1% level {r = (-) 0.4247} and attitude (X_{13}) at 5% level {r = (-) 0.2863}. Table 8 reveals the details of correlation amongst the independent variables and pond management practices.

DISCUSSION

The farmers of the district in general live under uncertain, harsh social and environmental conditions. People are basically small holders (80.12%) and 75.25% people live below the poverty line (Anon, 2009). GINI co-efficient of the district is 0.488 (AHDR, 2003). They operate their farms with little access to land, water, extension service and credit. Farming in the district itself is fraught with the uncertainties of floods, drought and anthropo-political conflicts. The fish farmers are normally repelled to high input farming technologies owing to (i) adoption does not sustain due to high cost involved; (ii) low access of the household to technology extension and credit; and (iii) vulnerability of the households to risk involved such as floods, drought and societal problems (Lightfoot et al., 1992).

Scientific and technological revolution (STR) is characterized by deep interconnection and interaction of processes and fundamental changes in all the areas of science, technology and

Table 7. Correlations amongst the independent variables.

Variable	Pearson correlation							
	Localiteness-cosmopoliteness (X_6)	Economic motivation (X_7)	Decision-making ability (X_8)	Scientific orientation (X_9)	Interest (X_{10})	Information seeking behaviour (X_{11})	Knowledge (X_{12})	Attitude (X_{13})
Age (X_1)	0.1051	0.3990**	0.3479*	-0.0284	0.1718	0.0232	0.0570	-0.1523
Education (X_2)	-0.0275	-0.0837	-0.1881	-0.0125	-0.1304	0.0994	-0.0888	0.2829*
Main occupation (X_3)	-0.1695	0.0838	0.0058	-0.0711	-0.2314	0.1740	0.1675	0.1598
Annual income (X_4)	-0.0984	0.6909**	0.6048**	0.0506	0.1498	0.4573*	0.2107	-0.0299
Operational holding (X_5)	0.2105	0.2000	0.1316	0.0749	0.0976	0.1193	0.1467	0.0204
Localiteness-cosmopoliteness (X_6)	1.0000	-0.0947	-0.1853	0.2279	0.0130	-0.2706	-0.1290	-0.1043
Economic motivation (X_7)	-0.0947	1.0000	0.5769**	-0.0894	0.1907	0.1977	0.1212	-0.2159
Decision-making ability (X_8)	-0.1853	0.5769**	1.0000	-0.1175	0.2699	0.4341**	0.1174	-0.2401
Scientific orientation (X_9)	0.2279	-0.0894	-0.1175	1.0000	-0.0835	-0.1945	-0.0956	0.1653
Interest (X_{10})	0.0130	0.1907	0.2699	-0.0835	1.0000	0.0072	0.0589	-0.3393*
Information seeking behaviour (X_{11})	-0.2706	0.1977	0.4341**	-0.1945	0.0072	1.0000	0.3268*	0.0321
Knowledge (X_{12})	-0.1290	0.1212	0.1174	-0.0956	0.0589	0.3268*	1.0000	0.1743
Attitude (X_{13})	-0.1043	-0.2159	-0.2401	0.1653	-0.3393*	0.0321	0.1743	1.0000

** Correlation is significant at the 0.01 level (2-tailed); *Correlation is significant at the 0.05 level (2-tailed).

production, with science playing the leading role as the productive force(Marinko, 1989). Science embodies Francis Bacon's postulate 'Knowledge is power'. The present study established the Bacon's postulate exhibiting positive correlation of farmers' knowledge with farmers' attitude, information seeking behaviour, decision making ability, economic motivation and income. Science and technology constitute the means of enhancing men's strength and the potentiality of his hands and brain. Production as such is unthinkable without economics and the science of management. Drawing on the actor oriented perspectives in rural sociology (Long and Long, 1992), it was advocated that success of adoption of a technology at higher level are not merely a function of the technology, nor of the research and extension methodology, but result from a complex conjunction of people and events with outcomes. According to Wilkening (1953) adoption of a specific practice is not the result of a single decision to act but series of action and meaningful decisions. Rogers (2003) explains that the adoption decision and its timing depend on decision maker's perception and inherent characteristics, with innovators at one extreme and laggards at the others. Farmers in the same environment have different objectives and livelihood strategies and therefore respond differently to a given technology. Only 18% of the respondents have adopted fish farming as major occupation, rest 82% were engaged in agriculture, other business and government jobs. Biot et al., (1995) suggested that 'different behaviour is as much a function of different opportunities and constraints as of different perception'. Even within the farm households, the ability to make decision on resource use and technology adoption varies according to age, gender and other category and actual decision can depend on a complex bargaining process amongst the members (Ellis, 1993; Jackson, 1995 and Biot et al., 1995). Beyond the household group processes and ability to harness them can play a crucial role in adoption decision (Chamala and Mortiss, 1990; Frank and Chamala, 1992; Pretty and Shah, 1994). While Wozniak (1984) opined that education increases ones' ability to receive decode and understand information relevant to making innovative decisions; Clay et al. (1998) found that education is an insignificant

Table 8. Correlations amongst independent variables and pond management practices.

Pearson correlation	Variable	Pond management
	Age (X_1)	0.2781
	Education (X_2)	-0.4247**
	Main occupation (X_3)	-0.0930
	Income (X_4)	0.0540
	Operational holding (X_5)	0.0879
	Localiteness-cosmopoliteness (X_6)	0.1885
	Economic motivation (X_7)	0.0381
	Decision-making ability (X_8)	0.2215
	Scientific orientation (X_9)	0.1095
	Interest (X_{10})	-0.0306
	Information seeking behaviour (X_{11})	-0.1369
	Knowledge (X_{12})	-0.0325
	Attitude (X_{13})	-0.2863*

** Correlation is significant at the 0.01 level (2-tailed); *Correlation is significant at the 0.05 level (2-tailed).

determinant of adoption decision. In this study it was hypothesized that high level of institutional education increases the probability of adopting new technology. The present study revealed negative correlation of educational qualification with decision making ability. Au and Enderwick (2000) explained that six beliefs *viz.*, compatibility, enhanced value, perceived benefits, adaptive experiences, perceived difficulties and suppliers' commitments; affect the cognitive process that determines the farmers' attitude towards technology adoption. The present study revealed positive correlation with main occupation, education, scientific orientation, information seeking behaviour, knowledge and operational holdings. The present study suggests a change in farmers' attitude for development of aquaculture in the district.

Conclusion

Fisheries have been a caste based activity in Assam. Flood plain lakes, which were once unmanaged natural water bodies, were the main source of fish. Baruah et.al. (2000) opined that historically there have been three distinct groups of people involved in fisheries activity: (i) those who catch fish for their own daily consumption; (ii) those belonging to the fisher community and (iii) the rural fisher entrepreneurs (leaseholders). Ordinary people usually catch fish daily for food, while fishers are full-time operators.

Aquaculture is comparatively new sector of food production and it is undergoing continuous change in Assam. During last the 20 years it has been mastering a driving force that has propelled aquaculture to the forefront. However, pond productivity is limited to 2800 kg/ha in Goalpara, Assam although much higher yields (5000 kg/ha) have been recorded by Luu et al. (2003) in China and Vietnam. Biophysical potential of aquaculture in Assam reveals that same production could be achieved if recommended pond management practices are adopted by the farmers. This study on farmer's adoption behaviour suggests that they should be substantially trained on the latest technological innovations in aquaculture for a paradigm shift from the current perception on aquaculture as a poverty alleviation programme to a prestigious income generating enterprise.

ACKNOWLEDGEMENTS

Authors are grateful to Dr. Anubrata Das, Director, National Research Centre on Pig (ICAR), Guwahati-781 131, Assam, for his inspiration, encouragements and supervision. They are also thankful to the Department of Fisheries and Department of Agriculture, Government of Assam for providing the secondary data incorporated in this paper. The authors acknowledge the helps received from their fellow colleagues working in the KVK Goalpara, Assam.

REFERENCES

AAU (Assam Agricultural University) (1997). Package of Practices for Fish Farming in Assam. Assam Agricultural University, Jorhat, Assam, India.

AHDR (2003). Assam Human Development Report, Planning and Development Department, Govt. of Assam, India. (http://planningcommission.nic.in/plans/stateplan/sdr_pdf/shdr_assam03.pdf).

SHBA (2009). Statistical Hand Book Assam, 2009. Department of Economics & Statistics, Govt. of Assam, India.

ARDB (2011). Online Agricultural Research Data Book (http://www.iasri.res.in/agridata/11data/HOME_11.HTML.)

Au AK, Enderwick P (2000). A cognitive model on attitude towards technology adoption. J. Managerial Psychol., 15(4): 266-282.
Baruah UK, Bhagowati AK, Talukdar RK, Saharia PK (2000). Beel Fisheries of Assam: Community based Co-management Imperative. Naga, ICLARM Q.,23(2): 36-41.
Biot Y, Blaikie P, Jackson C, Palmer-Jones R (1995). Rethinking Research on Land Degradation in Developing Countries. World Bank Discussion Papers 289, Washington DC, The World Bank.
Case A (1992). Neighbourhood influence and technological change. Reg. Sci. Urban Econ., 22(3): 491-508.
Chamala S, Mortiss P (1990). Working Together for Landcare: Group Management Skills and Strategies. Academic Press, Brisbane, pp. 369.
Clay D, Reardon T, Kangasniemi J (1998). Sustainable intensification in the highland tropics: Rwandan farmers' investments in land conservation and soil fertility. Economic Development and Cultural Change, 46(2): 351-378.
Ellis F (1993). Peasant Economics: Farm Households and Agrarian Development. 2nd Edition, Cambridge, Cambridge University Press, England, pp. 309.
Feder G, Just RE, Zilberman D (1985). Adoption of agricultural innovations in developing countries: A survey. Economic Development and Cultural Change, 33(2): 225-298.
Frank BR, Chamala S (1992). Effectiveness of extension strategies. In: Lawrence, G., Vanclay, F. and Furze, B. (Eds.). Agriculture, Environment and Society: Contemporary Issues for Australia, Macmillan, Melbourne. pp 122-140.
Jackson C (1995). Environmental reproduction and gender in the Third World. In: Morse, S. and Stocking, M. (Eds.). People and Environment, UCL Press, London. pp. 109-130.
Lightfoot C, Gupta MV, Ahmed M (1992). Low external input sustainable aquaculture for Bangladesh – An operational framework. Naga, ICLARM Quarterly July 1992, pp. 9-12.
Long N, Long A (1992). Battlefields of Knowledge: The Interlocking of Theory and Practice in Social Research and Development. London, Routledge, pp. 306.
Luu LT, Trang PV, Cuong NX, Demaine H, Edwards P (2003). Promotion of small-scale aquaculture in the Red River Delta, Vietnam. In: Edwards, P., Little, D.C. and Demaine H. (Eds.). Rural Aquaculture, CAB Publishing, Oxford, pp. 55-75.
Mandal SC, Singh P, Borthakur BC, Mahanta K, Pande HK (1981). Report on ICAR Research Review Committee on National Agricultural Research Project. NARP Document No. 39. ICAR, New Delhi, pp. 167
Marinko G (1989). What is the scientific and technological revolution? Progress Publishers, Moscow, USSR, pp. 318.
Moser CM, Barrett CB (2003). The disappointing adoption dynamics of a yield-increasing, low external-input technology: the case of SRI in Madagascar. Agric. Syst., 76(3): 1085-1100.
Munshi K (2004). Social learning in a heterogeneous population: Technology diffusion in the Indian Green Revolution. J. Dev. Econ., 73(1): 185-213.
National Research Centre on Pig NRCP (Annual Report) (2006-2007). Annual report of Krishi Vigyan Kendra Goalpara, National Research Centre on Pig, ICAR, Dudhnoi, Assam, India. Report submitted to the Zonal Coordinating Unit, Zone-III, ICAR, Umiam, Meghalaya, India (http://icarzcu3.gov.in).
Panse VG, Sukhatme PV (1985). Statistical Methods for Agricultural Workers. Indian Council of Agricultural Research, New Delhi. India, p. 359.
Pareek U, Trivedi G (1964). Manual of Socio-Economic Status Scale (Rural). Mansayan, Delhi, India, p. 32.
Paulraj R (1997). Aquaculture feed- Handbook of Aquafarming. The Marine Product Export Development Authority (Ministry of Commerce, Govt. Of India), Kochi, India.
Pingali PL, Rozelle SD, Gerpacio RV (2001). The farmers' voice in priority setting: a cross-country experiment in eliciting technological preferences. Economic Development and Cultural Change, 49(3): 591-610.
Pomp M, Burger K (1995). Innovation and imitation - adoption of cocoa by Indonesian smallholders. World Dev., 23(3): 423-431.
Pretty JN, Shah P (1994). Soil and Water Conservation in the Twentieth Century: A History of Coercion and Control. Research Series No. 1, Rural History Centre, University of Reading.
Rogers EM (2003). The Diffusion of Innovations. 4th Edition, New York, Free Press, p. 512.
Singha AK (1991). A study on the Determinants of Entrepreneurial, Behaviour in the Adoption of Selected Farm Practices by the Farmers of Cachar district of Assam; M.Sc. (Agric. Ext.) Thesis; Assam Agricultural University, Jorhat, Assam, India.
Sugunan VV, Sinha M (2001). Sustainable capture and culture-based fisheries in freshwaters of India. In: Pandian, T. J. (Ed.). Sustainable Indian Fisheries, National Academy of Agricultural Sciences, New Delhi, pp. 43-70.
Talukdar RK, Baruah UK, Bhagowati AK, Borua S (1999). Agri-ecosystem analysis for subsistence aquaculture. In: Baruah, U. K., Talukdar, R. K., Sharma, R. C., Das, P. K., Sharma, J. K., and Borua, S. (Eds.). Compendium on Participatory Rural Appraisal and Aquaculture, prospects and perspectives, Assam Agricultural University, Jorhat, Assam.
Wilkening EA (1953). Adoption of improved farm practices as related to family factors. Wisconsin Experiment Station Research Bulletin. 183 Wisconsin, USA.
Wozniak GD (1984). The adoption of interrelated innovations: A human capital approach. Review of Economics and Statistics 66: 70-79.
Zhang X, Fan S, Cai X (2002). The path of technology diffusion: which neighbors to learn from? Contemporary Economic Policy, 20(4): 470-479.

What factors determine the time allocation of agricultural extension agents on farmers' agricultural fields? Evidence form rural Ethiopia

[1]Mekbib Gebretsadik Haile and [2]Degnet Abebaw

[1]Center for Development Research (ZEF), Bonn University, 53113 Bonn, Germany
[2]Ethiopian Economics Association, P.O. Box 34282, Addis Ababa, Ethiopia

In Ethiopia, agricultural extension (AE) agents are expected to play a key role in linking smallholder farmers to agricultural scientists and researchers. AE workers assist farmers in the adoption of improved agricultural technologies and facilitate the adjustment of the new technologies to the local agro-ecological conditions. Using data collected from a total of 236 randomly selected extension agents working in eight purposefully selected weredas in Ethiopia, this study examines factors that affect the time allocation of extension agents on farmers' agricultural fields. The analyses reveal that perceptions of extension workers about the fairness of performance appraisal, equality of male and female agents, job security, resource availability, and workload manageability significantly affected the time that agents worked on farmers' plots. While off- and on-the-job trainings and participation of extension workers in the agricultural planning process encouraged agents to work more on farms. Better years of schooling and larger number of villages to work with had adverse effect on agents' working time. It is therefore suggested that agricultural policy should ensure that extension agents are not only available but also adequately accessible to farmers. More specifically, agricultural policy should bring improvement in the perception of extension agents about their working environment.

Key words: Agricultural extension, agricultural technology, time allocation, perception, Ethiopia.

INTRODUCTION

Ethiopia is one of the developing countries that is acknowledged in implementing agricultural development programs. The country bases its economic development on the agricultural development led industrialization (ADLI), a strategy which prioritizes the agricultural sector to lead the economy. Since more than a decade ago, ADLI has been used intensively to enhance the country's economy in the long run. It is considered as a long-term strategy to achieve faster economic development by adopting labor intensive improved agricultural technologies. The importance of the sector in enhancing the economy cannot be denied. Ethiopia's economy is based on agriculture which accounts for about 45% of GDP, 85% of exports, and 80% of total employment (MoFED, 2006). Nevertheless, the sector suffers from frequent drought, land degradation, traditional crop varieties and poor cultivation practices, and hence is

*Corresponding author. E-mail: haymakg@yahoo.com.

Abbreviations: ADLI, Agricultural Development Led Industrialization; **AE,** Agricultural Extension; **CSA,** Central Statistical Authority; **EEA,** Ethiopian Economic Association; **FDRE,** Federal Democratic Republic of Ethiopia; **GTP,** Growth and Transformation Plan; **IAR,** Institute of Agricultural Research; **IFPRI,** International Food Policy Research Institute; **MOFED,** Ministry of Finance and Economic Development; **PADETES,** Participatory Demonstration and Training Extension System; **PASDEP,** Plan for Accelerated and Sustainable Development to End Poverty; **WoARD,** *Wereda* office of Agriculture and Rural Development.

characterized by low productivity.

In order to address these inherent problems the government of Ethiopia (GoE) developed the Plan for accelerated and sustainable development to end poverty (PASDEP) in 2005/2006. The main objective of the PASDEP plan is to accelerate the transformation of agriculture from subsistence to a more market-oriented sector (FDRE, 2005). The government has emphasized the vital role of agricultural extension (AE) workers during the implementation period. The new five-year plan, the growth and transformation plan (GTP), which was launched in 2010 does also emphasize that agriculture will continue to lead the country's economy in the coming five years. The plan forecasts that agriculture will contribute about 35% of GDP in 2015 (FDRE, 2010).

The GTP clearly specifies that agricultural research and extension will play key role in the agricultural development strategy. Dissemination of existing and new improved agricultural technologies to all farmers, expansion of irrigation, transformation of smallholders to higher-value agricultural producers are some of the components of the GTP that focus on the agricultural sector. The *wereda* office of agriculture and rural development (WoARD) is responsible for the implementation of these agricultural policies at community levels via AE workers assigned to every *kebele*[1] throughout the country. Consequently, AE agents bear a significant amount of work in the sector. These workers should have, among other things, proper perception of their working environment, job satisfaction, adequate training related to their specialization, proper remuneration and other incentives to discharge their responsibilities appropriately.

According to Picciotto and Anderson (1997) the focus of AE service in many countries since the early 1960s has been to increase agricultural production and productivity. AE helps to reduce technology gaps between farmers practicing traditional techniques and those who adopted improved agricultural technologies by providing technical advice, information and training (Oladele and Mabe, 2010). Moreover, AE service contributes to increase the actual yields of farmers' plots towards their potential levels (Anderson and Feder, 2002). Thus AE has a dual purpose in linking scientists and farmers: it assists the adoption of improved agricultural technologies; and it facilitates the adjustment conditions (Kassa and Degnet, 2004). More precisely AE agents contribute to the development of the agricultural sector, among other things, by disseminating new information and technologies, training and mobilizing farmers, supplying inputs, and administering farmers' credits and loans. This is especially important in a country where only less than 5% of the nation's agricultural produce is generated by farmers who adopt improved and modern agricultural tools and farming practices (CSA, 2010).

As is the case for any other professional workers, working environment and the workers' perception are crucial factors in the decision making of AE agents regarding their time allocation and performance on agricultural fields. Nevertheless, agricultural agents in many developing countries work under disadvantageous conditions and with insufficient working facilities (Agunga et al., 1997). Inappropriate perception of extension agents about their working environment creates a disincentive for them to effectively work on farms. The amount of time that extension agents work on farmers' plots directly affects their overall job performance, which in turn has an effect on agricultural productivity. Hence, understanding the factors that affect the time allocation of extension agents on farmers' plots is crucial for policy makers. Moreover, identifying and appropriately addressing these factors is critical for the overall agricultural productivity and towards poverty alleviation in the end. The purpose of this paper is, therefore, to investigate the factors that affect AE workers' decision on their time allocation on agricultural farms, and hence their job performance. The paper makes use of data collected from 236 extension agents working in 8 *weredas* from 7 regions in the country.

The rest of the paper is organized as follow: A brief overview of AE in Ethiopia and the theoretical framework; Explanation of data collection process and research methodology; Interpretation of results; conclusion.

LITERATURE REVIEW

Agricultural extension in Ethiopia

It has been about six decades since agricultural research and extension service started in Ethiopia. Since the establishment of the Institute of Agricultural Research (IAR) in 1966, national agricultural research system was organized with autonomous management. The national agricultural research system had several stations addressing the main agro-ecological zones, and the major commodities throughout the country. However, since the early 1990s some independent agricultural research centers were launched in different regions and they were administered by the regional bureau of agriculture. In 1995 the GoE launched the participatory demonstration and training extension system (PADETES)

---

[1] *Kebele* is the lowest administrative unit in the country. There are three agricultural extension agents assigned to every *kebele*: these are extension agents specialized in crop, in livestock, and in natural resource.

as its AE system. PADETES aims, among other things, at increasing production and productivity of smallholder farmers through improved technologies; and by making extension workers play a great part in the development process.

With ADLI development strategy, Ethiopia benefits from the strong political determination of the government to enhance smallholder farming in the country. This determination is reaffirmed by the recent development plans of the country that give priority to agricultural comercialization and transformation of subsistence smallholder farming. This is paramount importance since the country has about 12 million rural households of which about 84% are smallholder farmers who own farmlands less than 2 hectares (CSA, 2008). According to Hazel et al. (2007) in countries with equitable agricultural distribution and with good agricultural potential, which is the case in Ethiopia, smallholder farming can play a significant role in the economic development of the country. In such conditions AE programs can be significantly pro-poor (Gebremedhin et al., 2009).

There have been significant changes undergoing in the AE service in the country in the last few decades. The country has made significant efforts to increase both the quantity and quality of extension service among rural farmers. Currently, there are more than 60,000 AE agents sharing their expertise to farmers according to their specialization. The AE workers are not limited to serve farmers about resource management and farming techniques per se. Besides supporting and training farmers on resource management and improved farming techniques, AE workers provide more services which directly or indirectly influence agricultural production and productivity. For instance, agents provide a service of information and skill development, input supply, credit and saving, marketing of agricultural produce, etc (Gebremedhin et al., 2009).

Theoretical framework

Several studies showed that the amount of time that instead spent on several other activities has never been significantly lower than that spent at work. According to Becker (1965) and Gronau (1977), individuals make decisions about the optimal allocation of their available time, T towards leisure (T_L), work at home (T_H), and work in the market (T_M). Home consumption time, that is leisure, differs from home production time since the latter is regarded as a time used to produce goods and services that can be close substitutes to goods in the market. Leisure, however, has only poor substitutes in the market. In the extreme, home production time and working time in the market are perfect substitutes if the individual generates the same utility from the consumption of goods and services whether they are home produced or purchased from the market.

Considering a single-person household, the individual makes decision on his time allocation in order to maximize a utility function derived from the consumption of a composite commodity, Z and Leisure, L:

$$U = U(Z, L) \tag{1}$$

The commodity Z can be either purchased from the market or produced at home and this does not affect the utility. Let Z_M and Z_H refer to the market expenditure for purchased goods and the market-value of home produced commodities, then the total consumption commodity Z can be given by

$$Z = Z_H + Z_M \tag{2}$$

The individual maximizes his utility function in Equation (1) subject to two constraints: budget and time constraints. Thus, the individual maximizes his utility subject to the two resource constraints: the endogenous budget constraint

$$Z_M = wM + V - \gamma C \tag{3}$$

where w is constant wage rate, M denotes field or market work, V denotes other sources of income, C refers to costs associated to working in the market (for eample, transportation cost, drudgery costs, etc), and γ is a dummy variable that takes the value 1 if the individual works in the market and 0 otherwise; and the time constraint

$$T = T_M + T_H + T_L + \gamma t \tag{4}$$

Where t is the costs of working in the market (field) in terms of time (for example walking time to and from work place).

A necessary condition for an optimal time allocation is that the marginal rate of substitution between goods and leisure should be equal to the marginal product of working at home, which in turn is equal to the shadow price of time (w^*). For $\gamma > 0$ (if the individual works in the market), the marginal product of work at home is also equal to the real wage rate (w).

$$\frac{\partial Z}{\partial T_H} = \frac{\partial U/\partial Z}{\partial U/\partial L} = w^* (= w \ if \ \gamma > 0) \tag{5}$$

The above outlined model yields a set of reduced-form equations for the endogenous variables as a function of different factors including individual, market (working environment), and other constraints. Of particular interest for our study is the reduced-form equation for the individual's time allocation for activities in the market (field). This equation for an individual working in the market (farm) could be given in the following form

$$T_M = T_M\ (w, C, V, \gamma, W, X) \qquad (6)$$

where W denotes variables that explain the working environment, X refers to all other relevant variables that influence the time allocation of the individual for work. However, in developing countries where markets are imperfect and thin, non-wage factors are more important to influence the working time of employees.

Several factors influence the time allocation of employees at work. This study gives great emphasis to the effects on time allocation for work of the perceptions of AE workers' towards their working environment. Workers may have unique perceptions towards their working environment. These may include workers' perceptions towards assigned tasks, individuals, interpersonal interactions, organizational standards, goals and regulations, physical location, and prospects-rewards-incentives. "The entire organizational environment (particularly as perceived by the organizational member) is considered to have a very important impact on an individual's motivations, satisfactions, and task performance" (Newman, 1977, P. 520). Bennel and Zuidema (1989) indicated that the most important role of agricultural managers is to ensure proper utilization of agricultural workers with appropriate attitude, motivation and perception. Several studies investigated ways by which employees could be encouraged to apply greater effort to their tasks (Welsh, 1980; Herzberg, 1959 as cited in Oloruntoba and Ajayi, 2003; Vernon, 1972). According to Herzberg (1959; cited in Oloruntoba and Ajayi, 2003) negative factors such as physical working conditions, poor pay, organizational policies, interpersonal relations; and positive factors such and recognition by others substantially influence workers' job satisfaction and hence affect their working time. This in turn influences their job performance and agricultural productivity. An empirical study by Oloruntoba and Ajayi as minimal supervision, promotion, growth opportunities, (2003) on private farms in Nigeria showed that most of these propositions hold true. They found that high remuneration, job security, recognition by staff and promotion are significant motivating factors for the private farm employees to work longer hours.

Darlene and Borman (1989) indicated that working environment such as physical surroundings, job satisfaction and management supervision, and workers' perceptions about these conditions can influence job performance. Job security and safety aspects such as pension scheme and workers' association, and organization policy and management style can produce different perceptions of workers about their working environment (Azril et al., 2010). For instance, institutional policies and management styles that put so much pressure and supervision on their workers should be discouraged as they could create negative perception about the organizational policy among the employees.

The purpose of this study is, therefore, to examine the effect of different factors on the amount of time that extension agents work on farmers' agricultural fields. According to PASDEP, even though extension agents are available in *weredas* and *kebeles,* they are not adequately accessible to farmers (FDRE, 2005). The document reported that close to 50% of the farmers in rural Ethiopia did not find extension service adequate. Thus, the results of this study will have crucial significance for the country's agricultural policy makers while making reviews to ensure that extension services are accessible for farmers on their fields. The study will have importance among other things in terms of pinpointing problem areas that policy makers should give emphasis to.

RESEARCH METHOD

Data

The data used for this study is obtained from a survey that was undertaken jointly by the Ethiopian Economic Association (EEA) and International Food Policy Research Institute (IFPRI) as part of a study on gender and governance in rural service in Ethiopia. The data was collected from eight purposefully selected *wereda*s from seven of the administrative regions: namely Tigray, Amhara, Afar, Oromia, Benishangul, SNNPR and Gambella (Table 1). One *wereda* was selected from each of these regions with the exception of Amhara region, from which two *wereda*s were selected. All the *kebeles* were selected from these eight *wereda*s for the survey, resulting in a total of 156 *kebeles*. Two AE agents, one specializing in crop and one specializing in livestock were interviewed from each kebele. However, in cases where a crop or livestock extension agent was not available, agents specializing in natural extension agents were interviewed in their places. We also interviewed 14 extension agents who have not specialized in any of these three but who work on agriculture in general. The survey questionnaire includes, among other things, information about their demography concerning their field activities, experience, training, performance, evaluation, professional networks, workload, availability of working resources and perception of their working environment. The survey was conductedbetween April and June 2009.

Descriptive statistics

From Table 2, one can observe that the extension agents are, on

Table 1. Sampling distribution of extension agents by district and by region.

District (region)	Sample size				
	Extension agents specialized in				
	Crop	Livestock	Natural resource	General	Total
Sekota (Amhara)	22	18	7	0	47
Bati (Amhara)	10	17	15	1	43
Ofla (Tigray)	19	17	3	0	39
Ibantu (Oromiya)	16	19	2	0	37
Sheko (SNNPR)	15	5	7	6	33
Gog (Gambella)	6	4	3	7	20
Telalak (Afar)	4	5	2	0	11
Yaso (Ben. Gumuz)	3	2	2	0	7
Total	95	87	41	14	237

Source: Survey data, 2009.

In terms of the different activities extension workers are engaged in, Table 3 shows the average percentage of the agents' time allocation for different tasks. One can see from the table that, on average, about 64% of the total time available to all extension agents is spent on farmers' agricultural fields or farmers' homes. Extension workers might go to farmers' homes for different purposes including for advice, holidays, weddings, funerals, and for other social and work related reasons. Office activities such as report writing, meetings and discussion forums take about 18% of the total time available to agents. However, the lion's share of the AE agents does not have access to transportation facilities. About 98% of the agents walk to work on farmers' agricultural fields (Table 4); and only less than 1% of the extension agents do have access to public transport. This has a negative effect on the time that the extension workers spend on farmers' plots of land. Almost none of the extension workers are provided with any means of transportation such as vehicle, bicycle or motorbike from their offices.

Table 5 shows that extension agents in Sekota *wereda* worked the highest aggregate number of hours on farmers agricultural fields during the past planting, harvesting and slack seasons followed by agents in Ibantu and Sheko *weredas*. On the contrary, extension workers in Telalak and Gog *weredas* worked the lowest hours on farmlands. This might be partially because sedentary farming is very common in the former *weredas* but not in the latter two. *Wereda* dummies are used in the empirical analysis to capture such *wereda* specific heterogeneities.

Variables

Dependent variable

We initially considered agricultural productivity to be the dependent variable. However, since there are up to three extension workers in each peasant association who could potentially work on the same agricultural land, it was difficult to disentangle the productivity of an individual extension agent. Thus, the total weekly time (T_M) that each agent worked on farm lands during the last planting, harvesting and slack seasons is taken to be a proxy for agents' productivity. We will, therefore, have multiple regression equations to capture variation of workers' time allocation for farm work across the three seasons. Thus, the dependent variables T_M^p, T_M^h, and T_M^s are the number of hours that each extension agent worked per week during the farmers' last planting, harvesting and slack seasons respectively.

Independent variables

The explanatory variables used in this study include agents' individual characteristics, salary, perception about work environment including safety and security, interpersonal relationship, availability of job related resources, organizational policies and management style, remuneration, manageability of workload and farm distances; organizational support; number of villages handled by agents; average farm size of households; clustering dummies; and other relevant variables. The choice of these variables is based on the above theoretical framework and previous literature that examined job satisfaction and job performance of AE agents (Long and Swortzel, 2007; Chyung 2005; Koustelios, 2001; Strong and Harder, 2009; Scott et al., 2005). Perceptions about working condition variables were measured in evaluative terms. Extension agents were presented with series of statements that describe their current job characteristics. They were then asked to indicate their level of agreement to the statements. Those who strongly and somewhat agreed with the given statement were coded 1 and those who strongly and somewhat disagreed were coded 0. Moreover, availability and/or access to transportation facilities to reach to working place affects the time that the agents spend on agricultural fields. For instance, in the absence of adequate transport facilities, travelling long distance by foot could affect the amount of their working time. However, since 98% of the extension workers do not have and/or access to any transportation means, this variable is not among our explanatory variables. Finally, since extension workers from the same *wereda* report to the same WoARD, we included district dummies to control for unobserved unique characteristics of each *wereda*. The descriptive statistics of potential explanatory variables included in this study are reported in Table I (Appendix A).

Empirical model

Following the theoretical model discussed, our particular interest i

Table 2. Demographic characteristics of the sampled extension workers.

Description		Extension agents
Family size, mean (SD)		2.47 (1.77)
Age, mean (SD)		26.66 (2.99)
Marital status (%)	Married	40.43
	Single	58.30
	divorced/widowed/other	1.28
	Total	100
Sex (%)	Male	92.98
	Female	8.02
	Total	100
Education Level (%)	TVET* 10+1 to 10+3	68.96
	TVET 12+1 to 12+3	27.15
	Other	3.89
	Total	100

Source: Survey data, 2009. * TVET= Technical and vocational education and training.

Table 3. Percentage of time AE workers spend on different tasks.

Task	Mean percentage of time spent
On farmers' fields or farmers' homes	63.96
Training farmers	15.70
Attending in-service training	5.02
Office work	18.41
Other	15.13
Total*	118.22

Source: Survey data, 2009. * Extension agents could work multiple tasks at the same time.

AE agent in our case. The time allocation of AE workers on farmers' plots of land may have seasonal variation. Accordingly to capture this seasonal variation, we disaggregate their working time in to three different seasons: planting, harvesting and slack seasons. Thus we have multiple independent reduced-form equations. Since the dependent variable (time at work in each season, T_M) has a continuous linear form and is always non-negative, a multivariate log-linear regression model is employed for this study. We will use an ordinary least square (OLS) estimation method. The log-linear regression equations can be given as follows:

$$\log T_{Mi}^{p} = (X;\beta) + \varepsilon_i^{p}$$
$$\log T_{Mi}^{h} = (X;\beta) + \varepsilon_i^{h}$$
$$\log T_{Mi}^{s} = (X;\beta) + \varepsilon_i^{s}$$
$$\log T_{Mi} = (X;\beta) + \varepsilon_i \quad (7)$$

where T_{Mi}^{p}, T_{Mi}^{h}, and T_{Mi}^{s} denote the amount of time that the ith extension worker worked per week on farmers' agricultural fields during the past planting, harvesting and slack seasons respectively, and T_{Mi} is the average of the aggregate time that extension agents worked per week in the three seasons; ***X*** is a matrix of explanatory variables that refer to individual characteristics, perceptions about working environment, specialization dummies and *weredas* dummies for the ith extension worker; ***β*** is a vector of parameters to be estimated; and ε_i are extension worker specific error terms.

RESULTS AND DISCUSSION

The final parsimonious regression models were chosen by the 'general-to-specific' modelling strategy (Hendry, 1995; Davidson et al., 1978). All the potential explanatory variables listed in Table I were used in the fully specified log-linear models for all multivariate regression models. We used OLS in the 'general-to-specific' procedure to drop variables that consistently have insignificant t-

Table 4. Common transportation means for extension agents in the survey area.

Transportation means	Percentage of extension agents
Horse/mule/camel	0.43
Foot	97.87
Public transport	0.85
Combination of different mode	0.43
Other	0.43
Total	100

Source: survey data, 2009.

Table 5. Aggregate weekly hours that agents worked on farmers fields in the past season, by *wereda*.

Wereda	Average weekly hours agents worked on farm			
	Planting season	Harvesting season	Slack season	Average
Sekota	41.94	37.62	30.02	36.53
Bati	37.55	27.90	24.07	29.84
Ofla	40.64	31.97	22.44	31.68
Ibantu	39.43	36.92	25.27	33.87
Sheko	42.30	33.79	24.55	33.55
Gog	27.45	20.5	16.85	21.6
Telalak	13.25	10.63	6.25	10.04
Yaso	34.33	34.33	16.67	31.79

Source: survey data, 2009.

statistic values. In the following discussion we rely on the results from the final parsimonious model where the average of the aggregate time that extension agents worked per week is the dependent variable[2].

Among the personal characteristics of extension workers included in the multivariate linear model above, number of trainings that the agents had in the last five years and significantly affected the time that they worked on farmers' agricultural fields at the 5% level. Giving on-the-job training to agents was found to have a positive effect on their time allocation on farm lands. Extension workers who have sufficient trainings will be well equipped with essential tools, knowledge and confidence to show farmers better farming practices right on the farm. However, extension agents who handled larger number of villages worked shorter hours on agricultural fields in the past season. The larger the number of villages an extension agent has to handle, the larger time that is wasted by walking to/from one village to another. *Ceteris paribus*, years of schooling negatively affected the time that agents worked on agricultural fields. This might be because more educated agents have higher expectations and consider working on a farm not a decent job. Such agents might spend most of their time in cities and towns looking for a better job which is as per their expectations. Average farm size of female households the agent worked with significantly positively affected his/her working time on farmers' agricultural fields. The larger the farm size of female farmers the more time extension workers spend on the farm, all other factors remaining constant. All other factors remaining constant, extension agents who participated in the number of villages that agents handled strategically *kebele's* agricultural planning process had a positive effect on the hours they worked on farmers' agricultural lands. One explanation may be that extension workers who involve in the *kebele's* agricultural planning feel that they own the plan and work hard to achieve it.

Perceptions of extension workers about the fairness of performance appraisal in the *wereda*, equal opportunity of male and female agents, performance appraisal, resource availability, and workload manageability are among the working environment perception variables that statistically significantly affected the amount of time the agents worked on farmers' fields. Even though they are not statistically significant, perceptions of extension workers about the timeliness of their salary, the fairness of hiring/recruitment, availability of supervisors whenever

Table 6. Econometric results.

Variables	Coefficient	T-statistic (Rob. SEs)
Personal characteristics		
Intercept	3.92	9.25 (0.42)***
Gender	0.03	0.23 (0.14)
Age	-0.01	-0.44 (0.01)
Marital status	-0.01	-0.12 (0.06)
Number of children below 14	0.03	0.73 (0.04)
Education	-0.03	-1.63 (0.02)
Work experience	0.003	0.22 (0.01)
Number of training	0.02	4.02 (0.004)***
Involved in the kebele's agricultural planning	0.16	1.46 (0.11)
Number of villages handled	-0.03	-4.26 (0.01)***
Average farm size of female households worked with	0.05	1.76 (0.03)*
Working environment perception variable		
Timeliness of salary	0.07	0.69 (0.11)
Merit based recruitment	0.04	0.58 (0.07)
Equal opportunities for male and female	-0.19	-2.62 (0.07)***
Performance appraisal	0.13	2.29 (0.06)**
Job insecurity	0.09	1.47 (0.06)
Resource availability	0.18	2.24 (0.08)**
Corruption in other weredas	0.13	1.83 (0.07)*
Disagreed with organizational policies	-0.02	-0.23 (0.07)
Supervisor is around	0.05	0.73 (0.06)
Workload manageability	-0.13	-1.88 (0.07)*
Specialization dummies		
Agents specialized in livestock	-0.004	-0.06 (0.07)
Agents specialized in natural resource management	-0.02	-0.24 (0.08)
General extension agents	-0.12	-0.98 (0.12)
District dummies		
Bati	-0.46	-4.41 (0.10)***
Gog	-0.30	-1.64 (0.19)
Ibantu	-0.14	-1.17 (0.12)
Ofla	-0.43	-3.60 (0.12)***
Sheko	-0.01	-0.05 (0.15)
Telalak	-1.06	-4.24 (0.25)***
Yaso	-0.08	-0.56 (0.15)
VIF	2.10	
Number of observations	236	
R-squared	0.43	

***, **, and* are significant at 1%, 5%, and 10% levels, respectively. The dependent variable is the logarithm of the average aggregate time that agents worked per week during the past planting, harvesting and planting seasons.

needed, and organizational policies/programs are other working environment perception variables with expected signs. *Ceteris paribus*, those AE workers who have a perception that there is fair performance appraisal in their wereda, on average, worked more time on farmers' agricultural fields compared to those who feel otherwise. Another variable that was found to affect the time that agents worked on farmers' agricultural fields is their perception about the availability of enough resources to carry out their responsibilities. *Ceteris Paribus*, on average, extension agents who feel that the resources they need are available were more likely to work more hours on farmers' plots. Surprisingly the perception of extension workers about their workload manageability statistically significantly and negatively affected the time they worked on agricultural fields. One explanation might be the fact that these extension agents underestimate their job and consider their task to be easy and they spend most of their time in office, with their family, in cities and other places.

Dummy variables for *weredas* Bati, Ofla and Telalak, Sekota being the reference *wereda*, were found to be statistically significant. The result implies that extension agents working in these three *weredas* were likely to work fewer hours on farmers' agricultural lands than those working in Sekota *wereda, ceteris paribus*. For instance, extension workers in Telalak (Afar) worked, on average, the lowest number of hours on farmers' agricultural fields per week in the past season. Consistent to the descriptive analysis the explanation may be that sedentary farming is not common in this *weredas* and the agents might not spend much time on the farm (Table 6).

CONCLUSIONS AND RECOMMENDATIONS

Agriculture is the main stay of the country's economy. The current government has emphasized in developing and investing in the sector. AE workers are given the lion's share of the responsibility for this agricultural development strategy. Accordingly, this study investigated the factors affecting the agricultural productivity of extension agents working in rural Ethiopia. Data from a survey of 236 AE agents working throughout the country were used for our empirical analysis.

The results showed that many variables that explain the perception of agents about their working environment significantly influenced their working time on farmers' agricultural fields. For instance, those agents who feel that there is fair performance appraisal in the *wereda*, sufficient resources to work with, and those who feel that they should work harder to secure their job in the future worked more hours on agricultural fields. It is therefore, paramount importance to frequently follow up the fairness of the performance appraisal and the availability of enough resources for the extension agents to work with in the *weredas.* There must also be some way of finding out, fining and firing extension workers who shirk on their work. On the other hand, those extension agents who feel that their task is manageable worked fewer hours on farmers' plots. It is therefore important to set a standard for the extension agents to achieve, and make the target level of performance not to be so easily achievable.

The number of villages handled by agents had negative effect on the agricultural productivity of the extension workers in the past season. It is therefore important to reduce the time wasted by travelling to and from villages by allocating an optimal number of villages to each agent. However, participating extension agents in the *kebele's* agricultural planning process increases their motivation and willingness to work more hours on farmers' plots. One reason is that agents feel that they own the agricultural plan and strife to achieve it. Moreover, providing on-the-job training also increased the motivation of extension agents to work more hours on farms. Finally, extension agents in Sekota *wereda* worked significantly more hours on farmers' agricultural fields than those in most other *weredas.* It might be of importance for each other WoRADs to prepare ways that their extension agents learn from those in Sekota *wereda.* Experience sharing schemes among these *weredas* may have a crucial role on this aspect.

REFERENCES

Agunga R, Ojomo C, Na S (1997). Burnout, job satisfaction and work situations as perceived by district extension officers, Ondo State, Nigeria. J. int. Agric. Ext. Edu., 4: 47-56.

Anderson JR, Feder G (2002). Rural Extension Services, World Bank, Agriculture and Rural Development Department, Washington DC.

Azril MSH, Jegak U, Asiah M, Azman AN, Bahaman, AS, Jamilah O Thomas K (2010). Can Quality of Work Life Affect Work Performance among Government Agriculture Extension Officers? A Case from Malaysia. J. Soc. Sci., 6: 64-73.

Becker GS (1965). A Theory of the Allocation of Time. Econ. J., 75: 493-517.

Bennel P, Zuidema L (1989). Human resource management and agricultural research: Overview and issues. Proceedings of the regional workshop on human resource management in NARS, Harare, Zimbabwe.

Chyung SY (2005). Analyze Motivation-Hygiene Factors to Improve Satisfaction Level of Your Online Training Program. 18th Annual Conference on Distance Teaching and Learning.

CSA (2008). Central Statistical Authority (CSA). Statistical Abstract, 2007/08. Addis Ababa, Ethiopia.

CSA (2010). The Federal Democratic Republic of Ethiopia Central Statistical Agency Agricultural Sample Survey. Report on farm management practices, June 2010. Addis Ababa, Ethiopia.

Darlene MO, Borman WC (1989). More Evidence on Relationships between the Work environment and Job Performance. Human Perform., 2: 113-130.

Davidson J, Hendry D, Srba F, Yeo S (1978). Econometric modeling of the aggregate time-series relationship between consumers' expenditures and income in the United Kingdom. Econ. J., 88: 661-692.

FDRE (2005). Plan for accelerated and sustained development to end poverty. Federal Democratic Republic of Ethiopia (FDRE), Addis Ababa, Ethiopia.

FDRE (2010). Growth and Transformation Plan. Federal Democratic Republic of Ethiopia (FDRE), Addis Ababa, Ethiopia.

Gebremedhin B, Jaleta M, Hoekstra D (2009). Smallholders, institutional services, and commercial transformation in Ethiopia. Agric. Econ., 40 (s1), 773–787.

Gronau R (1977). Leisure, Home Production, and Work--the Theory of the Allocation of Time Revisited. J. Pol. Econ., 85: 1099-1123.

Hazel P, Poulton C, Wiggins S, Dorward A (2007). The future of small farms for poverty reduction and growth. 2020 Discussion Paper No. 42. International Food Policy Research Institute (IFPRI), Washington, DC.

Hendry D (1995). Dynamic Econometrics. Oxford Universiy Press: Oxford.

Kassa B, Degnet A (2004). Challenges Facing Agricultural Extension Agents: A Case Study from South-western Ethiopia. African Development Bank, Blackwell Publishing Ltd. Oxford, UK.

Koustelios A (2001). Personal characteristics and job satisfaction of Greek teachers. Int. J. Edu. Manag., *15:* 354-358.

Long JL, Swortzel KA (2007). Factors influencing individual job *performance of Extension agents in the Mississippi State University Extension Service.* Proceedings of the 2007 American Association for Agricultural Education Research Conference, Volume 34. *Minneapolis, Minnesota.*

MoFED (2006). Ministry of Finance and Economic Development (MoFED), Survey of the Ethiopian Economy. Addis Ababa, Ethiopia.

Newman JE (1977). Development of a Measure of Perceived Work Environment (PWE). Acad. Manage J., *20*: 520-534.

Oladele OI, Mabe LK (2010). Job burnout and coping strategies among extension officers in North west province, South Africa. Afri. J. Agric. Res., *5*(17): 2321-2325.

Oloruntoba A, Ajayi MT (2003). Motivational factors and employees' job satisfaction in large-scale private farming in Ogun state, Nigeria. J. int . Agric. Ext. Edu., *10*: 67-72.

Picciotto R, Anderson JR (1997). Reconsidering Agricultural Extension. World Bank Res. Obs., 12: 249-259.

Scott M, Swortzel KA, Taylor WN (2005). Extension Agents' Perceptions of Fundamental Job Characteristics and Their Level of Job Satisfaction. J. Southern Agric. Edu. Res., 55: 88-101.

Strong R, Harder, A (2009). Implications of Maintenance and Motivation Factors on Extension Agent Turnover. J. Ext., 47(1) Article 1FEA2. Available at: www.joe.org/joe/2009february/a2.php

Vernon VH (1972). Working under pressure. Staple Press. London.

Welsh AN (1980). Managing people. Coles Publishing Co. Ltd., Toronto.

Appendix

Appendix A. Descriptive statistics of variables in this study

Table I. Descriptive statistics of dependent and potential explanatory variables.

Variable		Mean	Std Dev.
Dependent Variable			
Amount of time (hours) agents worked per week on farmers' agricultural fields during	planting season (T_M^p)	38.15	16.68
	harvesting season(T_M^h)	31.76	16.22
	slack season(T_M^s)	23.76	12.85
Independent variables			
Personal characteristics and other variables			
Age		26.66	4.99
Gender (1= Male, 0 = Female)		0.92	0.27
Marital status (1 = Married , 0 = Otherwise)		0.40	0.49
Number of children of age 14 and under (Children14)		0.61	1.05
Years of schooling		15.56	1.81
Experience in years		5.33	4.21
Number of trainings obtained in the last 5 years		6.63	8.32
Salary		939.44	162.38
Number of villages the agent handles		5.57	4.56
Received award? (Yes = 1, 0 = No)		0.17	0.38
Total time required to reach to *kebele* centre from the agent home (mts)		29.44	46.86
Involved in the kebele's agricultural planning process (Yes = 1, 0 = No)		0.69	0.46
Had experience Sharing-visits? (Yes = 1, 0 = No)		0.63	0.48
Average households' farm size that the agent works on	Male-headed households	2.17	2.49
	Female-headed households	1.16	1.13
Working environment perception variables			
I feel recognized by farmers and co-workers as a hard worker (1= agree, 0= disagree)		0.93	0.25
My salary and other benefits I receive allow me to maintain a decent standard of living (1=agree, 0=disagree)		0.39	0.49
DAs are paid equal to service providers in other wereda sector offices who do comparable tasks (1= agree, 0= disagree)		0.41	0.49
DAs always receive their salaries on time (1= agree, 0 = disagree)		0.74	0.44
The pay scales in WoARD reflect differences in workload and responsibility between different grades (1= agree, 0 = disagree)		0.46	0.50
DAs are hired purely on the basis of merit (1= agree, 0 = disagree)		0.51	0.50
DAs are promoted purely on the basis of merit (1= agree, 0 = disagree)		0.44	0.50
There are good opportunities for promotion in the WoARD (1= agree, 0 = disagree)		0.37	0.49
Performance appraisals in this wereda's WoARD are carried out in a fair way (1= agree, 0 = disagree)		0.48	0.50
DAs have a reason to be worried about losing their jobs in the near future (1= agree, 0 = disagree)		0.38	0.49
DAs in this wereda have enough resources available to carry out their work as required by professional norms (1= agree, 0 = disagree)		0.17	0.38
Corruption or misuse of funds is a problem in other weredas' WoARDs (1= agree, 0 = disagree)		0.50	0.50
I often disagree with the policies or programs we are asked to implement (1= agree, 0 = disagree)		0.33	0.47
There is hardly any political interference in our work (1= agree, 0 = disagree)		0.45	0.50
The majority of people in this office are well-qualified to do their job (1= agree, 0 = disagree)		0.56	0.50
The majority of DAs in this *wereda* work the expected hours (1= agree, 0 = disagree)		0.66	0.48

It is hard to make my supervisor happy (1= agree, 0 = disagree)	0.59	0.49
My supervisor knows enough about my daily activities to know if I am doing good or poor work (1= agree, 0 = disagree)	0.53	0.50
My supervisor is always around when (s)he is needed (1= agree, 0 = disagree)	0.51	0.50
My workload is manageable (1= agree, 0 = disagree)	0.44	0.50
The farm distances from where I stay are manageable (1= agree, 0 = disagree)	0.42	0.49
Inputs I need for my work come on time (1= agree, 0 = disagree)	0.13	0.33
I am satisfied with my job (1= agree, 0 = disagree)	0.79	0.41
Specialization dummies		
Agents specialized in crop	0.37	0.48
Agents specialized in livestock	0.40	0.49
Agents specialized in natural resource management	0.17	0.38
General extension agents	0.06	0.24
Wereda dummies		
Bati	0.18	0.39
Gog	0.08	0.28
Ibantu	0.16	0.36
Ofla	0.16	0.37
Sekota	0.20	0.40
Sheko	0.14	0.35
Telalak	0.05	0.21
Yaso	0.03	0.17
Bati	0.18	0.39

Source: survey data, 2009.

18

Trends in agricultural extension services in Africa: Option for new approaches

J. O. Saliu[1]*, P. C. Obinne[2] and S. I. Audu[1]

[1]Department of Agricultural economics and Extension, Kogi State University, Anyigba, Nigeria.
[2]Department of Agricultural Extension and Communi`cation, University of Agriculture, Makurdi, Nigeria.

Many developing countries are now at various stages in the process of economic liberalization, privatization and decentralization. This implies that technical programming, personnel management and budget appropriations for extension delivery will no longer be the same as they used to be under centralized public extension service delivery. Liberalization has bred different actors in extension delivery such as the public agencies, private service providers, produce organizations and Non-Governmental Organizations (NGOs). It is therefore important to find out the methods utilized by the current facilitators of extension activities and how the activities are funded. This paper thus examined the current approaches to agricultural extension in South Africa, Nigeria, Uganda and Madagascar. Findings revealed the following approaches: (1) partnership extension services, (2) people centered participation, (3) farmers organizations and (4) fund mobilization through shared responsibility among all the stakeholders. These approaches covered a significant part of each of the countries examined. However, time must be giving if the approaches are to stand the test of time. Effort must be intensified to see how the benefit from the approaches in the enclave coverage can be extended to every part of the nation. Enabling environment like development of viable markets, communication and other infrastructural facilities can be provided for the exploration of these approaches.

Key words: Actors, private service providers, non-governmental organization, public agencies.

INTRODUCTION

Sub-Sahara Africa consists of farmers who mostly reside in rural communities. The rural population is expanding rapidly while accessibility to health, education and even food is becoming increasingly difficult. For instance, about 800 million people in the developing world do not have enough to eat and out of this figure, about 180 million live in Sub-Saharan Africa (Sofi, 2001). This gloomy picture is emerging in a fast changing environment where most developing countries are at various stages in the process of economic liberalization. Economic liberalization means removal of all sort of economic restrictions such as price control on farm produce, government withdrawal from economic activities such as input supply, marketing and even direct agricultural production while decentralization is a means of transferring selected public responsibilities to regional, state or local level institutions by the national government. This implies that technical programming, personal management and budget appropriations for extension delivery can be delegated to other stakeholders like farmers cooperatives, private service providers, non governmental organizations, public agencies and so on (Neuchattel Group, 1999). The main public actors in most of the African countries are the National Agricultural Extension Services while private actors such as input suppliers, purchasers of agricultural products, private trainer-adviser, outreach agencies and private media are now participating in activities geared towards agricultural programme (Neuchattel Group, 1999). This development is springing up at a time when many developing countries are less

*Corresponding author. E-mail: josaliu@yahoo.com.

able to continue providing all the extension services expected of them as a result of financial limitations. For instance, Arokoyo (2003) asserted that the decentralization of agricultural farm radio programmes in Nigeria took place at the same time as the privatization of state radio broadcasting stations. These stations then started to charge commercial prices for radio broadcasting time slots. Many of the decentralized farm radio programmes, with much smaller budgets than those of the former national programmes, could simply not afford those prices. As a result, most Nigerian farm radio programmes have been discontinued. Since the inception of this wind of change, extension delivery has undergone various permutations informed by the experiences on ground in each country. It is in this circumstance that agricultural extension is expected to provide farmers with necessary education, skill and technical information to enable them take effective farm management decision for accelerated agricultural development. It is therefore important to know the various coping strategies adopted by these developing countries. We may also ask what effect has this change on the agricultural policy of the said countries? Who are the current facilitators of extension activities in these countries? What role have the farmers assumed in the planning, managing and implementing extension activities? Who provide fund for extension activities in African countries? This becomes essential to address the issue of inadequate farmers' involvement and high cost of Training and Visit (T & V) Extension System, which the national government in some of the developing countries like Nigeria can no longer afford as a result of the withdrawal of donor agencies, economic liberalization and decentralization. As such, this paper attempts to analyze the current trend and approaches to agricultural extension delivery in Africa to meet the interest and needs of the farmers as it has become obvious that T & V Extension System no longer meet the diverse interest and needs of farmers (Charles and Williams, 2005).

For the purpose of this paper, Africa is divided into two groups. Well developed agricultural service institutions and less developed agricultural services. Among the well developed ones are Nigeria, South Africa and Egypt, while the less developed agricultural service institutions include Uganda, Ghana, Zimbabwe and Madagascar. South Africa and Nigeria were picked from the well-developed group while Uganda and Madagascar were picked from the less developed group. In essence, this paper examines current trend and approaches to agricultural extension delivery in South Africa, Nigeria, Uganda and Madagascar. The specific issues discussed are:

A review of the activities of the actors involved in agricultural extension in the said countries, participatory approaches adopted by the various agricultural extension stakeholders in the said countries, fund mobilization practices for managing agricultural extension services in the selected countries.

AGRICULTURAL EXTENSION IN SOUTH-AFRICA

Actors Involved in Agricultural Extension

Prior to 1994, Department of Agricultural Development of the House of Assembly, which was the largest Department, was responsible for servicing while agricultural services for the white in South Africa during apartheid in seven agro-ecological zones. The 'Black agriculture;' was serviced by the departments of agriculture of the Homelands' and independent states. Following the change to a democratic South Africa in April 1994, all agricultural activities fell under one department (Van-Nickerk, 1995). However, the agricultural development remains uneven between the whites who own commercial farms and the black who are subsistence farmers. As such, South African agriculture has a strong commercial production sector which provides a foundation for national food security and even international trade, thus making a significant contribution to the national economy. This success was due to the agricultural policies and extension approaches adopted to date in South Africa (Worth, 2002). Agricultural extension officers or advisers are employed by departments of agriculture, cooperatives and some marketing organizations. A whole host of private firms and non-governmental development agencies also provide advisory services. However, information dissemination through extension services is yet to get to all farmers who will need the information.

Participatory approaches

National and provincial agricultural policies are formulated to address issues of food insecurity and rural poverty. These policies have direct effect on the agricultural extension approach in South Africa. Duvel (2001) observed that the model for extension in South Africa is a technology-centred one. He, however, noted that occasionally technology is transferred in a people-centred way, using people centred participatory (Participatory Rural Appraisal – PRA) methodologies, aimed at reducing barriers to technology adoption and also to adapt technologies and practices to local conditions. It is a popular opinion that PRA being used in South Africa has helped to address the needs of the farmers.

In essence a participatory diagnosis is the farm participatory technique of involving farmers in South Africa. It is therefore, a proclaimed priority of South Africa that development be people-centered (Crase et al., in Worth, 2002). In this wise, extensions activities are carried out by service provides (Government and Private) and farmers through joint planning and programme implementation. The people-centred policy is enshrined in the ethos of the South African constitution.

Fund mobilization for extension work

Funding for extension work is popularly based on assets and partnership with private sectors and farmers' organizations. This method ensures availability of fund for extension work throughout the year.

CURRENT AGRICULTURAL EXTENSION SERVICES IN NIGERIA

Actors involved in extension work

Nigeria probably has the most elaborate research and extension institution in sub-Saharan Africa, with a population of over 140 million and 71 million ha of arable land. The system comprises of 17 commodity – based research institutes and a special national extension institute, over 45 faculties of agriculture in conventional federal, state and private universities, three universities of agriculture, several colleges of agriculture/polytechnics and three international agricultural research centers viz; International Institute of Tropical Agriculture (IITA), a substation of International Livestock Research Institute (ILRI), (Okwu and Ejembi, 2001). All these institutions collectively and individually serve as the fountain of agricultural innovation for both public and private agricultural extension service providers. In Nigeria, agricultural extension service has been mainly public. Currently the major provider of public sector agricultural extension services, is the agricultural Development programmes (ADPs) in each of the 36 states of Nigeria.

The pre-ADP Extension Services (The conventional extension system) were presumed to be ineffective and could only provide limited services to the majority of farmers in the basic farming enterprises due to bureaucratic bottlenecks. This formed the basic fact underlying the removal of the extension service from the civil service system and the introduction of ADP in the early 70s has now became the only functional public extension service provider in the country (Ekpere, 1991). The World Bank-ADP approach has been used to re-organize the management of agricultural extension system for effective performance. The main feature is the training and visit extension system which include a single line of command, a well-defined geographical boundary of operation for each extension worker, a supervisor to supervisee ratio of not more than 1:8. Other characteristics include systematic programme of short training courses, removal of all non-agricultural extension functions from the responsibility of the extension worker, effective monitoring and provision of adequate transport facilities. These features ensure flexibility and prompt decision-making devoid of regular civil service bureaucracy. The current public extension service is now known as Unified Agricultural Extension Services. Which means the enlargement of the areas of focus to include service in fishery, livestock and women in agriculture. Oyebanji (1994) reported that the performance of the extension services since the inception of this professional system in Nigeria has been encouraging. Extension contacts with farmers have increased, several technologies have been disseminated and adopted by farmers. Idachaba (2005) also asserted that Anyigba ADP (Kogi State, Nigeria) greatly transformed the lives of Igala people through the revolution in extension services delivery. However, some prevailing problems existed and limited the effectiveness of the system. The problems include, insufficient funding at state level, inadequate or non-availability of inputs, poor logistic supports and inadequate staffing.

The issue of poor funding becomes critical after the withdrawal of outside funding as revealed in a study by Agbamu and Okagbare (2005) who explained that the effects of expiration of World Bank funding were noticeable in the inability of Ogun State Agricultural Development Project (OGADEP) to provide enough vehicles and motorcycles for its agricultural extension work, irregular payment of traveling claims, ill-motivated field-staff, reduced training session for village extension workers and reduced monthly technology review meetings. For instance 51.2% farmers were given extension services fortnightly during World Bank assistance when funding sources were 71.5% from World Bank, 15.9% from state government, 6.8% from International Fund for Agricultural Development (IFAD) and 15.9% from Federal Government of Nigeria. But this fortnight extension services fell to 40% after ceasation of World Bank assistance. However, OGADEP was still able to make significant patronage by looking inward for survival. The state government increased financial support to 43% IFAD, 44% and FGN 13.1%. The financial inadequacy coupled with the inability of the public extension services to meet the yearning and aspirations of the farmers in a deregulated economy made a number of private organizations to venture into agricultural extension.

Private extension/participatory approaches

The foremost participants in private extension services are United African Company (UAC), John Holt, Nigerian Tobacco Company (NTC) and Diocesan Agricultural Development Programme of the Catholic Diocese of Ijebu-Ode, among several others who became involved in agricultural production, processing and marketing some decades ago Adedoyin, 1995). About ten years ago, Green River Project of the Agip Oil Company, Ciba Geigy Agro-Chemical extension outfit, Olam Nigeria Limited, formally Agro-Millers at Makurdi, have been found to inject positive changes in the life of the communities where the programmes are located (Akele and Chukwu, 2004; Isife and Madukwe, 1999).

Informal Private Sector (IPS) is another dimension to extension services. IPS denotes private organizations that provide various extension services in the areas of agro seed, agro-chemicals, agro processing, micro-financing, farm tools and agro-consultancy with the aim of making profit from the ventures. A research conducted by Okoro et al. (2006), in Abia State comparing the effectiveness of agricultural extension services between IPS and Public Extension Services, revealed that most of the IPS practitioners are retired agricultural officials and or self employed graduates with few staff strength. Inputs are sold along with extension services.

Relationship between the IPS operators and farmers is mostly on individual basis. Ninety eight point five seven percent (98.57%) of the farmers got gift like crate of eggs from the IPS as a kind of public relation services. About 85% of the farmers got extension information from IPS while 71.4% of the respondents attributed growth in farm size to the effort of the IPS.

In summary private extension services appear to provide timely and appropriate services in terms of farmers' need. This is in agreement with the whys and wherefores of extension privatization.

Advocates of private extension services belief that it improves efficiency, improves public finance, encourages competition and private sector participation.

However, poor road network, inadequate finance on the part of the private extension practitioners, farmers inability to buy inputs and poor educational status of farmers resulting in slow adoption of technologies are the constraints affecting private extension services.

It is important to point out, that, privatization of extension services in Nigeria is still in the form of increased in private sector participation in provision of agricultural extension services and not a transfer of state owned asserts to the private sectors. One of the popular farmers organization that could be used to promote extension ctivities is Apex Farmers' Association of Nigeria (AFAN).

The type of partnership formed by this organization with other stakeholders is not very strong. Nigerian government has no guidelines regulating activities of private extension service providers at the moment.

Fund mobilization

Generally, fund for agricultural extension services are from donor agencies and the three tiers of government (Federal, State and Local Government). Private organizations have not been actively involved in sharing funding for extension programmes in Nigeria.

NEW APPROACHES TO AGRICULTURAL EXTENSION SERVICE IN UGANDA

Actors involved in extension services

Uganda was strongly involved in the use of pubic extension services to improve agricultural production. Ministry of agriculture and government agencies adopted the training and visit extension system that was sponsored by donor agencies. Donors were usually inflexible in the way in which they provide funds. In this wise, research and training programmes fell short of meeting the farmers need, efforts were duplicated due to poor coordination between public extension outfit and some Non Governmental Organizations involved in the provision of extension services. Poor road network, unattractive market prices for agricultural produce continued to serve as barriers to effective extension services. With the introduction of liberalization and globalization, a policy that stimulated decentralization and privatization of extension services became popular in Uganda (Mundy and Sultan, 2001).

Participatory approach

Ugandan government has put in place policies that promote decentralization, privatization, liberalization and developed a new plan for modernization of agriculture in the late 90s, (Oryokot, 2003). This policy has stimulated a popular participatory approach called participation through organization. We now have Uganda National

Farmer's Association (UNFA) that has established a "demand-driven, cost-recovery" extension system as an alternative to public sector extension in a number of districts (Carney, 1998). Using this typology of farmers participation, the extension system becomes more farmer led and as such farmers' needs and interests receive more attention.

The new plan developed to modernize agriculture in Uganda is based on two strategic pillars, raising overall agricultural yields and diversifying small holder production patterns into a mix of higher-value, export oriented commodities along with lower-value food staples. This plan was developed by a broad constituency of stakeholders officials and politicians, farmers, Non-Governmental Organizations (NGOs), Civil Society and the Community of donors (SAA, 2001). The new plan places a high priority on agricultural research and extension with particular attention on technology generation and transfer through decentralization of activities. The current position of Ugandan agriculture is that the country has regained its position as the largest coffee producer on the continent, the tea industry has been revitalized, a small export-oriented horticulture industry is emerging and maize exports to Kenya continue to grow (NSAA, 2001). This implies that the agricultural policy of decentralization and participation through organization is working well for Uganda.

Fund mobilization

Fund for extension services in Uganda is now based on partnership with government agencies, farmers organizations and some non governmental organizations. Farmers now mobilize fund to render information services, advisory, training marketing and credit services. This situation assists to guide research and makes the farmers to be active participants rather than passive recipients of new technologies. However farmers that cannot pay for the annual fees are not likely to benefit from the farmers organizations extension programmes.

NEW APPROACH TO AGRICULTURAL EXTENSION IN MADAGASCAR

Actors involved and participatory approaches

The experience of Madagascar in previous extension approaches (that is teaching and visit extension approach to technology transfer) has not yielded the expected results. Many farmers were left on their own for information on farm inputs. Insufficient funding and inaccessibility to credit by farmers constituted a major problem to extension services in Madagascar. This has led to the current popular support for farmers' organizations, collaboration and diversification strategies embarked upon by some private agencies. For instance, 11 farmers' organizations totaling 132 farmers had collaborated extension work with Project South West (PSW) for production of peapods and another 45 producers' organization benefited from bean seeds production in the mid-west Region.

Various projects are currently implemented through partnerships with the private sector and NGOs in Madagascar which is believed to increase confidence in the extension service. The partnership style is being supervised by the Ministries of Agriculture, Livestock and Applied Development Research. A number of initiatives have being generated by these relationships. The mode of operation is "tight regulation by contracts" with the different institutions concerned. As many as 20 agents from private sectors have participated in the implementation of the project's activities while 66 NGOs now maintain close partnerships according to their areas of competence and comparative advantage with respect to the project technical consultations.

Fund mobilization

Fund is now being mobilized through the communities participating in extension services and supported with funds from government agencies. The local communities are becoming more committed to project executed in their localities. For instance, 32 communities in 8 agricultural regions recruited out of their own resources and made available for extension services 87 agricultural counselors on a two-year renewable term. In effect, both private and public extension systems are made to compete for extension services (Schorosca and Seck, 2003). Under this approach, provision is made for regular communication and consultation on fund mobilization and extension activities with all the actors involved in the extension process. That is, farmers' organizations' NGOs, Private sector organization, local decentralized community, governments and donors. As a result of this approach, the linkages between farmers, extension agents and researchers have improved and more effective operations have been witnessed in Madagascar. However, many communities are left out of this extension system because the system could not provide for those who could not afford to pay their annual membership fees.

SUMMARY

In summary, Agricultural Extension Service is taking a new dimension in Africa, from complete public services to more private sector involvement. Certain features are common among the operators of this new extension focus despite differences in circumstances and environment.

The four (4) countries reviewed operate:

i) Partnership extension services which involves so many stakeholders, like the farmers, Non-Governmental Organizations, Private Sectors and even the Government. Though partnership of the above description is not very popular in Nigeria.
ii) They also decentralize their programmes for more popular involvement of people who were hitherto less involved in extension work.
iii) The nature of participation is now more people-centred. That is, participation through farmers' organization which is more pronounced among all the four countries reviewed except Nigeria where organized farmers groups are still few and less active in accepting the responsibility of recruiting agricultural extension agents out of their own resources.
iv) The kind of participation adopted by these countries permitted involvement of stakeholders' right from programme planning to implementation of action plan.
v) Fund mobilization common to all the nations under review is that of shared responsibility among all the stakeholders. However, Nigeria still needs to mobilize individuals and private organizations to embrace public – private partnership strategy of funding agricultural extension services. If by so doing, it will enable the country to harness more fund for their exten-sion activities.

Conclusion

Economic liberalization has influenced the mode of providing extension services in Africa. The style of decentralization and popular involvement of partners in extension services seems to be new and limited in coverage. Effort must be made to under study these approaches to see how it can be sustained and extended to every part of the countries in Africa. The flexibility in the ability of any approach to suite different problems exploited by their nation should be taken into consideration.

Recommendation

1) From the review, countries where partnership with various stakeholders such as Non-Governmental Organization, Farmers Cooperatives, private sector service providers and so on is being practiced seems to enhance availability of extension service to farmers who would have not had the opportunity. It is therefore recommended that effort be made by each country in Africa to liaise with every available stakeholder that will enhance availability and use of agricultural extension services.
2) A buttom-up approach that is people-centred which will enhance popular participation understanding and recognition of farmers' indigenous knowledge and experience as observed in South Africa should be put in place and use for identification, planning and implementation of agricultural extension projects.
3) Based on the findings in the review, unavailability of fund was identified to have disrupted the free flow of extension information to farmers. Fund mobilization through shared responsibility among all the stakeholders will enhance continuous and sustainable agricultural extension programmes. Conscious attempt should be made to mobilize fund from all stakeholders such that farmers can show interest and even sponsor extension programmes as found in Madagascar and Uganda.
4) A policy that would stimulate formation of farmers organization which will encourage self reliance, demand driven and cost-recovery should be put in place by each country. The similitude of this organization is the Uganda National Farmers Association (UNFA) which represents about 90,000 farmers throughout Uganda and serve members with extension information through annual membership fees of 1500 shillings. Another example is federation des organization non governmentales du Senegal (FONGS) with about 29 sub groups and 150,000 members in nearly 3000 village groups. FONGS extends services to about 900,000 rural people or about 0% of Senegal's population (Mundy and Sultan,2000).

REFERENCES

Adedoyin SF (1995). Non Governmental Organizations in Agricultural Development: The Experience of the Diocesan Agricultural Development Programme of the Catholic Diocese of Ijebu-Ode, Nigeria. Paper Presented at the Second Annual National Conference of the Agricultural Extension Society of Nigeria. (AESON), held at NAERLS/ABU, Zaria Nigeria pp. 97-109.

Agbamu JU, Okagbare GO (2005). Institutional Strengthening of Ogun State ADP upon Expiration of World Bank Assistance. In: Orheruata AM, Nwokoro SO, Ajayi MT, Adekunle AT, Asumugha GN (eds). Agricultural Rebirth for Improved Production in Nigeria. Proceedings of the 39th Annual conference of the Agricultural society of Nigeria, held at the University of Benin, Benin City pp. 322-325.

Akele SA, Chukwu GO (2004). Poverty Alleviation through Sustainable Root and Tuber Crop Production. Proceedings of 8th Triemual Symposium-International Society for Tropical Root Crops, African Branch (ISTRC AB) Ibadan. Nigeria pp. 116-119

Arokoyo T (2003). ICTs for Agricultural Extension Transformation ICTs – transforming Agricultural extension? CTA's observatory on ICTs 6th Consultative Expert Meeting, Wageningen 6p.

Carney D (1998). Changing Public and Private Roles in Agricultural Service Provision Overseas Development Institution, London pp. 110-115.

Charles A, Williams Z (2005). "Participation in Agricultural Extension" Accessible on the internet on www.agri-ext.org/appendix_participationin_agricultural_extension_htm. (Accessed and retrieved on June 2007).

Duvel G (2001). Participatory Programmed Agricultural Extnesion. Paper delivered at Mid-Term Review Workshop on Extension Service, Midrand, South Africa pp. 63-67.

Ekpere JA (1991). Agricultural Extension, Searchlight of the Nigerian Small Farmers. 1990 University Lectures. Ibadan: University of Ibadan Press 3p.

Idachaba FS (2005). Agricultural and Rural Development in Nigeria. The policy perspective. A text of convocation Lecture Delivered at KSU. Anyigba. pp. 3 - 4.

Isife BI, Madukwe MC (1999). Non-governmental Extension Services in Nigeria. Feature activities and policy issues. Proceeding of the 5th annual National conference of the Agricultural Extension society of Nigeria. (AESON) University of Nsukka pp. 164 -168.

Mundy P, Sultan J (2000). Farmers Groups and Markets, Ugandan National Farmers' Association, Giving Farmers a Voice Information Revolution, How Information and Communication Management is changing the lives of Rural People. Technical Centre for Agricultural ad Rural Cooperation (ACP – EC, Cotonou Agreement) Pp. 59-61.

Neuchatel G (1999). Common Framework on Agricultural Extension, Neuchatel Group, Paris. Accessible on the Internet on www.odi.org.uk/rpeg/efe.pdf (Accessed and retrieved on June 2007).

Okoro BO, Chukwu GO, Onumadu T, Okoye BC (2006). Sustaining Agricultural Extension through informal private sector Participation in Abia State. In. Asumugha GN, Olojede AO, Ikeorgu JG, Ano AO, Herbert U (eds). Repositioning Agriculture for Sustainable Millennium Development Goals in Nigeria. Proceedings of the 40th Annual Conference of the Agricultural Society of Nigeria (ASN), held at National Root Crops Research institute, Umudike, Abia State, Nigeria 297p.

Okwu OJ, Ejembi EP (2001). The Historical Development of Agricultural Extension in Nigeria. J. Sust. Trop Agric. Res. 2(3): 93 -99.

Oryokot J (2003). Increasing Farmer Access to Information, Knowledge and Technologies. The role of ICTs. ICTs - Transforming Agricultural Extension? CTA's observatory on ICTs 6th Consultative Export Meeting, Wagenigen 6: 23-25.

Oyebanji OO (1994). The Development of Agricultural Extension in Nigeria, Agricultural Extension in Africa. Proceedings of an international Workshops. Yaounde, Cameroun. Post Bus, AJ

Wageningen. The Netherlands: Technical Centre for Agric. Rural Dev. 2: 240.

Sasakawa Africa Association (SAA) (2001). Uganda Approaches to Agricultural Extension; Feeding the future, Newsletter, April edition p1.

Schorosch F, Seck O (2003). Agricultural Extension in Madagascar: Broadening Services, Leveraging Partnerships. Accessible on the Internet on Pmohan@worldbank.org (Accessed and retrieved on March 2007).

Sofi I (2001). The State of Food Insecurity in the World 2001, FAO, Rome. Accessible on the internet on www.fao.org/news/199img/sofiepdfdowloaded (Accessed and retrieved on March 2007).

Van NRV (1995). The State of Agricultural Information Services in South Africa. In Ania LO, Kaniki AM, Ojiambo JB (eds). Agricultural Information in Africa Blackhorse Publisher Lagos p. 95.

Worth S (2002). Sustainable Extension: Not Transformation but Renewal. Proceeding of the 18th Annual Conference of Association for International Agricultural and Extension Education (AIAEE), Durban, South Africa. pp. 118-120.

The role of extension in agricultural adaptation to climate change in Enugu State, Nigeria

Nicholas Ozor[1]* and Nnaji Cynthia[2]

[1]African Technology Policy Studies Network (ATPS), 3rd Floor, the Chancery, Valley Road, P. O. Box 10081-00100 Nairobi, Kenya.
[2]Department of Agricultural Extension, Faculty of Agriculture, University of Nigeria, Nsukka, Enugu State, Nigeria.

The clear evidence that climate change is already a reality calls for actions not just to slow down the process or enhance the sinks of greenhouse gases (mitigation) but also to assist those affected or threatened already to adjust in natural and human systems to a new or changing environment (adaptation). The paper examined the role of extension in agricultural adaptation to climate change in Enugu State. A multi-stage random sampling technique was used in selecting 120 respondents. Descriptive statistics were used in analyzing data from the study. Results show that the most significant effect of climate change in the area was intense weed growth ($\bar{X}$= 4.52) while the most popular adaptive measure adopted by respondents was the use of resistant crop and animal varieties/species (95.43%). The study proposed that the key roles of extension in agricultural adaptation to climate change were in the re-training of extension staff to acquire new capacity in climate risk management ($\bar{X}$= 3.93), setting up of emergency management units in extension agencies ($\bar{X}$= 3.91), dissemination of innovations on best practices and building resilience capacities of vulnerable people in climate risk management ($\bar{X}$= 3.90), and providing feedbacks to governments and interested agencies on climate change issues ($\bar{X}$= 3.9). The paper recommends a change in the extension service agenda to accommodate the challenges currently posed by climate change by adopting new roles identified in the study and the need for increased research and innovation for sustainable adaptation to climate change.

Key words: Role, extension, agricultural adaptation, climate change.

INTRODUCTION

Climate change refers to any change in climate overtime, whether due to natural variability or as a result of human activity (inter-governmental panel on climate change, IPCC, 2001). It can also be seen as change in climate which is attributed directly or indirectly to human activities that alter the composition of the global atmosphere and which are in addition to natural variability observed over comparable time periods (IPCC, 2007). Climate change has become a global issue in recent times manifesting in variations of different climate parameters including cloud cover, precipitation, temperature ranges, sea levels and vapour pressure (Ministry of Environment of the Federal Republic of Nigeria, MOE FRN, 2003). The variations in climate parameters affect different sectors of the economy such as agriculture, health, water resources, energy etc. The main cause of climate change has been attributed to anthropogenic (human) activities. For example, the increased industrialization in the developed nations has led to the introduction of large quantities of greenhouse gases (GHGs), including carbon (IV) oxide (CO_2), methane (CH_4) and nitrous oxide (N_2O) into the atmosphere. These GHGs are the primary causes of global warming (IPCC, 2007). The global increases in CO_2 concentration are due primarily to fossil fuel use and land use change, while those of CH_4 and N_2O are primarily due to agriculture (IPCC, 2007). Agriculture is therefore the main culprit of climate change producing

*Corresponding author. E-mail: sunny_ozor@yahoo.com.

significant effects through the production and release of GHGs.

Clearing of forests for agricultural production replaces forests with crops thereby reducing the rate at which carbon sequestration (trapping and absorbing carbon (IV) oxide gas) occurs. There is no doubt therefore that the earth is getting warmer and human beings are mainly to be blamed (Spore, 2008). With reference to sub-Saharan Africa, there is growing interest on the likely impacts of climate change on agriculture, economic growth and sustainable development. Incidences of climate change include changes in soil moisture, soil quality, crop resilience, timing/length of growing seasons, yield of crops and animals, atmospheric temperatures, weed insurgence, flooding, unprecedented droughts, sea level rises and many more. More than 70% of those living in African, Caribbean and Pacific (ACP) countries work in the agricultural sector and climate change for these people are not a theoretic discussion; it is the difference between life and death (Spore, 2008). There are projecttions of increases in rainfall in the humid regions of southern Nigeria, which are accompanied by increases in cloudiness and rainfall intensity particularly during severe storms (IPCC, 2007). Similarly, the savannah areas of northern Nigeria were projected to experience less rainfall, which coupled with temperature increases, reduces soil moisture availability. Changes in climate are severely affecting agricultural production in many African countries (UNEP, 2007). Increased temperatures and accompanying decrease in water availability reduce the length of growing seasons and yield potential and hence the areas suitable for agriculture, further adversely affecting food security over the continent (Thornton et al., 2006).

An increase in temperature is also conducive for a proliferation of pests and diseases, which are detrimental to crop production. Climate change affects livestock productivity directly through effects on the availability of feed and fodder. In coastal regions that have major lagoons or lake systems, changes in fresh water flows and greater intrusion of salt water into lagoons, as a result of climate change, affect inland fisheries or aquaculture (Cury and Shannon, 2004). Sea levels are rising because of high temperatures, glacial retreat and increased rainfall in some areas. A rise in the sea level results in an agricultural land loss. Erosion, sub-mergence of shorelines, salinity of the water table due to the increased sea level mainly affects agriculture through inundation of low-lying lands (MOE FRN, 2003). The rise in sea level in Atlantic Ocean for instance may have catastrophic impacts on large coastal cities such as Banjul (Gambia), Lagos and the Niger Delta regions (Nigeria) and Alexandria (Egypt) (Nicholls and Tol, 2006). In order to sustain the agricultural sector that plays pivotal roles in human existence in terms of the provision of food, fibre, fun, fuel and income, strategies of change need to be urgently initiated to cope with the changing climate. Agricultural extension has key roles to play in initiating this change. This is because adaptations to climate change impacts require changes in knowledge, attitudes, resilience capacities, and skills of the people and agricultural extension can bring this change.

Agricultural extension according to Leeuwis (2006) is a series of embedded communicative interventions that are meant, among other things, to develop and/or induce innovations which supposedly help to resolve (usually multi-actor) problematic situations. It has been observed that agricultural extension is involved in public information and education programmes that could assist farmers in mitigating the effects of climate change (MOE FRN, 2003). According to them, such involvements include awareness creation and knowledge brokerage on the issues of climate change; building resilience capacities among vulnerable individuals, communities and regions; encouragement of wide participation of all stakeholders in addressing climate change issues; and developing appropriate frameworks for coping/adapting to climate change effects/impacts. Besides, there is very little documentation on local adaptation to climate change (Pandey, 2006) especially when it is known that some African communities have developed traditional adaptation strategies to cope with climate variability and extreme events (Osman-Elasha, 2007). It is pertinent therefore to ascertain what effects or impacts climate change has on agricultural productivity in the study area?

What adaptive measures have farmers adopted in cushioning the effects or impacts of climate change? And what roles can agricultural extension play in building the adaptive capacities of vulnerable people in the area?

Purpose and objectives

The overall purpose of the study was to ascertain the role which agricultural extension can play in climate change adaptation in Enugu State. Specifically, the objectives were to:

i) Ascertain the different effects or impacts of climate change on agriculture in the study area;
ii) Identify the adaptive measures adopted by respondents in cushioning the effects of climate change in the area; and
iii) Ascertain the role of agricultural extension in building adaptive/resilience capacities of vulnerable people to climate change effects/impacts.

METHODOLOGY

The study was carried out in Enugu State of Nigeria. Enugu State is one of the 36 states in Nigeria and is located between latitude 5° 56'N and 7° 06'N and longitude 6° 53'E and 7° 55'E (Ezike, 1998).

The State has a population of 2,452,996 (NPC, 2006). The vegetation of the state is mainly forest type but stretches out into derived savannah in the northern fringes. Enugu State experiences distinct wet and dry seasons with a total annual rainfall of about 1,700 mm (Enugu State Government Official Gazette, No. 25, 1997). The major occupation of people in the state is farming. Major crops cultivated include, cassava, yam, cocoyam, vegetables, oil palm etc, while major livestock reared are poultry, goat, sheep and cattle. The State has seventeen local government areas (LGAs) and is divided into three agricultural zones namely:

1) Enugu North Zone, comprising Igbo-Etiti, Uzo Uwani, Nsukka, Udenu, Igbo-Eze North and Igbo-Eze South LGAs;
2) Enugu East Zone, comprising Nkanu West, Nkanu East, Enugu North, Enugu South, Enugu East and Isi Uzo LGAs;
3) Enugu West Zone comprising Udi, Ezeagu, Oji River, Awgu and Ani LGAs (ENADEP, 2007).

Farmers in the State constituted the population for the study. A multi-stage random sampling technique was used to select respondents, thus two agricultural zones (Enugu North and Enugu West) were randomly selected. Two LGAs were randomly selected from each of the two zones. For Enugu North Zone, Nsukka and Udenu LGAs were randomly selected while Udi and Oji River LGAs were selected from Enugu West Zone. Two town communities were also randomly selected from each LGA to give a total of eight town communities. They include Ibagwa-Ani and Okpuje (Nsukka), Obollo Eke and Amala (Udenu), Agbala-Enyi and Ugwuoba (Oji River), and Awhum and Nsude (Udi). Fifteen farmers with long years of farm experience were purposively selected from each of the town community. The preference for farmers with long years of farm experience (not less than 20 years) was made so as to gain the most useful information from them on their experiences of changes in climate as it affects their farming practices and production. This brought the total number of respondents that participated in the study to 120. Data for the study were collected through a semi-structured interview schedule that was validated by experts in agricultural extension and climatology.

A focus group discussion (FGD) was held in each zone to obtain more in-depth information from farmers on the subject matter. A pilot test was also conducted as part of the instrument validation and to test for reliability. Data were collected between June and October 2008. The interview schedule had three sections. Section A ascertained the different effects/impacts of climate change on agriculture in the area (Objective 1). A five-point Likert-type scale of 'to a very great extent, to a great extent, to some extent, to little extent and to no extent', with nominal values of 5, 4, 3, 2 and 1, respectively was used to determine the extent of effects/impacts of climate change on farmers' agricultural production. Relevant information on the possible effects/impacts of climate change on agriculture obtained from literature, observations and focus group discussions was utilized. The respondents' mean scores were obtained for each response option, such that items with $\overline{X} \geq 3.00$ were regarded as having significant effects/impacts on agriculture in the study area while items with $\overline{X} < 3.00$ were regarded as not having significant effect /impact on agriculture. Section B dealt with identification of the adaptive measures adopted by farmers in cushioning the effects of climate change (Objective 2). In order to obtain a quantitative measure of respondents' agreement on the use of each technique in cushioning the effects of climate change, a checklist of various adaptive measures which were obtained from literature, personal interviews, observations and the FGD was utilized. Respondents were requested to tick the measures used in adapting to the climate change effects/impacts. Any response option (adaptive measure) with score ≥ 50% was regarded as being useful and important while items with scores < 50% was regarded as not being useful and important in cushioning the effects of climate change by farmers.

Section C sought to ascertain the roles that agricultural extension could play in building adaptive/resilience capacities of vulnerable people to climate change effects/impacts (Objective 3). These roles were obtained from relevant literature, extension professionals and the FGD. A four-point Likert-type scale of strongly agree, agree, disagree and strongly disagree with nominal values of 4, 3, 2 and 1, respectively was used to obtain a quantitative measure of the extent to which the identified roles of extension were perceived by the respondents as effective. The respondents' mean scores were obtained for each response option such that items with $\overline{X} \geq 2.50$ were regarded as effective role of extension, while items with $\overline{X} <$ 2.50 were regarded as not an effective role of extension in building adaptive capacities against climate change. Objectives 1 and 3 were analyzed using mean scores and standard deviations, while percentage scores were used to analyze Objective 2.

RESULTS AND DISCUSSION

Climate change effects/impacts experienced by farmers

Data in Table 1 shows that the most significant effects/ impacts of climate change experienced by farmers in the study area were intense weed growth ($\overline{X}$ = 4.52), incidence of pests and diseases ($\overline{X}$ = 4.28), soil erosion ($\overline{X}$ = 4.27), lack of portable water for human consumption and livestock use ($\overline{X}$ = 4.15), loss of vegetation/pastures ($\overline{X}$ = 4.10), distortion and destruction of wildlife ecosystems ($\overline{X}$ = 4.06), decrease in soil fertility ($\overline{X}$ = 4.06) and health related issues of climate change which can affect production for example hunger and famine, drudgery and stress from heat, etc ($\overline{X}$ = 4.01). Others include items 9 to 23 as shown in Table 1. On the other hand, respondents did not perceive items 24 to 27 as significant effects/impacts of climate change on agriculture. Data in Table 1 also shows that the standard deviations for all the issues considered were less than 1.5. This indicates that farmers' individual scores as regards their experience of climate change effects/ impacts on agriculture did not differ much from their mean scores. The high incidence of weeds in farmers' farms can be attributed to favourable climatic conditions that support weed growth in the area. This may include high moisture conditions experienced during the greater part of a farm year, poor agricultural practices and bad timing of weeding. Flooding and wind erosion may also contribute to intense weed growth because they help in weed dispersion especially to new areas where they have never existed before.

Farmers reported in all the FGDs conducted that they had to weed their farm plots more than three times if they were to obtain any meaningful harvest from the farm. This was against one or two times that they had previously weeded a typical farm plot some two decades

Table 1. Distribution of respondents based on effects of climate change on agriculture (n = 120).

Effects/impacts of climate change on agriculture	Mean scores	Std. deviation
Intense weed growth.	4.52*	0.85
Incidence of pests and diseases.	4.28*	1.17
Soil erosion.	4.27*	1.21
Lack of portable water supply for human consumption and livestock use.	4.15*	1.24
Loss of vegetation/pastures.	4.10*	1.36
Distortion and destruction of wildlife ecosystem.	4.06*	1.36
Decrease in soil fertility.	4.06*	1.30
Health related issues of climate change which can affect production for example hunger and famine, illness, drudgery and stress from heat, etc.	4.01*	1.35
Uncertainties in seasonal climate for example changes in rainfall pattern, temperature, etc.	3.97*	1.25
Decrease in yields of crops and animals.	3.96*	1.15
Storage losses in roots and tubers.	3.94*	1.20
Post harvest losses due to climate variability.	3.90*	1.38
Flooding..	3.83*	1.49
Premature ripening.	3.76*	1.39
High rainfall intensity.	3.58*	0.89
Heat from high temperature.	3.53*	1.24
Shortening of crop cycle.	3.52*	1.45
Drying of rivers, lakes and surface water bodies.	3.48*	1.43
Crop destruction in the field by rain and wind.	3.40*	1.12
Drought.	3.35*	1.42
Decrease in soil moistures.	3.32*	1.09
Loss of farm land/households to floods and erosion.	3.25*	1.35
Change in storage quality of fruits and vegetable..	3.16*	1.29
Changes in taste of fruits and vegetables.	2.83	1.27
Low rainfall intensity.	2.31	1.11
Increased salinity/water pollution due to climate variability.	2.29	1.33
Decrease in fish population due to salinity, water level, ocean currents, speed, etc.	1.87	1.49

Source: Field data 2008; *Significant effect/impacts; SD = Standard deviation; Cut off point = 3.00.

ago. They equally noted that there were new weed emergence in their farm plots that appeared to be recalcitrant. These symptoms really support the fact that climatic conditions have changed tremendously. It had been reported that heavier than normal rainfall will lead to increased growth of weeds, increase in erosion due to high rainfall and the indirect effect which include increase in pest and disease build up (MOE FRN, 2003). The implication of high incidence of weeds on farmlands is that it increases the cost of production through payments for labour or use of herbicides. Increased pest and disease infestation, which is also considered by farmers as significant in the area, was seen as an indirect effect of increased rainfall intensity, high humidity and warmer environment. Farmers noted in the FGDs that losses could be so high that nothing is harvested at times in a farm plot due to the incidence of pests and diseases. Efforts to control them could also be costly and frustrating, they observed. Research suggests that higher mean temperatures increase pest developmental rates and fecundity and frequency of outbreaks, and lead to expansion in the range of insect pests diseases and weed species (Spore, 2008). It was further noted that altered wind patterns are expected to change the spread of wind-borne pests and of bacteria and fungi that are crop disease agents. Climate change also has repercussions for human and livestock health, by shifting the distribution of certain disease vectors.

The World Health Organization established a clear link between heavy rainfalls over much of eastern Africa in 2008 and major outbreaks of malaria (Spore, 2008). These challenges will put serious strains on agricultural development and livelihood systems especially in the rural economy. High intensity of rainfall expectedly leads to flooding and causes serious soil erosion. In certain cases, farmers have lost their farmlands, crops, livestock, and even their homesteads to the devastating effects of flooding and erosion. In extreme cases, farmers have lost their lives to flooding and erosion. A farmer narrated how he lost all his property to flooding in one of the many occasions of heavy rainstorm that swept his building in the night. Similar stories abound in many parts of Nigeria and Africa as a whole. In 2007, many parts of Africa suffered severe drought while floods on much of the continent destroyed roads and buildings and wiped out millions of hectares of farmlands (Spore, 2008). These effects will not only affect the rural agricultural economy but the entire nation, continent and the world as a whole. Urgent action is therefore needed to mitigate and build resilience capacities of vulnerable people especially farmers so as to save the developing world from imminent danger. With increasing consequences of climate change, the availability of water for both human consumption and livestock use is put at great risk. This is because climate change sometimes leads to the drying up of rivers and streams, which human beings and livestock depend upon. The case could be worse in many African countries where the systems of portable water supply by governments have failed.

According to Ozor (2009), the drying up of streams and rivers in some communities due to climate change ultimately lead to their search for water in neighbouring communities with its attendant man hour losses, and the propensity to trigger conflicts and hardships on the people. Prolonged incidence of drought, urbanization, overgrazing and crop production lead to loss of vegetation/pasture that enhances carbon sequestration. This invariably distorts the wildlife and endangers some species. Again, global warming causes soil degradation and erosion as a result of increased precipitation in the areas thereby leading to decline in soil fertility. The aspects of health that will be exacerbated by climate change include: increased cases of cataracts (eye disease) especially in the northern parts of Nigeria due to low cloud cover and greater intensity of solar radiation; increased cases of malaria and typhoid due to increased rainfall and temperature in certain parts of the country; and increased cases of water-borne diseases such as cholera and dysentery due to urban flooding and improper disposal of wastes (Anyadike, 2009).

Adaptive measures adopted by respondents in cushioning the effects of climate change

The most significant adaptive measures adopted by farmers in coping with climate change effects in the study area are presented in Table 2. These include; use of resistant crop and animal varieties/species (95.43%), use of organic manure (94.16%), mixed farming (93.09%) and diversification in crop enterprise (91.81%). Other useful adaptive measures adopted include items 5 to 26. On the other hand, items 27 to 33 were not considered as significant adaptive measures in the area. The use of resistant crop and animal varieties/species provides useful adaptations and resilience to the effects of climate change. Such crops or animals are known to survive and complete their life cycles normally even when the environment will not allow others to thrive. The increased adoption of resistant varieties/species may not be unconnected with the fact that farmers encounter serious pest and disease infestations and weed growth, which have been recognized as the most significant effects of climate change in the area. It has already been reported that one of the many adaptations to climate change involves the use of resistant varieties such as early maturing varieties or drought resistant ones (Maddison, 2006). However, farmers noted that the issue of availability of the resistant varieties at the appropriate time is a problem. This has led them to once in a while fall back to local species that might fall victim of the environmental changes. It was reported also that the non-availability of seeds may be a significant impediment to adaptation (Maddison, 2006). This underpins the need for sustainable research and innovation in agriculture.

Increased application of organic manure was adopted by farmers not only to increase the fertility of the soil in the most sustainable way but also to increase the moisture content of the soil. Farmers noted that with the increase in temperature and evapo-transpiration resulting from climate change, measures to conserve moisture for crop use is necessary hence they preferred the use of organic manure. They equally pointed out that it is cheaper for them and most readily available unlike inorganic fertilizers. Mixed farming practice is adopted by farmers for many reasons such as to ensure food security, increased income, reduced incidence of pests and diseases, among others. In the FGDs conducted, almost all farmers confirmed that they engaged in mixed farming practice. In response to higher temperatures, farmers have resorted to using heat tolerant crop varieties, crop varieties with high water use efficiency, early maturing crop varieties, and increased crop and livestock farming, (mixed farming) (Benhin, 2006). Diversification in crop production is aimed to achieve similar objectives like mixed farming. The only difference is that it concentrates on diversifying crop production alone. It is worrying to note that some modern and scientific coping strategies such as irrigation schemes, increased mechanization of agricultural production and use of weather forecasts were not significant in the study area. Farmers noted that they do not know how to use the modern technologies and besides, most of them were not aware of the facilities.

Few farmers that reported that they were aware of the modern technologies observed that they are expensive and not accessible. This circumstance poses great challenge for agricultural extension.

The role of extension in agricultural adaptation to climate change

Data in Table 3 shows that farmers perceived most of the

Table 2. Distribution of respondents based on the adaptive measures adopted by respondents in cushioning the effects of climate change (n = 120).

Adaptive measures	Percentage score*
Use of resistant crop and animal varieties/species.	95.43
Use of organic manure.	94.16
Mixed farming.	93.09
Diversification in crop enterprise.	91.81
Changes in planting dates.	87.92
Changes in harvesting dates.	87.94
Processing of crops to reduce post harvest losses and enhance shelf life of crops.	87.25
Mulching/ use of cover crops.	86.81
Changes in the timing of land preparation activities.	86.30
Prompt weeding.	85.72
Water storage in ponds.	85.74
Construction of shelter for animals and human using palms, grasses and other non-conductors of heat.	83.51
Contour cropping across hill slopes.	81.32
Changes in planting depth of seeds and seedlings.	79.16
Afforestation/ reforestation initiatives.	76.94
Construction of drainage/dams.	74.45
Increased range land.	73.65
Contour and cut-off ditches.	73.61
Expansion of cultivated areas.	69.70
Adoption of zero or minimum tillage.	71.19
Road catchments.	68.17
Contour bund.	67.83
Flexibility in the location of watering points.	64.41
Cultivation of marginal lands.	63.37
Culling of infected animals.	62.23
Decrease in stocking rate of animals.	56.56
Half moon catchments.	51.62
Laws against deforestation activities.	49.81
Changing from production to marketing of agricultural products.	48.90
Abandonment of crop land or destruction of infected animals.	31.96
Lengthened fallow.	31.03
Use of weather forecasts.	28.67
Increased mechanization of agricultural production.	27.99
Irrigation schemes.	25.69

Source: Field data 2008; *Multiple responses; Cut off mark = 50%.

roles of extension identified in the study as effective in cushioning the effects of climate change. They include in order of importance the re-training of extension staff to acquire new knowledge and skills in climate risk management ($\bar{x}$= 3.93), setting up of emergency management units by extension agencies that will attend to victims of climate risks ($\bar{x}$= 3.91), dissemination of innovations on best practices and building resilience capacities of vulnerable people in climate risk management ($\bar{x}$= 3.90), and providing feedbacks to governments and interested agencies with situation reports on the causes of climate change, its effects, and the local knowledge and practices of the rural people ($\bar{x}$= 3.90). Other effective roles of extension include items 5 to 10. On the other hand, items 11 and 12 were not considered as effective roles of extension in agricultural adaptation to climate change. The standard deviations from the mean for all the perceived roles were less than 1.0, indicating that farmers' individual responses were not far apart from the mean. Climate change presents new challenges and threats to food security in most countries especially the developing ones. This demands that extension service brace up to the development by re-training its staff to acquire the capability (knowledge and skills) in

Table 3. Mean distribution of respondents' perception of the roles of extension in agricultural adaptation to climate change (n = 120).

Role of Extension	Mean score	Std. deviation
Re-training of extension staff to acquire the new knowledge and skills (capacity) in climate risk management.	3.93*	0.36
Setting up of emergency management unit by extension agencies that will attend to victims of climate risks.	3.91*	0.29
Dissemination of innovations on best practices and building resilience capacities of vulnerable people in climate risk management.	3.90*	0.40
Providing feedbacks to governments and interested agencies with situation reports on various causes of climate change, its effects, and the local knowledge and practices of the rural people.	3.90*	0.42
Use of demonstration methods in teaching farmers the measures used to mitigate or adapt to the effects of climate change.	3.84*	0.50
Organizing seminars, workshops, and field days to sensitize farmers and the public on climate risk management.	3.82*	0.49
Use of farmer-to-farmer extension strategy to promote awareness and adoption of best practices in climate risk management.	3.80*	0.48
Use of information communication technologies (ICTs) such as the internet, radio, television, media vans, leaflet, and posters etc, to create awareness on the climate change issues.	3.77*	0.45
Formation of Young Farmers Club (YFC) in schools to educate and encourage young farmers in learning about climate change issues with a view to reducing human causes and improving adaptation options.	3.70*	0.69
Use of farmer field schools (FFS) to promote faster learning by farmers on the measures used to mitigate and adapt to the effects of climate change.	3.37*	0.71
Denying farmers who indulge in poor agricultural practices that contribute to climate change such as bush burning access to extension services.	2.38	0.67
Use of law enforcement agents against persons that deliberately indulge in practices that contribute to climate change such as bush burning.	1.56	0.45

Source: Field data 2008; *Significant roles of extension; Cut off point = 2.5.

managing the risks that climate change poses especially in rural areas where the greater part of agricultural activities take place. The retraining program will mostly involve staff who are directly in contact with farmers and those that supervise and manage them. They will include extension agents, block extension agents, block extension supervisors, subject matter specialists, zonal extension officers, zonal managers, directors of extension services, directors of rural institutions development, and the programme managers of agricultural extension agencies such as the agricultural development programmes (ADPs).

According to Ozor (2009), there is need for change in roles and capacity in the extension system so as to accommodate the new dimensions brought about by climate change. In order to provide practical solutions to hazards and sudden uncertainties in agricultural production, there is need to establish emergency management units within every extension agency that will be trained to take charge of victims of disasters emanating from climate change risks such as flooding, erosion, acid rains, drought, submergence and poisoning from agricultural chemicals (insecticides, pesticides, herbicides, fertilizers, etc). The role of extension in emergency management is advocated because they work very closely and live with the rural farmers and can provide at least the best first aid assistance before the arrival of experts. It therefore makes it important for extension to be linked to the state emergency services, health workers, and the police. The primary role of extension is in the dissemination of innovations to targeted clientele ranging from farmers, pastoralists, fisher-folks, hunters, foresters, wine-tappers and other rural residents who depend on agriculture in one way or the other. Therefore, it behoves on extension to disseminate the best practices and innovations currently being developed by numerous research efforts across the globe on how to boost the adaptive capacity and resilience of vulnerable people to the effects of climate change. This is the only way to save the agricultural system that employs over 70% of Nigeria's population and contributes about 40% to the nations GDP.

Another significant role of extension in agricultural adaptation to climate change is its feedback role to government and other interested agencies on climate change issues. Extension agents live and work with the people in rural areas. This affords them the opportunity to be knowledgeable on issues that border on climate change in their areas of coverage. During the fortnightly meetings (FNTs) and block meetings (BMs), extension agents are expected to give situation reports on their duty stations. From this, government and non-governmental bodies can become aware of climate risk situations in the rural areas and then be able to render assistance, make policies, or execute programmes to manage the challenges identified by agents. In this circumstance, challenges such as flooding, drought, pest and diseases infestation, submergence by water and other catastrophic effects of climate change can be reported promptly. Similarly, local practices which have helped a particular individual or community to adapt to the effects of climate change can also be reported and advertised thereby creating the avenue for such practice(s) to be replicated and up scaled in other vulnerable areas. In a related study conducted in Australia, Cotching et al. (2009) recommended effective communication and feedback mechanisms in addressing systemic environmental impacts. Again, farmers perceive the use of demonstration methods as a significant role of extension in disseminating the coping and adaptive measures that could reduce climate change risks among vulnerable communities. It is common parlance that 'seeing is believing'.

Based on this, extension agents can use various extension teaching approaches including method demonstration, result demonstration, print media (for example, posters, leaflets, etc) and computer/ telecommunication media (for example internet, television, cinema, radio, computer, etc) to further inform and educate farmers on various issues of climate change. Farmers learn by doing and practices learnt during a demonstration session could lead to adoption of the technology. Other significant roles of extension in agricultural adaptation to climate change as perceived by farmers in the study area, border on the use of groups to achieve wide scale adoption of innovations. Areas identified include organizing seminars, conferences, field days and workshops; use of farmer field schools, use of farmer to farmer extension strategy; and the use of young farmers club. All these strategies centre on maximizing mass adoption of innovations through group approaches. According to Garforth (2005), the five common sets of arguments for focusing extension efforts on groups rather than individuals are for efficiency, effectiveness for learning and action, equity, demand-orientation and empowerment. Junge et al. (2009) emphasized the need for farmer education and training through capacity building. Using group efforts in communicating adaptive/resilience measures to climate change boosts understanding and knowledge on the issue as it achieves multiplier effects.

CONCLUSION AND RECOMMENDATIONS

The study sought to ascertain the roles of extension in agricultural adaptation to climate change in Enugu State, Nigeria. Specifically it examined the effects/impacts, adaptive measures and roles of extension in climate change adaptation. Results show that the most significant effect/impact of climate change in the area was intense weed growth, while the most significant adaptive measure adopted by respondents was the use of resistant crop and animal varieties/species. The key role of extension in agricultural adaptation to climate change was in retraining of extension staff to acquire the new knowledge and skills (capacity) in climate risk management. It is therefore recommended that extension should brace up to the new challenges posed by climate change by retraining its staff, mounting awareness programmes, and disseminating proven measures to boost the adaptive/resilience capacities of various stakeholders including farmers, foresters, pastoralists, hunters, etc to climate change in the area. These will of course reduce the anthropogenic causes and hence the effects/impacts of climate change. Also, there is need for increased research and innovation in agriculture to find out more sustainable ways of adaptation to climate change.

REFERENCES

Anyadike RNC (2009). Climate change and sustainable development in Nigeria: conceptual and empirical issues. *Enugu Forum Policy Paper 10*; African Institute for Applied Economics (AIAE); Enugu, Nigeria: 13-18.

Benhin JKA (2006). Climate change and South African agriculture. Centre for Environmental Economics and Policy in Africa (CEEPA), University of Pretoria, South Africa.

Cotching EW, Sheriff L, Kilpatrick S (2009). Integrating farm production and natural resource management in Tasmania, Australia. J. Agric. Educ. Ext. 15(3): 287-300.

Cury P, Shannon L (2004). Regime shifts in upwelling ecosystems: Observed changes and possible mechanisms in Northern and Southern Bengaels. Prog. Oceanogr., 60: 223 – 243.

ENADEP (2007). Enugu State Agricultural Development Programme (ENADEP) Annual Report, Enugu.

Enugu State Government Official Gazette (1997). Enugu State Official Gazette No. 25.

Ezike JO (1998). Delineation of old and new Enugu State. Bulletin of the Ministry of Works, Lands and Survey, Enugu State.

Garforth CJ (2005). Groups and organizations in extension. Comparative Extension Module – IDM005 (10). International and Rural Development Department, The University of Reading.

Intergovernmental Panel on Climate Change (IPCC) (2007). Impact, Adaptation and Vulnerability. Contribution of Working Group I of the Intergovernmental Panel on Climate Change to the Third Assessment Report of IPCC. Cambridge University Press. London.

Intergovernmental Panel on Climate Change (IPCC) (2001). Impact, Adaptation and Vulnerability. Contribution of Working Group II of the Intergovernmental Panel on Climate Change to the Third Assessment Report of IPCC. Cambridge University Press. London.

Junge B, Deji O, Abaidoo R, Chikoye D, Stahr K (2009). Farmers' adoption of soil conservation technologies: A case study of Osun State, Nigeria. J. Agric. Educ. Ext. 15(3): 257-274.

Maddison D (2006). The perception of and adaptation to climate change in Africa. Centre for Environmental Economics and Policy in Africa (CEEPA). Discussion Paper No. 10, CEEPA, University of Pretoria.

Ministry of Environment of the Federal Republic of Nigeria (2003). Nigeria's First National Communication under the United Nations Framework Convention on climate change. Abuja, Nigeria.

Nicholls RJ, Tol RSJ (2006). Impacts and Responses to Sea Level Rise: A Global Analysis of the GRES scenarios over the twenty-first century. Philosophical Trans. R. Soc. 364: 1073-1095.

Leeuwis C (2006). Communication for Rural Innovation: Rethinking Agricultural Extension; Third Edition. Blackwell Publishing. The Netherlands.

National Population Commission (NPC) (2006). Official Census Report. Abuja, Nigeria.

Osman-Elasha B (2007). Africa vulnerability. Tiem skills) in po 63: 7.

Ozor N (2009). Implications of Climate Change for National Development: The Way Forward. Debating Policy Options for National Development; Enugu Forum Policy Paper 10; African Institute for Applied Economics (AIAE); Enugu, Nigeria, pp. 25-42.

Pandey N (2006). Societal Adaptation to Abrupt Climate Change and Monsoon Variability: Implications for Sustainable Livelihoods of Rural Communities. Winrock International–India.

Spore (2008). Climate Change. A bi-monthly magazine of the Technical Centre for Agricultural and Rural Cooperation (CTA). Wageningen.

Thornton PK, Jones PG, Owiyo TM, Krusta RL, Herero M, Kristyanso P (2006). Mapping climate vulnerability and poverty in Africa. Report to the Department for Inter-national Development. International Livestock Research Institute, Nairobi, 200p.

United Nations Environmental Programme (UNEP) (2007). Global Environmental Outlook. Environment for Development. Kenya: 540p.

20

Appraisal of privatisation of agricultural extension services in Cameroon

Fonteh Athanasius Amungwa

Department of Sociology and Anthropology, Faculty of Social and Management Sciences, P. O. Box 63, University of Buea, Republic of Cameroon. E-mail: dramungwa@yahoo.com.

This paper examines the significance and impact of privatisation on the management and organisation of agricultural extension services in Cameroon. The focus is on how government action to implement new policies in agriculture is perceived by, and has affected farmers. The issues explored are the pressures for privatisation, the implementation of privatisation, influence on agricultural extension and farmers' responses to privatisation. Qualitative and quantitative research methods related to the Grounded Theory approach were used. 114 farmers, 26 extension agents and 14 senior extension officers were purposively sampled on the basis of their involvement in four agricultural extension projects in three regions, which typify different ecological and socio-cultural aspects of farming in Cameroon. The findings show that persistent economic constraints have increased the country's dependency on external sources of funding for agricultural extension programmes with pressure to reduce government spending, a high level of awareness among farmers with respect to new organisational changes in extension, and their willingness to accept innovations in their methods of working with support services that provide appropriate solutions to their farm problems. Privatisation of agricultural extension services has brought with it greater involvement with private companies and non-governmental organisations, and greater collaboration among them in delivering extension services to farmer clients. This increases the effectiveness and sustainability of the information systems available to farmers.

Key words: Agricultural extension and privatisation, liberalisation and organisational change.

INTRODUCTION

With the advent of political independence more than 40 years ago, most developing countries chose development models whereby the public sector controlled almost all aspects of the economy. In the agricultural sector, this included a domineering role of the state in the supply of physical inputs, credit, research and extension and marketing systems, either directly or through agricultural parastatals. In Cameroon, policy reforms in agricultural extension were particularly interventionist from the time of independence up to the end of the 1980s. This was a period during which production was partly in the hands of State-owned enterprises and there were numerous price controls and substantial input subsidies financed with the use of export taxes. An economic crisis which brought about radical changes in the management of agricultural extension services by the Ministry of Agriculture and Rural Development and that of Livestock, Fisheries and Animal Industries has characterized the period after the mid-1980s. The crisis resulted in major policy changes in the agricultural sector. Liberalization and privatisation policies were adopted aimed at disengaging the state from the production and commercialization of agricultural products. Between 1988 and 1992, the agricultural budget was slashed and subsidies for agricultural inputs like fertilizers, pesticides and insecticides were completely removed and farmers had to face the market price for these inputs. Furthermore, the state liquidated the National Produce Marketing Board, shifting respon-sibility for purchasing of the country's major export crops like coffee, cocoa and cotton from farmers and their international sale to the private sector. Thereafter, the government sought to attract private investment, particularly through privatization of some state-owned enterprises like the Cameroon Cotton Development Corporation and the Cameroon Development Corporation to increase production levels and rural incomes. The ratiotionale for privatisation was that the market performs better in the production and provision of certain goods and services. By shifting hitherto state-owned services to the private sector, it is assumed that resources would be allocated in a more judicious manner, leading to improved production and overall welfare. From the 1990s under the Rural Sector Development Strategy, prepared as a component of the Poverty Reduction Strategy Paper (PRSP), poverty, food insecurity, poor market integration, and unsustainable utilization of natural resources are identified as major challenges to rural sector growth. Agricultural extension thus represents a fundamental challenge to reaching the goals of Cameroon's PRSP of reducing poverty, increasing agricultural production, promoting food security, improving the quality of produce and ensuring agricultural and environmental sustainability. In the continuous search for more appropriate institutional arrangements for the delivery of agricultural extension services, questions are raised about whether certain public activities like agricultural extension could be performed more efficiently by the private sector (Gros, 1994). While it is generally agreed that privatisation leads to the delivery of more efficient and better quality of services and lowered government expenditure, it is necessary to emphasise the importance of situational analysis to avoid the temptation of applying formulae in developing countries simply because they have succeeded elsewhere. The purpose of this paper is to investigate the significance and contribution of privatisation to agricultural extension services in Cameroon and explore how government action to implement new policies in agriculture is perceived by, and has affected the actions of stakeholders actually engaged in farming and extension work. The pressures for privatisation, its implementation, influence on extension and farmers' responses in the Cameroon context are discussed. The specific objectives are to: define and describe the process of privatisation as this has occurred in Cameroon; identify the sources of pressures for privatisation; explore farmers' abilities and attitudes towards payment for advice; and the changes brought about in extension by priva-tisation. Agricultural extension privatisation as used in this paper refers to the different ways by which farmers can contribute towards the costs of running extension services. It is related to the view of information transfer in agriculture as a private good characterised by intellectual property, process licensing, and the use of paid con-sultants and differentiated production and marketing processes (Rivera and Cary, 1997). In a broad sense, privatisation of agricultural extension refers to various funding and service delivery arrangements that require farmers to contribute financially towards the costs of running the services in exchange for the advice they get to improve their productivity and income.

STATEMENT OF THE PROBLEM

The Cameroonian economy has been in crises since 1988 owing to changes in the international economic environment and poor internal management. The crises has resulted in acute balance-of-payments difficulties, which together with the need for external resource flows forced the government to adopt stringent Structural Adjustment policies[2] and stabilisation programmes to restore the economy on a growth path recommended by the Bretton Woods Institutions (Amin, 2008; IFAD, 1988). This programme partly involved a freeze in further employment into the civil service and a reduction of more than 60% in the salaries of state employees in 1993. This situation led to an increase in the rate of unemployment and a deterioration of living conditions, particularly in rural areas. Austerity measures adopted within these programmes included currency devaluation, reduction of public expenditure, and curtailment of imports demand.

The crisis has also led to the withdrawal of the state from production activities and its retreat from direct support to farmers, calling into question its role as the direct manager of agricultural extension services. The gradual disengagement of the State from agricultural activities after 1987 resulted in a substantial reduction in support for production by State-owned companies, the elimination of price and quality controls on coffee and cocoa, and liberalization of the marketing of these products as well as the inputs used to produce them (OECD and MINADER, 2006; FAO, 2006). The removal of subsidies and price controls, liberalisation of the economy, and limitation of the size of the public sector, have placed responsibilities on communities which had not been adequately equipped for the challenges. Involving the private sector in modernizing agricultural production and productivity is crucial, as demographic growth is sustained in Cameroon with the rate of 2.8% and the actual production of food crops is insufficient to cover the food needs of the population.

CONCEPTUAL FRAMEWORK

The privatisation of agricultural extension services is happening as an organisational response to pressures from an increasingly complex and rapidly changing environment (Lawrence and Lorsh, 1967). Privatisation is based on the assumption that there is relevant technology to disseminate because if there happens not to be any, a change in service provider can do nothing to increase the effectiveness of extension. Agricultural information has an economic value and is therefore an input in agricultural production. 'Relevant advice' has an attractive cost-benefit ratio and therefore should be saleable, in some shape or form (World Bank, 1994). Extension advice is increasingly being provided by commercial firms, banks, and private consultants, publishers of farm magazines and other computerised information services and data banks (Schwartz, 1992; Umali and Schwartz,

[2] The World Bank defines these programmes very broadly as policy reforms involving changes in relative prices and institutions designed to make the economy more efficient in using resources.

1994). Agro-input and food processing firms provide information to their clientele about the appro-priateness or range of uses of their products as part of their advertising strategy to expand or protect their market shares. Cameroon's adoption since 1990 of a more liberal agricultural policy has made it possible for private companies to be involved in fertiliser and seed supply, which were previously the exclusive domain of the public sector (MINAGRI, 1994). When farming becomes highly commercialised, the corresponding sup-port services will tend to become highly specific, providing adequate economic incentives for private sector involvement. Information on new technology can be considered a public good but as a certain level of tech-nology becomes widely accepted in agriculture, advice becomes a private good. At this stage, farmers would require more individually-tailored advice, and if it is of high quality, they should be willing to pay for it if the services provided are relevant to their needs The current trend in extension service provision is towards priva-tisation whereby farmers are expected to share the responsibility for this service and pay all or part of the costs (Van den Ban and Hawkins, 1996). This is happening because of budget deficits for the government to cover extension costs, and the need for extension agents to be accountable to farmers who contribute all or part of the costs of the services. Galal and Shirly (1994) observe that with sound investment policies and market-friendly practices, privatisation can provide substantial economic gains and increase a given country's GDP by an average of 2.5% with the divestiture of half of its public enterprise. The tendency for private extension to target mainly the more progressive farmers can be overcome by encouraging small farmers to contribute towards the cost of extension in 'non-cash' terms so that the accountability link is maintained. Extension services provided by the private sector, or even profit-oriented parastatals, can only upgrade the quality of overall extension support available to farmers (Harter, 1992; Wilson, 1991; Rolls, 1998). A conceptual orientation of agricultural extension services in which governments and other agencies like seed companies, implement dealers, fertiliser distributors, farmer co-operatives and school systems network in providing solutions to farmers' problems can only be advantageous (Rivera and Gustafson, 1991). Some Latin American countries have made conscious efforts to mobilise resources at national, regional or local levels and to contract private sector advisors, municipality staff or farmer organisations to deliver extension services (Zijp, 1998).

Privatisation offers governments the opportunity to restructure internally, reducing waste and seeking ways to improve their own capacity to identify problem areas and capitalise on opportunities for collaboration with the private sector, NGOs, and farmers' organisations (Rolls, 1998; 1998; Umali and Schwartz, 1994). It takes some of the financial pressure off the government, ensures financial sustainability and provides the basis for the estalishment of a more demand-driven responsive extension service. It involves increasing private sector participation in the delivery of services and promoting the adoption of various cost-recoveries, commercialisation, and other alternatives in increasing agricultural productivity (Antholt, 1994; Rivera and Cary, 1997). The participation of universities and agribusinesses, research and related service institutions in responding to farmers' demands tends to affect traditional perceptions of extension, contributing to install a demand-driven extension system. Key areas of intervention involving the promotion of innovative technology, capacity building and contractual relationships with local NGOs require better management of agricultural extension services to ensure that farmers get good advice from vendors of agricultural inputs and agribusinesses that transform and market products. The concepts, which now dominate the debate in agricultural extension, are effectiveness, efficiency and accountability.

MATERIALS AND METHODS

The methods used in collecting data for this study were related to the Grounded Theory approach (Strauss and Corbin, 1990). The approach lays emphasis on the active role and interrelationships of persons, groups and organisations in developing action strategies in the process of socio-economic change. The Grounded theory approach is qualitative and the investigation is conducted in an autonomous research area with its own social structure, boundaries and history. Questionnaires were administered to farmers, extension agents and senior extension officers sampled purposively from 4 extension organisations operating in 3 regions, covering 2 broad, distinct cultural and agro-ecological zones. These were, for government organisations: the Integrated Rural Development Poles Project (RDPP), Bafut, in the Northwest Region and the Société Camerounaise de Palmeraires (SOCAPALM), and Edea, in the Littoral Region. For non-governmental organisations: the African Centre for Social and Economic Development (INADES-Formation), Bamenda, in the Northwest Region; and the Centre de Dévelopement Autocentre (CeDAC*)*, Sangmelima, in the South Region.

Primary data was obtained from 114 key farmer informants selected on the basis of membership in farmers' group, involvement in the group's activities and their ability to read and write (32 from RDPP, 14 from INADES-Formation, 31 from CeDAC, and 37 from SOCAPALM). 26 extension agents responded to questionnaires. These included 13 from RDPP, 4 from INADES-Formation, 3 from CeDAC, and 6 from SOCAPALM. 14 senior extension officers, including 1 from each of the 4 case studies and 10 from the Ministry of Agriculture also responded to the questionnaires. In addition, 450 farmers and 52 members of staff of extension organisations were reached through focus group discussions and personal interviews. The data collected from questionnaires were analysed using the statistical programme for social scientists (SPSS). Although there were ecological and socio-cultural differences in the 4 case studies, the responses of farmers of the different projects showed no substantial variations.

RESULTS AND DISCUSSION

Pressures for privatization

The respondents' understanding of privatisation showed

Table 1. Involvement of private companies in extension (% responses).

Category of respondents	Yes	No
Farmers (N = 114)	37.2	62.8
Extension agents (N = 26)	59.1	40.9
Senior extension officers (N = 14)	100.0	0

that a majority perceived the concept as the selling of public-owned organisations to the private sector and the encouragement of private-sector management. Most of them identified the early 1990s as the period when the policy of privatisation was adopted in Cameroon. In effect, Cameroon's reform of state-owned enterprises was actually initiated in May 1989 under the economic adjustment programme and formalised in the Declaration on Economic Policy Recovery. Privatisation was encouraged through the sale of some state-owned organisations, retrenchments and the use of severance packages, performance contracts, training and advisory programmes and the suppression of state subventions to enterprises that performed poorly.

The awareness rate with respect to private companies' involvement in providing extension services was 100% for senior extension officers, 59.1% for the extension agents and 37.2% for the farmers. Considerable diversity now exists among the extension service providers. The differences were attributed to educational levels, the degree of exposure to modern sources of information and involvement in extension decision-making. The relatively high proportion of farmers (62.8%) and extension agents (40.9%) who were unaware of the involvement of private companies in extension, suggests an information gap between senior officials, the field staff and farmers. Farmers were unable to make a difference because of the existence of partnerships between private companies and Cameroon's National Agricultural Extension Programme which utilises public extension agents for testing the technologies of private organisations among farmers Table 1.

The types of extension activities undertaken by private companies as reported by respondents included: advice on farm-related matters, the sale of seeds and seedlings, fertilisers and agro-chemicals, education and training. Private input supply companies provided information with their products thus acting as a marketing device. Some of the private companies in Cameroon like the Farmers' House Ltd. (a seed firm) and PAMOL Estates Ltd. (an oil palm grove company), have introduced new technology for producing high quality maize, oil palms and horticultural products. Extension advice was provided not only to increase the farmers' product quality, but also as a way to promote partnership to ensure that the farmers sold to the companies that provided the advice.

Seventy-five percent of the farmers, 100% of the extension agents and 72.7% of the senior extension officers agreed that the World Bank, the IMF and the European Union were key promoters of privatisation in Cameroon. The methods used by the government to encourage privatisation, as identified by most of the three categories of respondents included the sale of poorly-managed state-owned companies, retrenchments and use of performance contracts, severance pay, sup-pression of subventions, training and advisory programmes to facilitate small and medium private initiatives. These initiatives were backed by political speeches to convince farmers and the general public about the prospects offered by privatisation.

Implementation of privatization

The study investigated perceptions regarding payment for extension advice by farmers during the past two years, the amounts involved, and respondents' attitudes to such commercialisation of services.

Table 2 shows clearly, that a large percentage of the respondents did not affirm farmers' payment for extension advice. It was still not widely recognised that privatisation leads to payment for extension advice. Smaller farmers are more likely than bigger farmers to be deterred from using a service for which they have to pay. Interestingly, 30% of the senior extension officers stated that farmers had paid for extension advice. This is explained by the fact that the State's policy of liberasation and disengagement from direct involvement in agricultural production activities, created an opportunity for some extension staff in active government employment to provide fee-for-service extension as independent consultants to supplement their incomes. Large-scale farmers capable of producing a large marketable surplus can afford to pay for extension advice, but to achieve this in traditional agriculture, farming must be modernised and made profitable. Oil palm production attracted a new class of farmers from the senior ranks of state and private sector employees who could afford to negotiate and pay a fee for extension services. Farmers of SOCAPALM and CeDAC's oil palm schemes and members of Common Initiative Groups associated with INADES-Formation and RDPP carrying out selective income generating activities could also afford to pay for extension services through contract arrangements.

A number of farm-level and organisational changes attributed to privatisation were recognised. The attitude of farmers towards seeking advice from diverse organisations, hiring extension agents with the hope of making gains was generally favourable and demonstrates a posi-

Table 2. Farmers' payment for extension advice (% responses).

Category of respondents	Yes	No
Farmers (N = 114)	5.3	94.7
Extension agents (N = 26)	13.6	86.4
Senior extension officers (N = 14)	30.0	70.0

Table 3. Payment for farm advice during past 2 years (% responses).

Amount	Farmers (N = 6)	Extension agents (N = 4)
Less than CFAF 10,000	0	33.3
CFAF 10,000 to 15,000	33.3	0
More than CFAF 15,000	66.7	66.7
Total	100	100

tive attitude towards the eventual emergence of consultancy services in extension. Demand for paid agricultural extension services depends upon the type of farmer and the expected benefits from investment in new information and technology. Market competition associated with highly marketable products will enhance effective demand for new information and technology. Large-scale farmers can spread the cost for extension information, resulting in lower per unit cost of extension thus making the service affordable. Small farmers will incur higher per unit cost of extension service, making it less profitable to pay for services. Unless such farmers have the desire and capacity to adopt innovations and start producing a larger surplus for the market, they will have limited incentive to pay for extension services.

Farmers in this study gave the impression that the involvement of diverse organisations in extension would enable them to seek advice from best sources that enable them to make gains. For privatised extension services to be more effective and sustainable, the economic, cultural, social and political environment must be favourable. The government has to provide essential infrastructure, enhance macroeconomic stability, a functioning legal system and competitive markets.

Among the 6 farmers and 4 extension agents who understood that payment was involved in privatisation of extension services (Table 3), 66.7% of both the farmers and extension agents affirmed the payment of more than CFAF 15,000 for farm advice during the past two years. Fee-based extension services to generate cost recovery appeared to be acceptable to this group of respondents. Demand for extension services would be influenced by factors such as the agricultural production cycle, market trends and the balance sheet of the farmer. The Cameroonian small farmer generally gets an income of between CFAF 150,000 and CFAF 300,000 bi-annually. CFAF 15,000 paid for farm advice in 2 years gives an annual average of CFAF 7,500 which is about 2.5% of the farmer's annual income. When self- consumption of some of the farm produce is considered, farmers evidently believe that they can afford to spend this percentage for extension services with the conviction that such investment will yield profit.

During the late 1990s, the privatisation of some of Cameroon's agricultural companies including SOCAPALM was an on-going process. Discussions with farmers and extension agents involved in the company's out-growers' scheme revealed uncertainty about the implications of privatisation and some of the farmers even feared losing their farms to foreign interests in the process.

Fifty-six percent of farmers, 66.7% of extension agents and 88.9% of senior extension officers were in favour that farmers' groups should hire the services of extension consultants. This finding demonstrates a positive attitude towards the eventual emergence of a 'fee-for-service' extension or consultancy services in Cameroon. However, 43.8% of the farmers and 33.3% of the extension agents did not see a need for hiring consultants, perhaps because of the severe financial difficulties of the state and farmers. A combination of public, private and voluntary extension efforts would be required to serve diverse target farmer populations. What matters most is to ensure that farmers have access to appropriate technology, information and advice through the most cost-effective approaches possible. Above all, the coexistence of several extension systems and approaches requires better co-ordination to fit diverse clientele rather than centralising (to create uniformity) and standardising one approach.

The willingness and ability to pay for extension advice may differ between the subsistence and commercial farmers. Subsistence crops (roots and tubers, cereals, fruits and vegetables) will require continued support from public extension services now and in the future. The implication of a change from providing agricultural advice free of charge to payment for advice is that the demand for paid services will come mainly from commercialised farming operations, particularly from large-scale farmers. Economic constraints make it difficult to deliver effective services to all farmers through direct contact using public extension services and serious thought needs to be given to the adoption of flexible approaches that are affordable in the short term and sustainable in the long run.

This study showed a higher proportion of farmers and extension agents in favour of payment for advisory services. Farmers in this category expected to get more regular extension advice upon payment, to improve their skills and increase farm production and income. Some of the extension agents argued that they do not receive much economic benefits from their work and farmers with the means should pay for their services. Respondents who differed on this issue argued that extension agents were state employees and it would be unfair to expect poor farmers to pay for advisory services. The perceptions of senior extension officers in the study were split between agreement (50%) and disagreement (50%) suggesting that farmers will be more selective when seeking advice for which payment is made.

These trends show that public sector extension agencies should not be the sole providers of information and advice to farmers. The private sector has a comparative advantage in areas where farmers demand a level of specialised services, generally for cash crops, not provided by public sector extension. Such services could have a spill-over effect on other crops if farmers improve their general farming skills, and apply the knowledge to the whole enterprise. Involvement of the private sector in providing extension services in some speciality areas is not sufficient argument for not retaining public extension services, particularly for crops and farmers who might not be reached by NGOs and the private sector. Public policy needs to encourage private sector extension wherever such initiatives are forthcoming without neglecting small farmers (mostly women) who, in resource-poor regions, cannot afford to pay for advice. Depending on the type of agricultural activity and level of wealth, farmers who demand personalised services should pay in return for good quality services while appropriate measures are taken to ensure a legal framework for such transactions.

With the government's commitment to the privatisation of public and para-public enterprises and the overhaul of the financial system (banks and insurance), many Cameroonians have lost their jobs. In SOCAPALM alone, 119 of the 2,266 workers lost their jobs (personal communication with General Manager). Two successive salary cuts in the first half of the 1990s and the effects of devaluation of the CFA Franc in 1994 by 50% have aggravated the poverty situation of state employees including agricultural extension staff who would like to supplement their incomes in any possible way including consultancies.

Influence of privatization on extension

Investigation of the influence of privatisation also focused on aspects of cost recovery. This provides a new source of financing extension services - the beneficiaries themselves. The degrees to which cost recovery pro-grammes are applied vary from one country to another (Bunney and Bowcutt, 1991). The results of this study indicated that the influence of privatisation on extension was still partial. Eighty-eight percent of the farmers, 80% of the senior extension officers and 77.3% of the extension agents reported that cost recovery was not being practiced. However, probing revealed some farmers and extension agents in favour of making financial contributions towards the running of agricultural extension services as long as these made their work to be profitable. Extension agents and farmers appear to be accepting the principle of payment for advice when it was requested. Senior extension staffs appear to be less supportive of the trend.

With the application of cost recovery measures, the more proficient farmers are more likely to pay for individual extension services that have a direct bearing on their income than the small farmers who generally have less capacity to pay for services. Cost- recovery implies that a choice of provider of extension services also exists and therefore policy makers and extension agents should be prepared to make the necessary attitude adjustments. These may include acquiring marketing skills and adopting a business orientation by the extension agents.

Farmers who affirmed that cost-recovery measures were being applied indicated costs incurred on fertilisers, farm chemicals, seeds, and private extension charges, indirect tax on production, and transportation fares to provide advice. Direct financial contributions were also made for participation in training programmes and the acquisition of printed extension material; extension agents mentioned fees, their transport to and from the venues of contact with farmers and foodstuff from farmers. Under SOCAPALM's contractual scheme with farmers, the corporation provided credit, inputs and technical advice. The costs of some of these services were recovered through the purchase of palm produce from the farmers. Initiatives of the Ministry of Agriculture and Rural Development to sub-contract certain extension activities to private partners, provided other opportunities for cost-sharing, particularly in respect of extension efforts for high value export crops.

Farmers associated with INADES-Formation and CeDAC's extension programmes, gave evidence that these organisations applied cost-recovery measures for delivering agricultural extension services. The market gardening and oil palm activities of farmers in Babanki Tungo and Sangmelima respectively, provided them with income for the purpose. Cost-recovery instils a sense of financial discipline and could be most efficiently achieved through indirect methods associated with the market economy, such as market fees. Extension information and advice are marketable commodities of individual benefit for which it is appropriate for the user to pay depending on the market demand and profitability of the commodity involved. However, public extension policy should ensure that small farmers have free access to information and advice until they can be well organised to contribute towards the cost of services provided. The operational implications of cost-recovery schemes, like defining the criteria upon which charges are made and actually collecting the payments and training extension agents in new skills, particularly in marketing techniques, needs to be considered by policy makers.

Farmers' responses to privatization

Respondents' perceptions regarding the degree of influence of privatisation on extension work were rated on

Table 4. Degree of influence of privatisation on extension work (% responses).

Category of respondents	No influence	Very little influence	Don't know	Some influence	Total influence
Farmers (N = 114)	17.4	32.1	15.6	22.0	12.8
Extension Agents (N = 26)	4.5	22.7	13.6	45.5	13.6
Senior Extension Officers (N = 14)	20.0	0	20.0	50.0	10.0

a five-point scale. Answers under 'no influence' and 'very little' influence were considered as no influence while 'some influence and 'total influence were considered as influence.

The results in Table 4 show 34.8% of the farmers, 60% of the senior extension officers and 59.1% of the extension agents agreeing that privatisation had influence on extension work. The proportion of farmers, extension agents and senior extension officers who perceived that privatisation had 'total influence' on extension work was closely related but contrasted with the proportion which thought it had 'no influence'. It is noticeable that 20% of senior extension officers and 13.6% of extension agents indicated that they did not know about the influence of privatisation on extension work, perhaps because of constraints of communication and the sharing of information on policy decisions and the need for concrete evidence. Senior extension officers revealed some scepticism about privatisation because of the poverty of a majority of farmers and the uncertainty of being faced with competition from the private sector. In 1996 the public sector and private organisations were initiating for the first time, a number of partnerships in extension and some of the senior public extension officials could not imagine the influence this would have on extension work. Changes in agricultural policy emphasising professionalism, development of private sector linkages between farmers and agricultural merchants and suppliers, NGOs and other stakeholders affected the attitude of public extension staff and farmers. Privatisation is contributing to a changing structure of the agricultural extension sector especially with regard to diversification in the supply of services, cost-sharing, input distribution and competitive marketing. However, privatisation was criticised as a major cause of retrenchments, reduced prices for agricultural commodities, poverty and the proliferation of traders in the agricultural inputs sector.

Eighty-six percent of the senior extension officers, 68.2% of the extension agents and 55.7% of the farmers agreed that privatisation leads to organisational changes in extension. The changes included: the involvement of NGOs and private companies in providing advisory services, better management and availability of more farm inputs, improved crop varieties but also, unemployment in some cases Table 5.

Table 5. Privatisation leads to changes in extension organisation (% responses).

Category of respondents	Yes	No
Farmers (N = 114)	55.7	44.3
Extension agents (N = 26)	68.2	31.8
Senior extension officers (N = 14)	85.7	14.3

The current change in the organisation of extension services was remarkable compared to the past, when the state subsidised farm inputs, and controlled their distribution. Liberalisation of the fertiliser sector had attracted some businessmen who imported and sold at very high prices, paying little attention to technical guidance of farmers. Senior extension officers saw the changes in terms of the re-organisation of rural communities and granting of more responsibilities to the farmers. Successful implementation of privatisation in extension requires not only the support of senior extension officers but also political initiative, financial and institutional support, investment promotion and capacity building.

IMPLICATIONS AND CONCLUSION

Privatisation was perceived as one aspect of reform of state-owned enterprises initiated under the economic adjustment programme aimed at reducing state spending. Private companies were involved in providing advisory services, though public sector funding and guidance remained essential to provide an enabling environment for all the stakeholders. Privatisation and cost-recovery programmes were introduced as a measure to relieve the financial pressure on the state, and also under external pressure from donors. The level of awareness among farmers about new organisational changes including privatisation and hence, the effects of government policy at the local level was high. Interestingly, the farmers showed a willingness to pay for extension services that were regular and relevant for improvement of their skills, farm production and income. A shift from receiving free advisory services to paying for such services implies that commercially-oriented farmers will be most capable of paying than subsistence farmers.

However, efficient farmer organisations may provide the opportunity for small farmers to co-operate and take advantage of the economies of scale with respect to payment for extension services. New government policies had brought about greater participation of private companies and non-governmental organisations in extension service delivery to farmer clients. This trend is increasing the effectiveness and sustainability of the information systems available to farmers, implying that farmer needs and demands will be reflected through market mechanisms including prices for information. Farmer attitudes would need adjustment from traditionally receiving free information and advice to payment for the services and therefore serious thought has to be given to ways of expanding their income-earning ability. The existence of several service providers implies that farmers will become more selective in seeking the advice they require with the possibility of influencing the direction of extension services and the type of information they need. Extension privatisation also implies that there will be a concentration on farming activities of direct benefit to farmers, with short-term goals, and a tendency to neglect long-term issues. These changes require that extension services be structured in ways that permit rapid interventions for the delivery of up-to-date and reliable advice at the farmers' requests, while ensuring that poor farmers are not left out. The training for extension staff will require adjustments in the level of specialisation to enable advisers cope with field problems and the changing needs of farmers. The conclusion is that privatisation is becoming a widely accepted organisational change and response that can offer alternative opportunities for efficiency, effectiveness and sustainability in extension service delivery. However, the economic efficiency which privatisation is supposed to bring about will depend more on how interested the new service providers and farmers perceive each other in the context of liberalisation.

REFERENCES

Amin AA (2008). Developing a Sustainable Economy in Cameroon. Council for the development of Social Science Research in Africa, Dakar.

Antholt CH (1994). Getting Ready for the twenty-first century. Technical change and institutional modernization in agriculture. World Bank paper No. 217, Asia Technical Dept. Series. Washington, D.C: World Bank.

Bakuzakundi M (1997). Evaluation de l'impact du Programme National de Vulgarisation Agricole par les bénéficiaires: Rapport final. Yaoundé: Ministry of Agriculture.

Bunney PMG, Bowcutt DE (1991). Making a business of an extension service. Agric. Prog. 66: 36-43.

Food and Agriculture Organization, FAO (2006). Food and Agriculture Indicators: Cameroon, November.

Galal A, Shirly M (1994). Does privatisation deliver? Highlights from a World Bank Conference. An EDI Study, Washington, DC: World Bank 2: 3.

Gros JC (1994). 'Of cattle, farmers, veterinarians and the World Bank: The political economy of veterinary services privatisation in Cameroon'. Pub. Admin. Dev. 14(1): 37-51.

Harter D (1992). Commercialization in Britain. Inperpaks Interchange, 9(1): 5-6.

International Fund for Agricultural Development, IFAD (1988). Structural Adjustment and the Smallholders in Sub-Saharan Africa: an IFAD Perspective. Paper for the JCGP high level meeting, New York.

Lawrence PR, Lorsh JW (1967). Organisation and Environment. Managing Differentiation and Integration. Boston, Massachusetts: Harvard Business School Press.

MINAGRI (Ministry of Agriculture), Republic of Cameroon (1994). Lettre de Politique Agricole, Avril.

OECD (Organization of Econonomic Cooperation and Development) and Ministry of Agriculture and Rural Development, MINADER (2006). Analysis of the interaction between macroeconomic policies and agricultural development, Yaoundé.

Rivera WM, Gustafson DJ (1991). Agricultural extension: Worldwide institutional evolution and forces for change. London: Elsevier.

Rivera WM, Cary JW (1997). Privatising agricultural extension. In B, E. Swanson, R. P. Bentz and A. J. Sofranko (Eds.), Improving agricultural extension: A reference manual . Rome: FAO. pp.203-211.

Rolls MJ (1998). The study of information transfer in agriculture. In Agriculture Progress pp. 24-28.

Schwartz L (1992). 'Private Technology Transfer in Sub-Saharan Africa', Washington, DC: Tropical Research and Development, Inc.

Strauss A, Corbin J (1990) Basics of qualitative research: Grounded Theory Procedures and Techniques. Newbury Park, Sage Publications.

Tchoungi, R. Steve Gartland, J. A. Mope Simo, Fondo Sikod, Augustin Youmbi, Michel Ndjatsana, James Mimpenny (1994). Structural Adjustment and Sustainable Development in Cameroon. A WWF for Nature Study, London: ODI.

Umali DL, Schwartz L (1994). Public and private agricultural extension: Beyond traditional frontiers. Discussion paper Washington, DC: World Bank Publication p. 236.

Van den Ban AW, Hawkins HS (1996). Agricultural Extension (2nd ed.), Oxford: Blackwell Science Ltd.

Wilson M (1991). Reducing the costs of public extension services: Initiatives in Latin America. In W.M. Rivera & D.J. (Eds.), Agricultural extension: Worldwide institutional evolution and forces for change. Amsterdam: Elsevier.

World Bank (1994). Privatisation: The lessons of experience. Country Economics.

Zijp W (1998). 'Extension: Empowerment through Communication. In Ian Wallace (Ed.) Rural Knowledge Systems for the 21st Century. AERDD, The University of Reading pp. 15-49.

Farmers training effectiveness in terms of changes in knowledge and attitude: The case of Holeta, Melkassa and Debre zeit Agricultural Research Centres, Ethiopia

Tsion Tesfaye[1]*, Ranjan S. Karippai[2] and Teklu Tesfaye[3]

[1]University of Swaziland, PO Luyengo, Hill Street, LA-8, Swaziland.
[2]Haramaya University, College of Agriculture, Department of Rural Development and Agricultural Extension, P. O. Box 138, Dire Dawa, Ethiopia.
[3]Ethiopian Institute of Agricultural Research, Research and Extension Division, P. O. Box 1234, Addis Ababa, Ethiopia.

Several agricultural technologies and high yielding varieties have been disseminated from the research centres operating under the umbrella of Ethiopian Institute of Agricultural Research such as: Holeta, Debre Zeit and Melkassa Agricultural Research centres. While the core functions of Ethiopian Institute of Agricultural Research (EIAR) are technology supply, popularization, national coordination and capacity building and policy development, the research and extension divisions of EIAR are responsible for transfer of technologies that are being developed in the respective research centres to farmers and other functionalities through training. There is, therefore, an alarming need to improve agricultural extension activities with the involved farmers through training. Therefore, the objective of this study was to the effectiveness of farmers' training in improving their knowledge and attitudes. One woreda was purposively selected from areas where each one of the three research centres is offering training to the farmers. Equal numbers of respondents of 40 trained and 40 untrained farmers from each of the three research centres were selected for the study. Quantitative data were collected from the respondents. Structured interview schedule was used for collecting the essential quantitative data from the sampled trained and untrained respondents. The quantitative data were analyzed using descriptive statistical tools and also chi-square t-test were employed. The major output of the study indicate that training offered by the three agricultural research centre significantly improved knowledge of potato, onion and durum wheat extension packages, attitude of farmers and level of practice of farmers compared to those of untrained sample farmers.

Key words: Farmers training, effectiveness, Ethiopia, research, onion, durum wheat, potato.

INTRODUCTION

Ethiopia is a country with a land mass 1.1 million km^2, an estimated population of about 77 million (Tsedeke, 2007). Agriculture is the most important enterprise, providing employment for more than 85% of the country's population and accounting for more than 40% of the total GDP and 90% of export earnings (Zeleke, 2000; Tsedeke, 2007). According to FAO data, close to 32 million ha of the total area is agricultural land but an average of approximately 12 million ha of this potential agricultural area is cultivated annually (CSA, 2006).

The country is agro-ecological, diverse and therefore produces wide range of crops and animals. Major crops include: cereals, roots and tubers, pulses, oilseeds, vegetables, fruits and cash crops such as: Coffee, cotton, tea, sugarcane and tobacco. This country ranks among the highest in Africa in its livestock resources. Cattle, sheep, goats, donkeys, chicken, camels, and honeybee are the major livestock species. The country also has rich water bodies and huge aquatic resources associated with

*Corresponding author. E-mail: tsdanatan@yahoo.com.

them (Tsedeke, 2007). Among several institutions in the country, Ethiopian

Institute of Agricultural Research (EIAR) is engaged in agricultural research and extension activities covering all activities associated with crops and livestock production. The vision of EIAR is to see that all Ethiopians engaged in agriculture, agro-pastoralist, and all agriculture-related business become beneficiaries of improved and appropriate agricultural technologies (Tsedeke, 2007). The mission is to conduct research that will provide improved and appropriate agricultural technologies that will contribute to increased agricultural productivity and nutrition quality, sustainable food security, economic development, and conservation of the integrity of natural resources and the environment. EIAR's role would also include finding innovative and effective means of technology dissemination, in collaboration with its partners in extension, higher learning institutions, advanced research institutes, International Agricultural Research Centres, Non-Governmental Organizations, and Community-Based Organizations (Tsedeke, 2007).

Several agricultural technologies and high yielding varieties were released from the research centres operating under the umbrella of EIAR such as: Holeta, Debre Zeit and Melkassa Agricultural Research centres. While the core functions of EIAR are technology supply, popularization, national coordination, capacity building and policy development, the research and extension units of EIAR are responsible for transfer of technologies that are being developed in the respective research centres to farmers and other functionalities through training.

The research and extension department is charged with the responsibility to support these needs through effective farmers' training. Thus, quite significant amount of time and money has been spent on agricultural technologies dissemination/transfer through farmers' and development agents' training programs which have been organized by each Agricultural Research centre of EIAR. To have a clear assessment of these efforts, the evaluation of training is also an important part in the training process cycle.

In evaluating an extension training program, one needs to consider that most training activities exist in a larger context of projects, programs, and plans. Raab et al. (1987) define training evaluation as a systematic process of collecting information for and about a training activity which can then be used for guiding decision making and for assessing the relevance and effectiveness of various training components. The objective of the study is to study the effectiveness of farmers' training in improving their knowledge and attitudes on the selected packages.

RESEARCH METHODOLOGY

A purposive sampling design was followed for the selection of the study research centres namely Holeta Agricultural Research centre, Debre Ziet Agricultural Research centre and Melkassa Agricultural Research centre. This judgmental sampling technique was followed by the investigator due to the reason that these three agricultural research centres were the most important ones currently involved on technology generation and dissemination under EIAR system. Around each research centre, the study was undertaken in one Woreda, which was selected purposively from where the Agricultural Research centres had been executing training for the past three years to farmers. This judgmental sampling method was chosen based on the preliminary study undertaken by the investigator. Woreda welmera, Ada'a woreda and Adama woreda were selected for this study from the woredas where Holeta, Debre Zeit and Melkassa Agricultural Research centres had been offering training for the last three years, respectively. The woredas which were sufficiently close to the Agricultural centres to enable a series of visits to selected farmers by the researchers over a limited period of time was another criterion.

Holeta Agricultural Research centre offered training for farmers from different woredas on agricultural technologies related to barley, wheat, tef, faba bean, field pea, chickpea, linseed and potato during the last three years. The number of farmers who accessed improved agricultural technologies associated with different crops was found to be higher in Welmera woreda when compared with the other woredas. Therefore, training of potato package being the important one was chosen in the case of Holeta Agricultural Research centre to measure training effectiveness in this study. The reason for this was also attributed to the fact that Holeta Agricultural Research centre was coordinating national potato research and extension programs and actively involved on dissemination of potato packages nation wide.

Similarly, Debre Zeit Agricultural Centre had been offering training on improved agricultural technology packages on several crop and livestock related areas of development. The major technology dissemination areas were associated with durum wheat, chickpea, lentils, tef, and poultry and beef production. Further, durum wheat agricultural technology adoptions as well as disseminations were being coordinated nationally by Debre Ziet Agricultural Research Centre. Thus, training of durum wheat package was purposively chosen to assess training effectiveness study in the case of Debre Zeit Agricultural Research Centre. Among the Woredas, Ada'a woreda was found to be highly associated with the dissemination of durum wheat package.

On the other hand, Melkassa Agricultural Research Centre was highly engaged on adoption of onion production technology and nationally coordinating onion research. Therefore, the onion training program was purposively selected for this centre in order to study the effectiveness of the trainings that were being offered to the farmers by the centre. Adama woreda was selected for the study due to the reason that the highest frequency of farmers had training from Melkassa Agricultural Research Centre on onion package.

Holeta Agricultural Research Centre is located at a distance of 45 km from the capital city of the country, Addis Ababa. The centre is located at $8^{0}30'$E latitude and $9^{0}00'$N longitude with 2400 m altitude. Major soil type of the area was nitosols and vertisols. The Welmera woreda consists of 61% and 39% dega and weyna dega, respectively. Figure 1 and 2.

The study related to Debre Zeit Research Centre was contacted at Ada'a woreda (around Debre Zeit town). Ada'a woreda contains 3, 3 and 94% dega, weyna dega and kola (hot climate), respectively. The area is found 47 km from Addis Ababa with an average geographic coordinate of 8°44'N latitude and 039°01.5'E longitude and an average altitude of 1900 m above sea level. The study area consists of almost entirely of Alfisol/Mollisol and Vertisols with high clay content.

The Melkassa Agricultural Research Centre (MARC) is found near Awash Melkassa ($8^{0}24'$N latitude and $39^{0}12'$E longitude) that is 17 km southeast of Nazareth town and 117 km away from Addis Ababa. The area is situated at an altitude of 1550 m als. The soil of MARC farm had a dominantly loam and clay loam texture.

The study related to Melkassa Agricultural Research Centre was

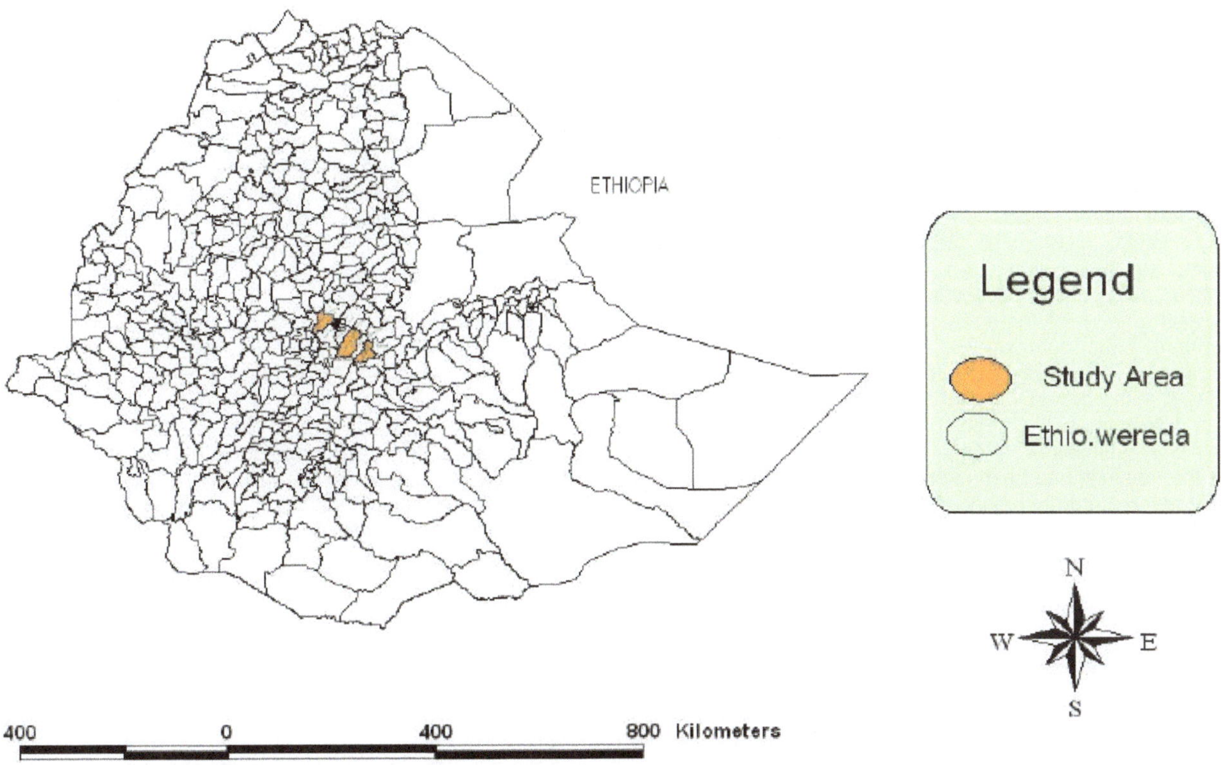

Figure 1. Location of study *woredas* in Ethiopia.

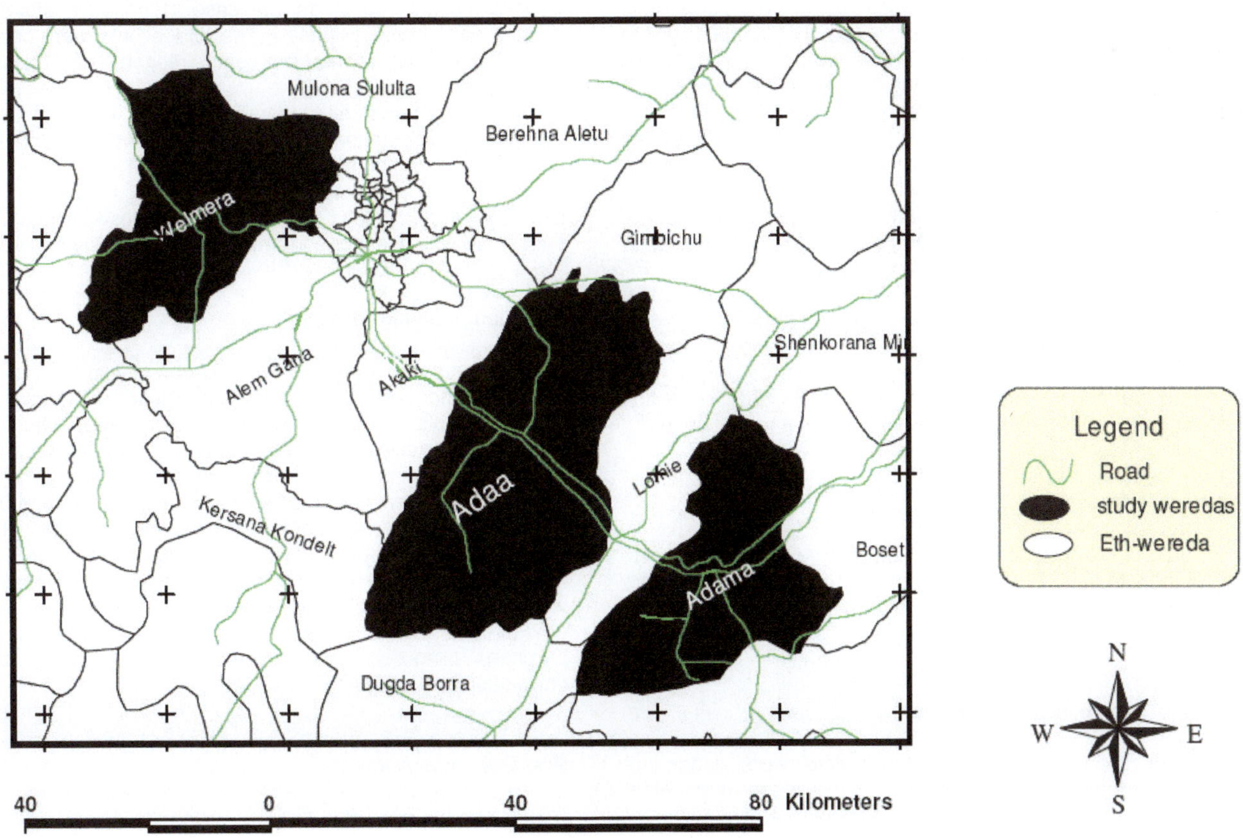

Figure 2. Detail view of location of study *woredas* in Ethiopia.

conducted at W. adama. The woreda consists of 31, 45 and 24% low land, mid-high land and high land, respectively.

Sampling, types and methods of data collection

A multistage sampling procedure was selected for the purpose of this study. From the 14 zones in Oromya region, East and West Shewa were chosen purposively. The research centres under EIAR namely Holeta, Debre Zeit and Melkassa Research Centres belonging to these zones were also chosen purposively. As described earlier, one woreda was purposively selected from areas where each one of the three research centres is offering training to the farmers. Equal numbers of trained and untrained respondents were used for this study. Therefore, from each one of the three woredas (Welmera, Ada'a and Adama) 40 trained and 40 untrained farmers were chosen for comparison purpose. The untrained farmers were selected from quite a significant distance away from where the trainings had been offered in order to avoid the cases of knowledge transfer from trained farmers to the untrained ones.

Welmera woreda was in the mandate area of Holeta Agricultural Research Centres and the farmers trained by that the centres were included in the sample from this area. Ada'a woreda was under the area of Debre Zeit Agricultural Research Centres and hence the farmers included from this woreda were those who were trained by Debre Zeit Research Centres. Melkassa Agricultural Research centre was located in Adama woreda, and the farmers trained by the centre were selected from this woreda. The sample included 40 trained and 40 untrained farmers that were selected randomly from the three woredas of the list of farmers under each centre. Since training effectiveness study through measurement before and after training sessions is impossible due to availability of time, the untrained farmers were used as control which takes care of immanent changes over time. Untrained farmers were selected randomly, from the sampling frame creating using the list of farmers growing the selected crop, with the help of development agents and peasant association (PA) leaders. Furthermore, farmers research group members were avoided from the sampling frame. Since the topics selected for the study for the centres were 'potato package' (for Holeta research centre), 'durum wheat package' (for Debre Zeit research centre) and 'onion package' (for Melkassa Agricultural Research Centre) , the farmers who had participated in the respective trainings were only included in the preliminary lists that were prepared as well as in the sample. Under each research centre one woreda was selected and under each woreda two PA were selected (one for trained and the other for untrained sample farmers) purposively. The PAs which consisted of the highest numbers of trained farmers in the woreda were selected for the purpose of this study. This training effectiveness study was intended to be carried out through quantitative data collection methodology. Quantitative data was collected from the respondents.

Methods of data analysis

All the data were processed and analyzed using appropriate statistical tools to fulfill the objectives of the study. The quantitative data was analyzed using descriptive statistics like mean, frequency and t-test.

Definition of variables and hypotheses

The important variables investigated in the research were, dependent and independent variables. Dependent variable was a variable that was affected or explained by another variable. An independent variable was a variable that causes changes in another (Sarantakose, 1998).

The general objective of this study was to assess training effectiveness. Any farmers' training was intended to bring about desirable changes in the behavioral dimensions of the participants such as: Knowledge and attitude leading to better on-the-job performance. For the purpose of this study, two major behavioral dimensions were considered such as: Knowledge and attitude to reflect the training effectiveness. Knowledge and attitude were treated as dependent variables in this study.

Attitude measurement

Attitude was defined as "the degree of positive or negative effect associated with psychological objects like symbol, phrase, slogan, person, institution, ideal or ideas towards which people can differ in varying degrees" (Thurstone, 1946).

The focus of this parameter was on the attitude of farmers towards the technology offered by the EIAR centres. "Attitude was defined in this study as the degree of positive or negative feeling of farmers' towards technology that were offered to them by the three Agricultural Research centres of EIAR.

Farmers' attitude towards the technology packages was measured using a summated rating (Likert type) scale. The scale was prepared with larger number of items initially and subjecting them to editing and screening in the light of pre-testing so as to include only the relevant items reflecting both positive and negative effect on a five point continuum. The items covered on all aspects related to the application of the given technology. Before administration, the scale was tested for its content validity and sufficient levels of reliability based on the pre-test results.

The attitude of a respondent was measured by adding the total scores obtained for ten item in the scale, by attributing 4 score for 'strongly agree', 3 score for 'agree', 2 score for 'undecided', 1 score for 'disagree' and 0 score for 'strongly disagree' responses in the case of positive items. In the case of negative statements the scoring pattern was reversed. The total scores were calculated by adding individual scores that each respondent obtained for all statements.

The total scores of attitude varied from 0 - 40. For the descriptive analysis three categories such as: Low, medium and high were employed. Since the score range was 0 - 40, the respondents were categorized into three such as: Low (0 - 13), medium (14 - 26) and high (27 - 40) for analysis with the help of descriptive statistics and total score was used for correlation and regression analyses.

Knowledge measurement

Knowledge of trained and untrained farmers was measured using a "Teacher-Made Test". The test items included 14 questions related to onion, potato and durum wheat technology package under Melkassa, Holeta and Debre Zeit Agricultural research centre respectively. Though 11 question, some 14 answers were expected. The scoring pattern was 1 score for correct answer and 0 score for wrong reply. The respondents were asked the question and the answers were recorded. Later these answers were evaluated and their total knowledge scores were calculated. Since the score range was 0 - 14 the respondents were categorized in to three such as: Low (0 - 4), medium (5 - 9), and high (10 - 14) for further analytical purposes using descriptive statistics and total score was used for correlation and regression analyses.

Practice measurement

Skill can be measured only by performance test. Data collection was done when the crop was not in the field. Skill test can not be done and if skill has to be measured it needs testing at each stage

Table 1. Knowledge of trained and untrained farmers under three agricultural research centers.

S/ No.	Research center	Trained		Untrained		t-value
		N	Mean	N	Mean	
1	Debre Zeit	40	10.57	40	6.08	8.83***
2	Melkassa	40	9.56	40	7.75	3.62***
3	Holeta	40	10.55	40	7.02	5.67***

***Significant at P ≤ 0.01 levels of significance
Source: own survey data (2007/8).

of farming which is beyond the scope of this study. However, practice was assessed as they responded but the accuracy of it can not be fully accepted to represent scale.

Thus, the practice of trained and untrained farmers under Melkassa, Holeta and Debre Zeit Agricultural Research centres was tested. Practice was not a dependent variable for final analysis, but was incorporated to generate some useful information. Practice was operationalized as the application of the knowledge in the real life situation. The practice of farmers was measured based on the recommended package. To test the practice of trained and untrained farmers, seven questions related to onion, potato and durum wheat technology package were used. These seven questions had seven answers. The scoring pattern was 1 score for correct answer and 0 score for wrong reply.

The respondents were asked the question and the answers were recorded. Later these answers were evaluated and their total practice scores were calculated. Since the score range was 0 - 7 the respondents were categorized in to three such as: Low (0 - 2), medium (3 - 5), and high (6 - 7) for further analytical purposes. This testing of practice was based on farmers' perception on their own practice, and hence it was not used as a valid measurement for further analysis. Practice/skill has to be observed for performance, but it was not possible to do it in the off season for crop cultivation.

RESULTS AND DISCUSSION

Effectiveness of training

To investigate the effectiveness of farmers training in improving their knowledge, attitude and practice of durum wheat, potato and onion extension packages 120 trained and 120 untrained farmers were in the study. From each research centres area 40 trained and 40 untrained farmers were used for this study. The advantage of assessing a knowledge, attitude and practice after training of farmers on a given agro-technology is one of the tools for information on the effectiveness of training (Adhikarya, 1997).

In this study, improvements in knowledge, attitude and practice of durum wheat, onion and potato extension packages that were disseminated by Debre Zeit, Melkassa and Holeta Agricultural Research centres, the frequency and percentage of respondents were obtained. The frequency and percentage of respondents ranged from low, medium to high categories. The differences between frequencies of respondents in the low, medium and high ranges were compared by using chi-square. This was to check the significance level of frequency of respondents that were classified in different categorizes (low, medium and high) within trained or untrained sample framers separately. Moreover, the significance differences between knowledge, attitude and practice of trained and untrained farmers were analyzed using t test.

Knowledge

A 'Teacher - made test' was prepared and administered to look at knowledge of the trained and untrained farmers, as discussed in methodology chapter. The answer of the farmers were evaluated and categorized into three levels such as: Low (1 - 4), medium (5 - 9) and high (10 - 14) based on the score ranges. The means of the knowledge of trained and untrained farmers were compared using paired t-test and are presented below in Table 1.

The t-test clearly showed that there was highly significant ($p \leq 0.01$) difference between mean score of knowledge of trained and untrained farmers who obtained training in potato, onion and durum wheat extension packages at Holeta, Melkassa and Debre Zeit Agricultural Research centres, respectively (Table 2). As can be seen from the Table 2, knowledge test indicated that the trained farmers had better level of knowledge when compared to the level of knowledge that untrained farmers had on the durum wheat, potato and onion extension packages that were provided by Debre Zeit, Holeta and Melkassa Agricultural Research centres, respectively. This finding is in agreement with the findings of Kefyalew (2006) that training kept the trained farmers more informed and updated.

Infact, this indicates that the untrained farmers also know something about these extension packages introduced into the area by the research centres. They can learn from the existing environment such as: Informal discussion with the trained farmers, by observing trained farmer's farm activity and from their life experience. However, from the result obtained, it could be seen that training kept the trained farmers more informed and updated on extension packages disseminated by Agricultural Research Centres.

The result, then, indicated that in terms of improving the knowledge of farmers the training organized by the Debre Zeit, Holeta and Melkassa was effective. Moreover,

Table 2. Knowledge of trained and untrained farmers under three agricultural research centres.

S/ No.	Research centre	Trained		Untrained		t-value
		N	Mean	N	Mean	
1	Debre Zeit	40	10.57	40	6.08	8.83***
2	Melkassa	40	9.56	40	7.75	3.62***
3	Holeta	40	10.55	40	7.02	5.67***

***Significant at p ≤ 0.01 levels of significance.
Source: own survey data (2007/8).

Table 3. Attitude of trained and untrained farmers under three Agricultural Research centres.

S/ No.	Research centre	Trained		Untrained		t-value
		N	Mean	N	Mean	
1.	Debre Zeit	40	27.60	40	22.92	3.62***
2.	Melkassa	40	28.87	40	22.85	3.72***
3.	Holeta	40	28.55	40	20.85	4.79***

*** Significant at p ≤ 0.01 levels of significance
Source: own survey data (2007/8).

the experts seemed to be able to transfer the required levels of knowledge for a specific technological package, which is the key factor to implement extension packages.

Attitude

The attitude of 120 trained and 120 untrained farmers was measured using a Likert type scale with 10 statements. The scale permits measurement of degree of positive or negative attitude towards durum wheat, potato and onion extension packages that was introduced by Debre Zeit, Holeta and Melkassa Agricultural Research centres, respectively. The differences between low, medium and high category of trained and untrained farmers' attitude of durum wheat, potato and onion extension packages was compared by chi-square. The mean scores trained and untrained farmers' attitudes were analyzed using paired samples t-test. The results of the analysis are presented below in Table 3.

The mean scores of attitude of trained farmers from Debre Zeit, Melkassa and Holeta Agricultural Research centre were significantly ($p < 0.001$) improved due to trainings offered on durum wheat, onion and potato extension packages (Table 3). Trained farmers had favorable attitude towards introduced durum wheat, onion and potato extension packages compared to the untrained farmers. Training rather enhances decision making by enabling the capacity to analyze information.

Under Debre Zeit Agricultural Research centre, the higher percentage of untrained respondents in the medium category indicates that the changes in attitude may not only be due to training but also due to information flow in the area. In MARC, although the larger number respondents were found to be categorized in high category for both trained and untrained farmers, the frequency of trained respondents was significantly higher than that of untrained ones in the same category. This could demonstrate that training that was offered by Melkassa Agricultural Research centre significantly improved the attitude of farmers towards improved onion extension package. This was in line with the findings of Kefyalew (2006) who stated that undergoing training by formal institutions and exposing oneself to the scientific information, it helps the individual to think rationally and seek new scientific information in all aspects of his/her life. Probably, the group situation in training and the group dynamics thereupon might have also influenced the participants to have an attitude change in the favorable direction.

Practice

The practice was operationalized as the application of the knowledge obtained from the training in the real life situation as stated by the respondents. The practice of farmers was measured based on their responses on application concerning the recommended durum wheat, potato and onion extension packages that were disseminated from Debre Zeit, Holeta and Melkassa Agricultural Research Centres, respectively. The differences between the low, medium and high category of trained and untrained farmers' practice was compared using chi-square and the means of practice of trained and untrained farmers was analyzed using paired samples t-test. The results of this test are displayed below in Table 4. The results presented in Table 4 clearly showed that

Table 4. Practice of trained and untrained farmers under three Agricultural Research centres.

S/ No.	Research centre	Trained		Untrained		t-value
		N	Mean	N	Mean	
1	Debre Zeit	40	5.90	40	3.12	10.40***
2	Melkassa	40	6.02	40	3.95	6.71***
3	Holeta	40	6.67	40	4.35	8.80***

***Significant at $p < 0.01$ levels of significance
Source: own survey data (2007/8)

Table 5. Knowledge, attitude and practice of trained and untrained farmers under three agricultural research centres (N=240).

Variables	Farmers	t-test					
		N	Mean	SD	SEM	df	T
Knowledge	Trained	120	10.229	2.797	0.255	119	9.967***
	Untrained	120	6.954	2.361	0.216		
Attitude	Trained	120	28.342	7.261	0.663	119	7.020***
	Untrained	120	22.208	6.841	0.625		
Practice	Trained	120	6.200	1.274	0.116	119	14.689***
	Untrained	120	3.808	1.451	0.133		

***,Significant at $p < 0.01$ levels of significance
Source: own survey data (2007/8)

the mean scores of practice of trained farmers on durum wheat, onion and potato extension packages were significantly ($P < 0.01$) higher under all the three agricultural research centres. The results presented showed that the training improved the levels of application of the scientific principles in durum wheat, onion and potato production due to trainings that were being offered by Debre Zeit, Melkassa and Holeta Agricultural Research centres.

Knowledge, attitude and practice test using pooled data

To further analyze the effectiveness of farmers training in improving their knowledge, attitude and practice over the centres of research operating under the umbrella of Ethiopian Institute of Agricultural Research the data were pooled together. The combined data which consists of three variables knowledge, attitude and practice of 120 trained farmers were analyzed using descriptive statistics such as: Frequency, percentage chi-square as well as paired t-test. The summary of paired comparison test is displayed below in Table 5.

The t-test result clearly showed that the mean score knowledge of trained farmers on extension package was significantly higher ($p < 0.01$) than the mean score knowledge of untrained farmers. This confirmed that the training offered by EIAR was effective in terms of improving knowledge of farmers. On the other hand the comparison between attitude of trained and untrained respondents using paired difference test indicated that the attitude of trained farmers significantly ($p < 0.01$) improved by the training offered by the centres. Similarly, the mean score of practice of trained farmers on extension packages was found to be highly improved when compared to untrained farmers practice of the same extension package. The paired comparison between the mean score of practice of trained and untrained sampled farmers showed that trained farmers are able to perform better than untrained ones.

In general, the knowledge, attitude and practice test clearly indicates that training significantly improved knowledge of farmers, improved attitude towards the packages and application of technology related to durum wheat, onion and potato production technology.

Conclusion

The knowledge, attitude and practice level of the sample trained farmers in Holeta, Debre Zeit and Melkassa Agricultural Research centres are presented in this study. The attitude of trained farmers was found to be significantly higher for farmers who obtained training from Melkassa, Holeta and Debre Zeit Agricultural Research centres and were classified as medium category, but significantly higher percentage of farmers' knowledge and practice which were classified in the highest category. The data presented in this study showed that untrained farmers knowledge, attitude and practice were found to be lower than the level of knowledge, attitude and practice that the trained farmers had acquired. It was observed that the significant numbers of untrained farmers' knowledge, attitude and practice level on extension packages were categorized in the medium range for all the three research centres. The survey data obtained from the three study area were combined together to look at the factors influencing the knowledge and attitude of the trained farmers who obtained trainings from the three research centres under EIAR. The result showed that education of farmers, wealth status, level of aspiration, information seeking behavior; extension contact and family size were the most importance independent variables which had significant influence on the knowledge of trained farmers. Whereas education and level of aspiration of trained farmers were the only two independent variables which had significant effect on the attitude of farmers for packages. The result clearly shows that the trainings offered by the Agricultural Research Centres were effective in terms of knowledge of technologies and attitude of trained farmers towards the extension packages. Trainings that had been conducted by the three research centres has shown to improve the knowledge, attitude, and practice of the trained farmers when compared to that of untrained farmers in the study area.

REFERENCES

Adhikarya Z (1997). Strategic extension campaign, increasing cost-effectiveness and farmers participation in applying agricultural technology. Rome. Sustainable Development Department (SD). Food and agriculture organization of the united nation (FAO).

Kefyalew BC (2006). Evaluation of farmers training programs the case of eastern Hararghe Babile and Hundene Woredas, M.Sc. Thesis, Haramaya University, Ethiopia.

Raab RT, Swanson BE, Wentling TL, Dark CD (1987). A trainer's guide to evaluation. Rome: FAO.

Thurstone LL (1946). Comment. Am. J. Soc. 52: 39-40.

Tsedeke A (2007). Focusing Agricultural Research to Address Development Deed. Direction for Agricultural Research in Ethiopia. ISBN: 99944-53-07-6, Pp 1-63. Addis Ababa, Ethiopia. Ethiopian Institute of Agricultural Research Publication.

Zeleke WM (2000). Study on functional literacy programme for agricultural and rural development in Ethiopia. Addis Ababa University, Ethiopia. Addis Ababa.

CSA (2006). Agricultural sample survey 2005/2006 (1998 E.C.) (September 2005-February 2006). Volume I: Report on area and production of crops (private peasant holdings, meher season). Statistical Bulletin 361, July 2006, Addis Ababa.

Sarantakos S (1998). Social Research and 2nd edition. Macmillan Press Ltd., London. p. 488.

Is PLAR (Participatory learning and action research) a sufficient approach for the purpose of supporting transitions for sustainable agriculture? A case study from Sweden

Karin Eksvärd* and Johanna Björklund

Centre for Sustainable Agriculture, Swedish University of Agricultural Sciences, Box 7047, SE 750 07 Uppsala, Sweden.

This article examines whether Participatory Learning and Action Research (PLAR) is a sufficient approach for the purpose of supporting transitions to improved sustainability in agriculture. Relevant outputs and outcomes are presented from a case study of Swedish organic greenhouse tomato growers researching their own views on what the concept 'organic' could comprise, and the sustainability of their production. The transitions to which this process gave rise were discussed. PLAR can create a learning environment where questions leading to systemic transitions are asked and acted on at the niche level (individual actors and groups). In this case, it facilitated learning from diversity and the development of a more sustainable basis for production. PLAR supported the participants to question their own practices and to find synergies among unavoidable goal conflicts. The article concludes that higher level transitions are needed also if transitions at the niche level are to drive systemic change in production regimes (rules and societal assumptions).

Key words: Sustainable development, organic agriculture, organic tomato production, participatory learning and action research, learning, transition.

INTRODUCTION

The need to align agricultural production systems with sustainable development, globally and locally, is no longer questioned. This implies a readiness to bring about changes in production systems and livelihoods, not just adjustments by means of technical solutions. This article presents how a group of organic greenhouse tomato growers, by researching the possibilities of improving their livelihoods as growers through Participatory Learning and Action Research (PLAR), came to question the level of sustainability in their production systems. Did their claim to be 'organic', the very basis for their income, meet their values on what sustainable 'organic' production should comprise? This included researching the increasing problem of earning sufficient income while also meeting social needs and ecological aims.

In Sweden, organic agriculture is one of the strategies that have been adopted for improving the sustainability of agriculture. As in most European countries, organic agriculture has been expanding in Sweden. During the period 2000 to 2005, the estimated percentage of the organic agriculture area utilised in EU-15 increased from 3.0 to 4.3% (European Commission, 2008). An increase in the number of organic producers and the current rapid increase in demand are driving an expansion of the market. There is an ongoing debate on how to make organic production a feasible option for a larger group of farmers. Two recognised strategies are making the

*Corresponding author. E-mail: karin@comlaetitia.se.

Abbreviation: G, Grower; **PLAR**, participatory learning and action research; **SLU**, Swedish University of Agricultural Sciences.

conversion to organic systems easier, and increasing the economic benefits of organic production. Some argue that making conversion easier may risk reducing the environmental gains from conversion and thereby losing important qualities of sustainability in organic production (Milestad et al., 2008).

For the Swedish organic vegetable market, the situation is different. Although the demand is continually growing, the number of growers is declining due to poor profitability and laborious modes of production (Nilsson, 2007). Retaining and increasing the number of organic vegetable growers and expanding the area managed organically, while at the same time maintaining and raising the environmental standards in production, is an important challenge for the sustainable development of the organic sector, as well as for the agricultural sector as a whole.

The demand on farming to deliver new services, products and ways of farming in order to improve sustainability is increasing in Sweden. These demands and concerns are implemented in the form of norms, laws or regulations, but sometimes also as incentives. Such institutional frameworks are structured as top-down processes where the farmers do not own the problems the regulations are designed to correct. Farmers are instead focused on how to deal with the regulations and to maximise the subsidies, rather than focusing on the problems they are best able to solve in the landscape structures and locations the institutional structures were designed to support (Ahnström et al., 2009). PLAR is a well-developed methodology for transdisciplinary learning and research processes for improving local situations such as farming through a 'bottom-up' approach (Pretty et al., 1995; Defoer, 2002). Working with group dynamics to reduce power asymmetries and avoid biases is a vital component (Pretty et al., 1995).

This article examines the following question: Is PLAR a sufficient approach for supporting the development of organic agriculture in order to maintain and increase sustainability? The case study used was the work of a PLAR group for organic tomato production in Sweden. The full PLAR process is described by Eksvärd (2010).

Background

The case material presented and analysed is based on a sub-component that emerged within the PLAR process for situational improvement in organic greenhouse tomato production. The process was initiated by the Swedish University of Agricultural Sciences (SLU), County Administrative Boards (Dalarna and Västmanland) and seven tomato growers in 1999 and is still continuing. The group was supported to identify and develop research question on topics of most interest to themselves. The sub-component presented in this article, addressed the specific question of 'what can actually be called organic?' That is, what the partners in the group actually considered to be sustainable agricultural practice; whether this was sufficiently sustainable to be called organic; and whether the production systems of group members met those criteria.

Swedish organic greenhouse tomato production consisted of only 23 200 m^2 in 2005 and is usually part of a larger production system. Although the individual production systems managed by group members were rather small at the time by European standards, overall they constituted almost 40% of the total organic tomato production area in their counties. The growers thus represent an important part of Swedish organic vegetable production.

METHODOLOGY

The entire project was conceived and implemented as a PLAR approach. The grower participants were identified over a process of several years of interaction between the advisors and the organic growers within the area. The scientific participants of the co-researching group in the sub-component presented here comprised two organic advisors, a researching facilitator and a researcher with expertise on resource sustainability. Participants active during the sub-component are presented in Table 1.

The sub-component was carried out actively between 2002 and 2004. It was supported by a variety of data gathering and data analysis activities. The results were fed into iterative cycles of reflection, planning and further co-learning activities. Data gathering tools included semi-structured interviews with individual participants and with the group as a whole, and measurements of resource use. Data analysis on resource use was carried out by means of estimation of 'emergy-based footprints' (Johansson, 2005). This footprint combines existing assessment methods such as emergy analysis (Odum, 1996) and ecological footprinting (Wackernagel and Rees, 1996). In the emergy-based footprint, the system boundaries are larger and the calculations are based on energy flows, while the Wackernagel and Rees method calculates biological production.

In short, emergy analysis accounts for the energy intensity embodied in products, including the environmental work and the work of humans in generating products and services, while ecological footprinting presents the resource use on an area basis. In an emergy analysis, all items are converted to a common basis of solar energy (the unit solar emjoules (sej)) using conversion coefficients (transformities). The transformity is the solar energy used to make one joule of a resource (Odum, 1996). The method had been comprehensively described by a number of authors (Brown and Herendeen, 1996; Odum, 1996; Björklund, 2000; Eksvärd and Rydberg, 2010).

In addition, numerous methods for supporting shared learning, reflection and mutual understanding were used, such as diagramming and visual representations (Pretty et al., 1995; Mikkelsen, 2005).

The process was facilitated and the facilitator kept written notes of all meetings. A review of the process was held in November 2004. The results of the semi-structured interviews and the quantitative data from the greenhouse resource study were presented and jointly interpreted. The results were assessed in the light of the initial research questions and related to what the group considered 'organic'. The most robust outputs were collectively

Table 1. Group members during the active period of pursuing the 'what is organic' question.

Code	Functional role	Durational involvement	Descriptors
R	Researcher	April 2002 – November 2004	Sustainable development and resources researcher married to F 10, female
A1	Advisor	Feb 1999 -	Close collaboration with facilitator until 2004, female
A2	Advisor	Feb 1999 -	Female
G1A	Grower	Feb 1999 -	Wife
G1B	Grower	Aug 1999 -	Husband
G2A	Grower	Feb 2002 -	Wife
G2B	Grower	Feb 2002 -	Husband
G3A	Grower	Jan 2001 -	Wife
G3B	Grower		Husband; occasional participation
G4A	Grower	May 2002 - ? 2005	Enterprise colleague, ended enterprise collaboration, male
G4B	Grower	(March 00) May 2002 -	Enterprise colleague, male
G5	Grower	Jan 2001- ? 2003	Sold farm, male
G6	Grower	Feb 1999 -	Male
G7	Grower	Aug 1999-	Male
G8	Grower	Feb 1999 -	Male
G9	Grower	Feb 1999 -	Male
G10	Grower	March 2000 -? 2003	Sold farm, male
G11	Grower	Feb 1999 -? 2003	Lacking time to continue.
RF	Facilitator	Feb 1999 – Jan 2004 (Nov 2004)	Action researching facilitator. Sporadic contacts after 2004. Visit at meeting Dec 2007. Female

Source: Group documentation.

identified for publication in a project report. Thereafter, follow-up information was collected sporadically by the researching facilitator, by means of informal interviews, up to December 2007. Further details on the methodology are given in the following sections.

Design of the sub-component: How would group members classify organic/sustainable practice?

At an early stage, the group identified a strategic issue of importance for the tomato growers and for development of the organic sector as a whole, as well as for the government's policy of encouraging a shift toward sustainable agriculture. This was 'What should the concept of 'organic' include to qualify as a guarantee that agriculture would assist in the transition to sustainable development?' The question had its roots in the first group meeting (17 February, 1999). It emerged clearly as a specific question in April 2002, and was discussed actively until November 2004.
The participants decided to begin with a brainstorming exercise in order to obtain a prior understanding of their interpretation of the concept. The process was then organised as follows:

1. Brainstorm – 'what is organic?'
2. Clarification of aims.
3. Semi-structured interview on views and values of organic production.
4. Collection and group resource data analysis of resource use, plant nutrients and how the greenhouse fits into the farm system.
5. Analysis and contextualisation of the material.
6. Drawing conclusions and decision on definition of organic.
7. Reporting.

Inspired by a study that described group interviews with organic farmers in Kansas, USA, on how farmers adopted agricultural strategies (Norman et al., 2000), the researcher and the researching facilitator suggested areas to be discussed during the group-based semi-structured interview (item 3 above) (24 March, 2003). The farmers then modified and approved these suggestions. The themes and topics prepared for the semi-structured group interview were:

1. Reasons given for growing organically.
2. Measures growers used for concluding how farm activities are functioning.
3. Visions and values that growers considered essential for sustainable production.
4. Difficulties with growing organic and what growers found to be unsustainable within their enterprises.

The group interview was taped and transcribed, and then analysed manually using a narrative analysis technique (Neuman, 2003). The resulting material was sent out to all group members for comment and then reported at a group meeting. This stimulated more discussions and elicited more inputs and clarifications on the issues that had been discussed.

As the growers realised they would not have the time to collect the resource use data (item 4), the researcher was asked to do this. Total resource consumption was analysed by measuring material flows of all construction materials, including ground

Table 2. Brainstorm on "What is organic?"

What is organic?
(a) Seed should be organic
(b) All fertilisation products should be organic
(c) We need to get more and more specialised; you do not want to work yourself to death.
(d) High quality and good natural status
(e) Economy
(f) Economising with resources.
(g) Tending to the microbial life in the soil. What is "easily soluble"? Should we give liquid fertilisers and grow plants in rock wool or apply nutrient through the soil?
(h) Improving local efficiency.
(i) Should we talk about "organic" or "sustainable"? What are ethical positions and what is organic?
(j) "Keeping the earth in the heart"
(k) Need to be ok about the fact that things change all the time.

Source: 29.04.02, Clarified for publication.

constructions and bought-in commodities. All transport of building materials, commodities and products for sale was documented. Working hours, costs and income were calculated. The aggregated resource use was then expressed per m^2 greenhouse area and year, and per kg tomatoes and year. In addition, quantitative analysis of resource use in four of the greenhouses was compiled as an 'emergy-based footprint' (Johansson, 2005).

Emergy analysis was used in this study to assess and compare resources of different kinds, such as the use of fossil fuels, iron and the energy in rain. With the emergy analysis, it was possible to put all different resources on a common basis, that is the amount of work done by the environment (the solar energy used) to generate them, and make a comparison that was consistent. By using emergy analysis, it was possible to include all resources in the same analysis.

RESULTS

Transitions during the PLAR process

Here, the findings on the question of whether PLAR is a sufficient approach for supporting the development of organic agriculture in order to maintain and increase sustainability were presented. The findings were presented in terms of three transitions: (1) The transition from factor trials to researching dynamic systems; (2) Transitions in the domain of knowledge creation, and (3) Transitions in actions for change. The subsequent analysis attempted to reveal how these transitions were catalysed by the co-researching PLAR experience.

Transition from factor trials to researching dynamic systems

The transition to researching dynamic systems for situational improvement, rather than factor trials as in problem-solving, is described below through following items 1 to 5 of 'Design of the sub-component'.

Brainstorm

At the time of initiation of this sub-component, the organic concept was perceived as being very complex. Table 2 illustrates the initial responses to what the group members thought should be included in the concept 'organic'.

From this exercise, the group derived two fundamental questions: 'What do we consider to be sufficiently sustainable to qualify as organic tomato production?' and 'What can be done to create more sustainable production systems and lifestyles?' A decision was made to examine how the group defined 'organic production' and how management options and choices of cultivation methods affected the sustainability of their production systems. From then on, the term 'organic' was used by the group as a synonym for 'sustainable practice', though recognising that sustainability is an evolving process.

Clarification on aims

The group agreed that the aims were:
'To be able to make choices that is in line with our values and worldviews, to clarify our standpoints and to have good grounds for choices of change. To improve the possibilities to do a job that is as 'organic' (sustainable) as possible and improve our capability of giving well-informed answers to questions from the world around and to discuss with the Swedish organic production

organisation and the European Union, for our own satisfaction' (6 November, 2002).

Two of the growers (G5, G10) subsequently stopped growing tomatoes in 2003 and sold their farms and G11 left the group due to lack of time. G10 explained the reason for his decision to be a high level of unease about being tied up in a lot of work in what was for him a questionable activity, namely producing tomatoes in greenhouses without an economic option to stop using fossil fuel.

Semi-structured group interview

During the semi-structured group interview (24 March, 2003), the respondents (G1A, G2B, G3A-B, G6, G7, G8) did not make any reference to estimates of whether their own production systems were sustainable or not, although the oil consumption issue and the stressful life were mentioned. Factors that promoted sustainability were named as: Using residues, cooperation and local circulation of nutrients. However, creating or using these factors in practice was considered messy and difficult. Growers asked how they could say what was more or less 'organic' (sustainable practice) among all the commodities and practices available to choose from.

G6*:* (… … Organic): "it is no pesticides, no fertilisers, but what are we for? […], there are several different ideas."
G3B: "A typical example is […], that you could get horse manure from a racetrack compared with buying Biovinass [organic certified liquid manure from yeast production], then I think the horse manure makes a lot more sense but then it is not organically certified and then you are caught up again."
G6: "But I may use the horse manure. But then you come to all that we have learnt, that it gives a lot of eutrophication of phosphorus and stuff. As I said, it [tomato production] was easier before but it is better now."

For all the growers, it was self-evident to follow the rules of the certification organisation, but they all accepted what G6 (24 March, 2003) called the 'grey zone', areas where they had to cope with how to put the rules into practice. They felt they were being forced to do things that they did not actually view as 'organic (sustainable)', because of the need to be efficient and save time. For instance, with reference to the need to support growing plants using strings hooked to a frame the growers commented:

> "Some of the stuff we use too, such as the hooks with string, they cost 1.20 to buy, with string and all. Should you buy a ball of string and wind on new string [… …] then you have 0.20 to do the winding. [… …] Well I [have started to] throw out the hooks and it is unheard of wastefulness!' (G8, 24 March 2003)
> […] 'at the rate that most production units become single person enterprises parts of the self-sufficiency thinking goes, there is lack of working hours and then you buy half ready products from outside [of your own system].' (G8, 24 March, 2003)"

The organic solutions they considered at the time to be more 'organic' were described as more time-consuming and as involving heavy work. This was especially true when it came to manuring strategies. The tight situation was also referred to in comments on time, workload and feelings about being able to cope.

"The big problem is how you are going to be able to cope with this in the future, it is so…, and it is heavy physically, you feel it in your arms and so… during the season there is no time for anything else, it is all green house…' (G2B, 24 March, 2003)

The will to have sustainable production, yet the personal conflicts this goal imposed, was clearly put by G1A:

"Sustainability: what is to be sustainable; me or the system? This brings conflicts. What is more important, the system or us, the individuals? It is not an easy choice when you are the individual."

The growers also clearly expressed frustration over prices, and politics that asked for more organic produce and lower food prices at the same time.

'Then there are powers that say there should be more organic food production in Sweden so that more have the possibility to buy, so that there is never any lack of products and that the prices come down to a more reasonable level and are more like the conventional. But that is not a product we want to produce and not what our customers ask for, they want products that are a bit better and we want to produce products that are a bit better' (G8, 24 March, 2003).

Group resource data analysis

The quantitative data on resource use for four of the greenhouses were aggregated and presented to the growers as an emergy-based footprint, that is, the theoretical area needed to produce the amount of natural resources used through local renewable resources. In Figure 1, the calculations were compared with two analyses of conventional and organic tomato production based on model data (Lagerberg, 1999). The emergy-based footprint for the four greenhouses (one shown both

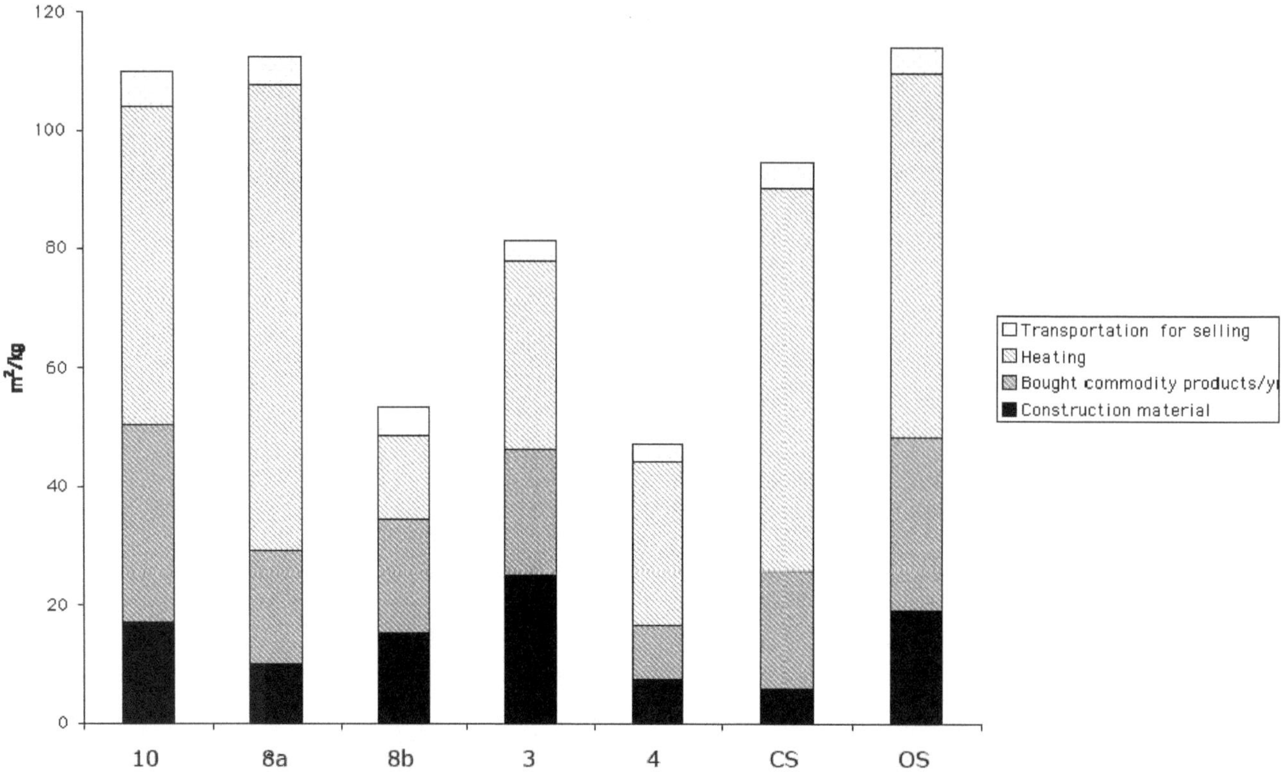

Figure 1. Ecological footprint area, that is, the theoretical area needed to produce the amount of natural resources used through local renewable resources, calculated through emergy analysis, in m^2 per kg produced tomato. Numbers on the x-axis are the ID-numbers given to the different production systems in the study. 'CS' conventional tomato production and 'OS' organic tomato production, both calculated from data taken from Lagerberg et al. (1999). Numbers on the x-axis are described in Table 3.

before and after conversion of heating system) varied greatly because of different modes of production, e.g. length of season, kind of greenhouse and yield. The greenhouses are described further in Table 3.

The largest footprint was caused by the oil-heated greenhouse with double polythene, extensive climate regulation and relatively high production (No 8a in Figure 1). When this greenhouse was converted to woodchip heating, the footprint nearly halved (No. 8b). All houses (except Nos. 8b and 3) (heated with wood pellets) were heated with oil, and the energy for heating was the largest item for all greenhouses except for these two. More construction material was needed for the heating systems in 8b and 3, which increased the footprint for the construction material. The low yields in No. 10 gave high impact values per kg and No. 4 had less intensive production. An overview of the greenhouse and production data is given in Table 3.

The grower with the highest yields and highest oil consumption used intensive ventilation to reduce humidity. Even in the warm seasons, improving the climate in the greenhouse by interchange of warm humid air for cooler and drier air that needed to be heated was profitable, as the oil was subsidised. For commercial growers with expensive greenhouses, investment in energy so as to obtain high and stable yields of good quality was a profitable option: G4B later commented that "you do not blaze away if you have a bad house... if you have a better house you go in for it more" (17 November, 2004). The results from the emergy analysis supported the ideas of the group that changing heating systems to a local and renewable energy source would have a significant influence on the sustainability of production.

Emergy-based ecological footprint calculations were also performed for a range of different nutrients used in production, and the application of heavy metals relating to this use was assessed (Björklund et al., 2005). The study on the amount and use of purchased nutrients highlighted the risk of importing heavy metals, especially high levels of cadmium in organic certified manure or in fertilisers originating from conventional systems. As the amount of nutrients applied per unit area is of an order of magnitude higher than in field crop production, it was shown that the given fertilisation recommendations are not possible if the heavy metal risks are to be avoided, as they are not adapted to organic greenhouse production.

Table 3. Description of the green houses and productions data.

Farm	Green house and production description
10	Double foil, 544 m^2, warm air heating system, oil, manuring and organic fertiliser, yield 18 kg/ m^2 from June to October. Marketing through local shops, farmers market and on farm. Other produce: Berries, cereals and poultry.
8 a	Double foil, 800 m^2, warm water heating system, oil, manuring and organic fertiliser, yield 32 kg/m^2 from May/June – October. Marketing through local shop, other produce: Cucumbers and tomato plants.
8 b	Double foil, 800 m^2, warm water heating system, woodchips, manuring and organic fertiliser, yield 32 kg/ m^2 from May/June – October. Marketing through local shop. Other produce: cucumbers and tomato plants.
3	Double foil, 506 m^2, warm water heating system, woodchips, manuring and organic fertiliser, yield 22 kg/m^2, season June – October. Marketing; pick your own. Other produce: Vegetables and flowers.
4	Single glass, 775 m^2, warm air heating system, oil, manuring and organic fertiliser, yield 20 kg/m^2, season June – October. Marketing through local shops. Other produce: Other greenhouse vegetables and flowers. Also place for work focused rehabilitation
Conventional system (CS)	Designed from best possible available data, glass house, 9 000 m^2, cropping in mineral wool, fertilization with nutrient solution, yield 42 kg/m^2, from March – October (Lagerberg and Brown, 1999).
Organic system (OS)	Designed from an enlarged research study, glass house, 9 000 m^2, cropping in delimited soil containers, warm water heating system, oil, Manuring and mulching, yield 23 kg/ m^2 from June to October (Lagerberg et al., 1999).

Source: Group documentation.

Simple reliance on the label of KRAV (Swedish organic production organisation) as a guarantee of good environmental practice was shown to be insufficient. The findings promoted discussion on the pros and cons of adding nutrients by use of farmyard manure just before planting only, or by adding doses of purchased manufactured fertiliser later in the season.

The discussion in the group led to a realisation that the choice of nutrients may have ethical implications, for instance if manure is imported from farming systems with lower animal welfare standards than the Swedish norms. The need to avoid the effects of avoiding nutrient loads to the environment in the form of heavy metals such as cadmium and the importance of building capacity in surrounding ecosystems to take care of nutrient leaching were other important findings.

Analysis and contextualisation of the material

In 2004, the researching facilitator and the researcher facilitated a carefully planned review of the process (17 November, 2004), from the first development of the question 'what can actually be called organic' to the accomplishments achieved by the end of 2004. The group recapitulated the aims, what had been done, decisions and findings. The semi-structured group interview on subsequent discussion was reviewed. All the growers present (G1A, G2A-B, G4A-B, G6, G9) then stressed the need to meet financial goals, not to live 'the fancy life' but simply to survive as tomato producers. The quantitative data from the greenhouses were presented and analysed and the material evaluated jointly. The questions participants considered to be satisfactorily answered were identified and it was decided what should be published.

With the whole procedure freshly in mind, the group held a brainstorming session based on the prompting question: "If you look back at the work, we have done and were asked to pick out key words that together would present a picture of the factors to be considered in deciding whether you think a cultivation system should be called organic or not, what would those key words be?' A matrix tool was used where all the key words generated were placed in a simple diagram with the y-axis representing 'more organic – less organic (not organic)' and values on the x-axis the 'knowledge-feeling' (Figure 2).

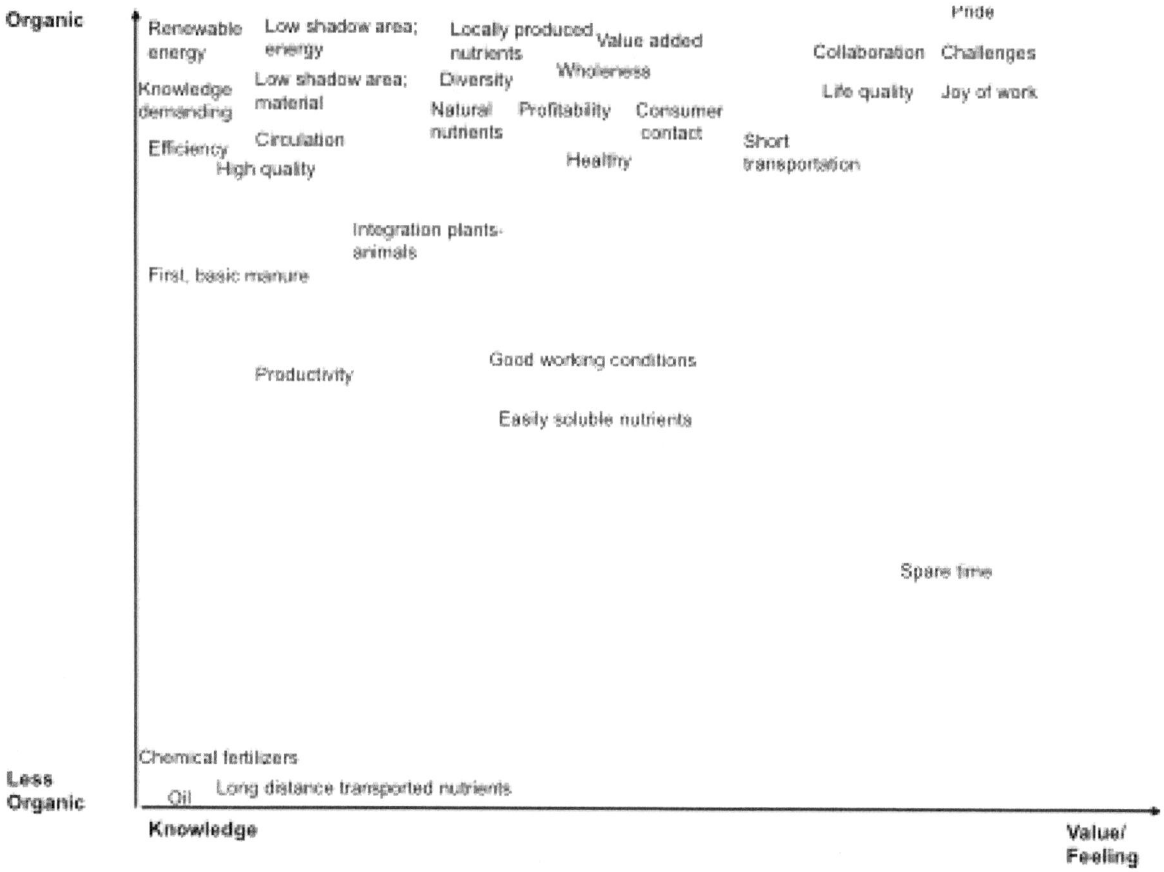

Figure 2. Tool to facilitate group discussion on whether key factors were more or less organic and whether this was something members knew as a fact or felt to be true (17 November, 2004).

The discussion generated from Figure 2 showed that local resource use, local nutrient circulation and a small ecological footprint were perceived as inherent in the participants' concept of organic. The products had to be of 'high quality'. 'Efficiency' was included in the organic concept. 'Productivity' in the sense of high yield was considered important but not a determining factor. Profitability was considered important to all kinds of production, as a condition of being able to continue, but there were doubts about whether this requirement was fulfilled. Easily soluble plant nutrients and long-distance transportation were clearly identified as being less organic. Moreover, the sense of being proud, well-being and feeling joy in one's work, meeting consumers and cooperation with others actors were perceived as being important for the concept of organic tomato production, although less easy to measure.

Transitions in the domain of knowledge creation

The findings on the transition in the domain of knowledge creation when researching dynamic systems were described through item 6 in the sub-component design process, drawing conclusions. Conclusions were drawn collectively on an overall level, but then also individually, as a person and for each farm respectively.

Drawing conclusions

After having worked through the diagram presented in Figure 2, the group phrased their own definition of organic production (sustainable practice) as:

'[...] the expression 'organic' strives towards production based on renewable natural resources with as low a shadow-area as possible, that is, a low consumption of natural resources. It is also important that the nutrients added are natural and based on locally circulated resources. The products should have a high quality and create communication with the consumers about added values to organic products and production, as well as organic consumption. Profit should be high enough to allow conditions for production that include high quality of life [not defined here], pleasure in work and a good working environment.'

Added to this was that organic production takes a high amount of knowledge, needs a holistic view of the production system and its environment, and often requires collaboration with others within the region.

For these growers, one of the aspects limiting their possibilities of creating as 'organic' (sustainable) a production system as they would like was shown to be time for themselves and their families. The entanglement of their financial, social and ecological situation became very clear. During recent years, all the growers had accepted practices they considered to be 'grey zones' organically speaking (that is, accepted by regulation but questionable in their view) in order to continue production. Examples given were buying in organic fertiliser instead of using farmyard or green manure, and becoming more specialised and giving up rearing animals.

After this discussion, arguments for a range of options grounded in group members' own individual situations were identified. The options considered included using imported, pelleted or liquid organic fertilisers, noting the ethical implications of importing nutrients from conventional pig production in neighbouring countries, but also noting the possibility of developing more accurate dosage and less leaching over time. Others preferred farm-made sheep manure compost, which may be heavy to handle, time-consuming to make and spread and may pose higher risks of nutrient leaching (depending on the soil and surrounding system), but is a locally produced and self-sustaining resource.

The articulated values were based on appreciation of the local supply of resources and a desire to be able to produce energy and nutrients on the farm. Some in the group saw the integration of crops and domestic animals as the most important option, whereas others thought that green manure had a place in organic tomato production. All agreed on the importance of taking a systems view, placing tomato growing in the context of the whole farm and the larger surrounding environment and society.

Finally, when asked by the researching facilitator how she was doing on the sustainability issue, F3A explained the recorded progress in these terms: "I do the best I can, I know it, [I] have better overview and have arguments" (December, 2007).

Transitions in actions for change

The sub-component and learning process on what 'organic' should include ended formally in 2004. However, some time after having carried out the group synthesis, the group decided to enter more deeply into questions regarding heating sources, and awareness of the issues also remained heightened in other areas. At this time, transitions in actions implementing the new knowledge and awareness started increasing.

By 2006 one couple had bought new bio-fuel heating equipment and were getting ready to install it and three other growers had decided to change their heating system (K. Sjöstedt, Hornudden, personal communication). Changing the heating system is a major and 'very costly affair'. By 2007, a further three enterprises had bought new equipment and installed it and another two had also decided to change their systems (E. Ögren, Västmanland County Administrative Board, personal communication). One grower explained (2006) that on their farm they would have to wait until 2007 for this change when they could get a 30% subsidy from the Swedish Board of Agriculture for the new heating system. "But we will never get our money back on this" the grower continued and explained that they had to decide "to go for it or quit". Continuing cultivation and keeping their oil heating was not a valid choice for them. The choice of bio-fuel added to their sense of security, as its supply and price is more predictable than that of oil and "it will make us feel better".

The group also decided the goals for their continuing work on nutrients. One is to 'work on adjusting nutrient levels to the needs of the plants for environmental reasons'. They are also continuing to work on a 'crib' tool to help calculate the nutrient requirements. This tool combines the scientific calculations of a researcher and data from their own practical experiences to improve the changes begun in nutrient management.

ANALYSIS AND DISCUSSION

The transition from research management to facilitation of learning was intentional, and initiated from the outset as a 'means to perform'. In this case, the means to perform entailed a transition from a hierarchical form of research management to the creation of conditions for co-researching. Starting off the enquiry process by agreeing on the methodological base and structure of the work turned out to be of real importance. Agreement on the principles of equality, collaboration and working systemically through learning processes made it easy to introduce group dynamic exercises and tools to build new relationships among the individuals in the group. This helped break the role-based status of who is to 'speak up to offer ideas, raise concerns or ask questions' and to create the psychological safety described by Nembhard and Edmondson (2006), which encourages participants to inquire beyond the boundaries of perception and practice. It provided the base for collegial participation that minimised power asymmetries and fostered a consensual decision-making process. The sharing of power to direct the process, to take a break, or actually quit the process, created a more secure environment in which difficult and sensitive questions could be posed.

The facilitator also worked to create a new kind of space for developing practical experience and socialising, though tight schedules of interaction with 'important' research and enterprise development activity. This served as a catalyst for motivation and joy, but also for the research in itself. The close contact between practice, social life, enterprise development and learning mirrored the lived experience of farm life and anchored the work in its context. It also fulfilled the growers' expressed need for developing a network of colleagues.

The growers themselves began to appreciate both the necessity and rewards of building new relationships within the group and between themselves and their enterprises:

> "I have learned things about practical production but the most important part is the contact with the growers and you others.' (G8, 15 November, 2000). 'It is the wholeness of it that is important. That we are a really broad group who work together to find the best for organic tomato production" (G11, 14 November, 2000).

Learning as systemic inquiry

The learning that has taken place has involved the participants improving their understanding of their own perceived system of interest, their understanding and belief in what is possible, and the ethical dimensions of their practices. This deep learning offered an opportunity to express the controversies they experience and tensions surrounding the concept 'organic'. The growers had their own individual systems, from single greenhouse production units run by each person to diversified family systems that included animals in the farm enterprise, and thus they experienced different possibilities and limits. The new solutions that were developed were based on investigation of interdependent components of an enterprise system in its biophysical and social environment. The group learned to let the solutions emerge as the learning unfolded. These processes have been identified as important to sustainable adaptation in complex situations (Kahane, 2004). The opportunities for learning were increased by the fact that the group displayed the required diversity necessary for systemic inquiry (Blackmore, 2005).

The learning process can be said to have also depended on an unlearning process. This unlearning is related to traditional role of responsibility aforementioned. However, unlearning of the initially fixed opinion that the energy question was not possible to handle went in parallel with learning that it was possible to change heating system – an example of unlearning old presumptions and relearning through action-based experience (Wang and Ahmed, 2003).

Accomplishing transitions

By clearly expressing the commonly felt conflict between saving one's own time and saving natural resources, and reducing environmental constraints, the decisions taken in the end became more informed. The existing economic drivers made increasing demands on production, meaning that more time was devoted to increasing yield and less time to the tasks that were resource saving, or that might lead to self-sufficient use of on-farm resources and local circulation of nutrients. The group members realised that the market system does not operate on the true costs of production. As a consequence, what society still considers a net benefit (cheap plentiful food) is experienced by those heading the transition toward sustainable agro-ecological production as negative pressures on their time, family and social relations, and on the viability of their enterprises.

By deciding to research the sustainability of their own farming practice, the group moved on from classical component research on questions associated with cultivation to questions of values, attitudes and performances, that is, they accomplished a transformation in the kinds of questions that could be asked. Growers needed to become confident and knowledgeable about their own standpoint and choices if this transition was to be enacted. The discussion diagram (Figure 2) encouraged the participants to start to express what organic 'is', and to leave the more negative terminology of what organic production *is* 'not'. Thereafter, phrasing definitions in terms of what organic should be empowered the participants to discuss the 'niched (micro level) problems' produced by trying to fit organic transitions into current market frameworks. According to Geels (2002) and Deuten (2003), it is useful in such cases to distinguish between experiments, niches, regimes and landscape, representing different levels of a system. 'Experiments' are local practices, which actors learn under local circumstances on the basis of research and practical inquiry. 'Niches' (meso level) and 'regimes' (macro level) are considered sets of rules that provide structure to these local practices, but at different levels of system stability. In niches, there is limited stability in the rules and uncertainty about future directions. Regimes are more stable than niches, since rules are replicated and shared among many different locations. Thus regimes offer more structuralisation of local practices (Raven, 2004). The landscape is even more stable, linking to the wider external development of a society. During processes of systemic innovation, the dominant regime tends to resist change. The initiative for change starts in isolated social environments, as a niche that is created by purposive experiment. Niches can succeed only when changes have been made at the regime level. The niche actors that cannot influence the regime will fail. The space for wider transformation emerges only when the dominant regime and landscape are removed, lifted or transformed.

In the case presented here, the growers' increased awareness and reflection on their own production systems brought about an increased confidence in presenting arguments concerning their decisions about how to manage their niche production and reduced the worry related to uncertainties concerning the judgements of customers and consumers. At the same time, participants' views of what organic (/sustainable practice) should include changed during the process, as they began to reflect on and challenge the implementation of the regulations imposed by the regime (organic certification) and to question the functioning of the market system (landscape). With widened and improved knowledge on these interdependent system hierarchies, it became easier for them to decide what was and what was not acceptable.

The group members demonstrated that although increased societal consciousness of the environmental impacts of oil consumption might affect their social relations with customers positively, from a short-term economic viewpoint they showed it would be more 'sustainable' for production to keep using subsidised oil. The resultant intensification of the 'urge to change' is exemplified in the transformation from oil to bio-fuels such as wood chips, which had a very clear outcome in terms of the environmental impacts. As the focus of the work widened to encompass the whole system, a new awareness grew within the group of the tight coupling between organic farming and conventional systems. Realising that national politics on food had an effect on, for instance, the time the growers spent with their families or whether they would accept fertiliser systems based on non-organic animal welfare, the growers acknowledged on a conscious level that to find solutions to these problems they had to pay attention to the interconnections between different areas of their production and between the enterprise and other system levels.

Conclusions

So, was PLAR sufficient for supporting the development of organic agriculture in order to maintain and increase sustainability?

If research can integrate efforts in a transdisciplinary research framework with farmers and growers as full members of the team, there is a much greater probability of working on questions relevant to the situation and reaching solutions that will be used. This is especially important as we approach natural limits of key resources such as fossil fuels, and it will require all of our ingenuity as teams to learn from natural systems and design our own agri-food systems for long-term sustainability.

In this case, PLAR shown to be useful for transitions in areas of interest, knowledge and actions taken, but was still not sufficient for support the development of an agro-ecological system. Goal conflicts in the growers' practical situation were found to be unavoidable. PLAR does not remove these, but does help participants better understand the reasons for these differences between social, environmental, enterprise profitability and production goals.

Higher levels of transitions than those at the niche level only are needed to drive systemic change in production regimes. PLAR can support the participants to search for synergies among these goals at different system levels. This way of researching is distinctively different to the predominant economic analysis of trade-offs between conflicting goals. This case created a context in which participants could learn from diversity and helped develop a more sustain-able basis for production at a niche level. There was no ambition to change the regime or the landscape in this case. PLAR's contribution to multilevel systemic transitions remains unclear, but the kind of collaborative environments developed in this case study supported the participants to ask the questions that challenge 'business as usual'. Higher-level transitions are needed if transitions at the niche level are to drive systemic change in production regimes.

It is important to note that the initial purpose of participants in the overall project was not to confront the sustainability issue. The challenge to sustainable practice emerged through engaging in a learning process for purposes more closely related to their own enterprises, values and lifestyles. The significant feature of participant engagement was that they dared to ask the questions that questioned their own production system. This kind of questioning can generate the systemic transitions necessary if farmers are to make a start in the journey towards achieving sustainability in a changing global climate.

ACKNOWLEDGEMENTS

Our gratitude to all the participants in the PLAR group: Olof Andersson, Karl-Gunnar Berglund, Ulf Engström, Bengt Eriksson, Kristina Homman, Anders Y Larsson, Adim Mizban, Britt-Inger Nillson, Hans Nilsson, Göran Pellas, Karin and Mats Sjöstedt, Lisbeth and Sven-Erik Wilhelmsson and Elisabeth Ögren. We also thank Janice Jiggins for her contribution in the editing process and Ekhagastiftelsen for their financial support.

REFERENCES

Ahnström J, Höckert J, Bergeå L H, Francis C A, Skelton P, Hallgren L

(2009). Farmers and nature conservation: What is known about attitudes, context factors and actions affecting conservation? Renewable Agric. Food Syst., 24: 38-47.

Björklund J (2000). Emergy analysis to assess ecological sustainability Strengths and weaknesses. Doctoral dissertation, Agraria 242. Swedish University of Agricultural Sciences, Uppsala.

Björklund J, Eksvärd K, Andersson O, Berglund KG, Engström U, Eriksson B, Johansson D, Larsson AY, Mizban A, Nilsson BI, Nilsson H, Pellas G, Sjöstedt K, Sjöstedt M, Wilhelmsson L, Wilhelmsson SE, Homman K, Ögren E (2005). Vad kan egentligen kallas ekologiska tomater? Ekologiskt lantbruk 42. Swedish University of Agricultural Sciences, Uppsala. 42 pp.

Brown MT, Herendeen R (1996). Embodied energy analysis and emergy analysis: a comparative view. Ecolog. Econ., 19: 219-236.

Defoer T (2002). Social learning for integrated soil fertility management in sub-Saharan Africa. In: Leeuwis C, and Pyburn R (eds) Wheelbarrows full of frogs social learning in rural resource management. Koninklijke van Gorcum, Assen. pp. 147-166.

Deuten JJ (2003). Cosmopolitanising Technology: A study of four emerging technological regimes. Twente University Press. Enschede.

Eksvärd K (2010). Facilitating systemic research and learning and the transition to agricultural sustainability. J. Agric. Educ. Ext., 16(3): 265-280.

Eksvärd K, Rydberg T (2010). Integrating Participatory Learning and Action Research and Systems Ecology – a Potential for Sustainable Agriculture Transitions. Syst. Practice Action Res., 25(1): 55–68

European Commission (2008). Eurostat homepage. http://epp.eurostat.ec.europa.eu/tgm/table.do?tab=table&init=1&plugin=0&language=en&pcode=tsdpc440 (last accessed 1 Dec 2008).

Geels F (2002). Understanding the Dynamics of Technological Transitions. A co-evolutionary and soci-technical analysis. Twente University Press, Enschede.

Johansson S (2005). The Swedish Foodprint: An Agroecological Study of Food Consumption, Doctoral thesis no 2005:56. Swedish University of Agricultural Sciences, Uppsala.

Kahane A (2004). Solving tough problems An Open Way of Talking, Listening, and Creating New Realities. Berett-Koehler Publishers. San Fransisco. 149 pp.

Lagerberg C, Brown MT (1999). Improving agricultural sustainability: the case of Swedish greenhouse tomatoes. J. Cleaner Prod., 7: 421 – 434.

Lagerberg C, Gertsson U, Larsen R, Gäredal L (1999). Emergy evaluation of five greenhouse tomato production systems. In: Lagerberg C (eds.). Emergy Analysis of the Resources Use in Greenhouse Crop Production and of the Resources Basis of the Swedish Economy. Doctoral dissertation, Agraria 191. Swedish University of Agricultural Sciences, Uppsala.

Mikkelsen B (2005). Methods for Development Work and Research. Sage Publications Ltd. London, 375 pp.

Milestad R, Wivstad M, Lund V, Geber U (2008). Goals and standards in Swedish organic farming: trading off between desirables. Int. J. Agricultural Resources, Governance Ecol., 7 (1/2): 23-39.

Nembhard IM, Edmondson AC (2006). Making it safe: the effects of leader inclusiveness and professional status on psychological safety and improvement efforts in health care teams. J. Organ. Behav., 27: 941-966.

Neuman WL (2003). Social Research Methods. Qualitative and quantitative approaches. Allyn and Bacon, Boston.

Nilsson U (2007). Ekologisk odling av grönsaker, frukt och bär – hinder och möjligheter för framtida utveckling. Ekologiskt lantbruk, 49. Swedish University of Agricultural Sciences, Uppsala. 51 pp.

Norman D, Bloomquist L, Janke R, Freyenberger S, Jost J, Schurle B, Kok H (2000). The meaning of sustainable agriculture: Reflections of some Kansas practitioners. Am. J. Alternative Agric., 15(3): 129-136.

Odum HT (1996). Environmental Accounting: Emergy and Environmental Decision Making. John Wiley & Sons Inc, New York. 370pp.

Ögren E, Homman K, Rämert B, Andersson O, Berglund KG, Eriksson B, Engström U, Johansson D, Mizban A, Nilsson BI, Lindström J, Lindström T, Pellas G, Sjöstedt K, Sjöstedt M, Wilhelmsson L, Wilhelmsson SE (2008). Metoder för att minska skadeverkningar av korkrot, *Pyrenochaeta lycopersici*, i ekologisk tomatodling. Lantbruks och fiskeenheten 2008:2. Västmanland County Administrative Board, Västerås.

Pretty J, Guilt I, Thompson J, Scoones I (1995). A Trainer's Guide for Participatory Learning and Action. International Institute for Environment and Development, London.

Raven R (2004). Conceptual Framework. Chapter two in: Strategic Niche Management for Biomass. A comparative study on the experimental introduction of bioenergy technologies in the Netherlands and Denmark. Published Ph.D. dissertation. Centre for Innovation Studies, Eindhoven, pp. 25 – 57.

Wackernagel M, Rees W (1996). Our ecological footprint. Reducing human impact on the earth. New Society Publisher, Gabriola Island. 160 pp.

Wang CL, Ahmed PK (2003). Organizational learning: a critical review. Learning Organ., 10: 8-17.

Extension agents access and utilization of information and communication technology (ICT) in extension service delivery in South East Nigeria

Ezeh Ann N.

Department of Agricultural Economics Management and Extension, Ebonyi State University, Abakaliki, Ebonyi State, Nigeria.

This study was carried out to assess field extension agents' access and utilization of information and communication technology (ICT) in extension service delivery in South East Nigeria. Multi-stage sampling procedure involving purposive and simple random sampling techniques were employed in selecting 120 Extension Agents used for the study. Primary data were collected with the aid of a well-structured questionnaire and analyzed using descriptive statistics according to the specific objectives of the study. The result of the socio-economic characteristics of the extension agents showed that 61.67% of the extension agents studied were males whose age bracket fall between 40-49 years and majority (78.33%) were married. Majority (85%) had postsecondary education with a household size of 1-3 persons and majority (51.67%) earned monthly income of ₦21, 000-₦40, 000. Further analysis revealed that although many ICT facilities exist, both contemporary and conventional ICTs, radio, television and phones sourced personally from open market were the most readily available ICT facilities owned, accessed, and utilized by most of the extension agents. Consequently, the level of access and utilization of these facilities were found to be generally low with a mean score responses of x = 1.8 and x = 2.4 respectively. The result of extension agents' opinion on the ICTs application in facilitating agricultural activities indicated; information on new varieties of seeds and seedling (x=2.6), acquire information from research institutions and farmers (x=2.7), best planting techniques (x=2.6) and use of fertilizer (x = 2.8) among others. Necessary recommendations such as strengthening the use of contemporary ICTs (phones, internet connected computer etc.), provision of adequate training on the use of ICTs for extension agents and general improvement in rural infrastructure were made among others.

Key words: Information and Communication Technology (ICTs), extension agents, access, utilization, facilities.

INTRODUCTION

Over the years, agricultural extension has been at the fore-front in the delivery of adequate information to farmers for increased productivity. According to Agbamu (2007), agricultural extension service delivery all over the world has been concerned with communicating research findings and improved agricultural practices to farmers.

The efficiency with which these information and practices are conveyed to farmers to a large extent would determine the level of agricultural productivity. Extension organisations have been concerned with what should be the appropriate means and approaches in getting the right agricultural information to the end-users (farmers). In recent times however, there has been revolution with regards to Information and Communication Technology (ICT) in agriculture and particularly in extension service delivery of Nigeria. This revolution is an intervention with the potential to ensure that knowledge and information on important agricultural technologies, methods and practices are put into right use by farmers. The use of ICT provides an important platform for actualizing this. ICTs consist of various collections of resources and technical tools that are used for connecting, spreading, storing and managing information (Pigato, 2004).

In other words, ICT represents the collection of hardware and software that is used for producing, preparing, transferring and storing data via devices such as computers, radios, televisions, etc., and it includes an extensive scope of traditional and modern media (Norad, 2002). In general, ICTs can be classified into three groups:

New ICTs

This group consists of computers, satellites, one-on-one connections, wireless phones (mobile), the internet, e-mail, the web, internet services, video conferences, CDROMs, personal computers (PC), distance control systems, informational-geographical systems, global positioning systems (GPS), electronic cameras, databases, etc. The hidden concept behind these technologies is that they are not automatically considered to be new, but their common and inexpensive availability has resulted in them being regarded as new.

Old ICTs

This group consists of radios, televisions, telephones, telegraphs, audio and video cassettes, films and slides. This group of technologies has been used for several decades.

Very old ICTs

This group of technologies has been used for several centuries and includes newspapers, books, photo albums, posters, theater, human interactions, markets and plays (Obayelu and Oyunlade, 2006).

Similarly, CTA (2003) 'information and communication technologies' are technologies which facilitate communication and the processing and transmission of information electronically. The use of ICT in agricultural extension and rural development is significant especially now that its use has witnessed an upsurge in almost all areas of rural life in several African countries; however, the persisting problems of connectivity, literacy, content and costs coupled with the issue of adequate access have continues to hinder expansive utilization of these facilities for agricultural information. In this respect, Omotayo (2005) observed that agricultural extension depends largely on information exchange between farmers and broad range of other actors who are the front line extension workers that are the direct link between farmers and other actors in the agricultural knowledge and information system (AKIS). In the past decade, a number of Sub-Saharan African countries have experienced slow agricultural development. Since then, the sector has not grown as fast as the population. Decline in agricultural development is attributed to a number of constraints that include inappropriate national agricultural development policies, lack of adequate information provision, low adoption of agricultural technologies and ineffective institutional frameworks (Asiabaka, 2009). However, inadequate provision of relevant, reliable and comprehensive information support to stakeholders in agricultural production has been identified as a major constraint (Kiplangot, 2003).

Information and communication technology have become an increasingly powerful tool for improving the delivery of basic services and enhancing local development opportunities. Today, a new paradigm of agricultural development is fast emerging in both developing and developed countries. The overall development of rural areas is expanding in new directions; old ways of delivering important information services to citizens are being challenged; and traditional societies are being transformed into knowledge societies all over the world (Meora et al., 2004). In Nigeria, the government having recognised the importance of ICT in extension service delivery. In this direction, she has established policies and projects among which are; liberalisation and privatisation policies to encourage private sector participation and attraction of foreign investment which consequently reduces the price of computers and other ICT equipments, making them more accessible. There was also the launching of the global system of mobile (GSM) communication, making phone lines more accessible nationwide and the establishment of internet cyber cafes in all the major urban areas in the country, though the rates are still high and service unsteady (SGF, 2002). Very important also was the establishment of information and communication support (ICS - Nigeria) in 2002, for agricultural growth (Oladimeji et al., 2004). This project though still on pilot phase in 8 states of the country (and in Abia State in the South-East) is aimed at establishing resource centres in rural areas, which makes information available to farmers and farm assistance organisations (ADPs, NGOs, Researchers,

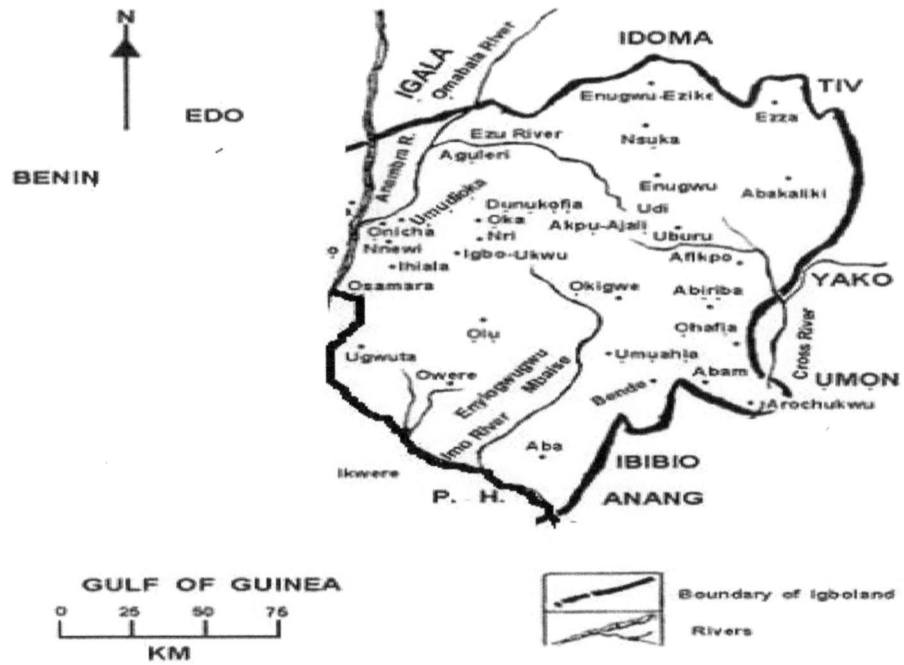

Figure 1. Map of south east Nigeria.

NAERLS etc.) at different centres and farm levels, as well as modernising their information base through the use of ICTs, such as telephone, radio, television, compact disk, read only memory (CD-ROM) internets etc. (Arokoyo, 2007). Apart from facilitating the acquisition and absorption of information/knowledge networking, codification, tele-working and science system, ICT could be used to access global knowledge and communication with other people.

Desta and Petta (2003) noted that access to more sophisticated and integrated technologies of biotechnology, natural resource management and material science has been made easier by ICT. However, what may be of paramount importance may not just be the existence of these facilities but how accessible they are to the change agents particularly the field extension agents who make use of them in reaching out to both zonal level extension personnel and the farmers. Equally relevant, is knowledge of the level of disposition of these extension agents to the utilization of information and communication technologies in carrying out their functions. Arokoyo (2007) asserted that the extent of application of ICT tools in extension service depends mostly on the targeted audiences, level of ICTs infrastructural development and its accessibility and affordability. The serious limitation of access to reliable telephone lines makes even ordinary intra-and inter-organisational networking for information exchange a frustrating experience (Alexander, 2004). Despite the importance of ICTs in extension service delivery, there seems to exist a gap in knowledge between what is existing and what needs to be done to ensure the optimal access and utilization of these ICT infrastructures so that it will contribute significantly to development of extension service delivery and agricultural development in the south-east Nigeria in particular and the country in general.

In the light of this, the study was to determine access and utilization of information communication technology (ICT) among extension agents in extension service delivery in South East, Nigeria. To accomplish this, the following specific objectives were addressed:

(i) describe the socio-economic and personal characteristics of the extension agents in Southeast Nigeria;
(ii) identify the existing and sources of ICT facilities for extension agents in the study area; and
(iii) assess the access and utilization level of ICT facilities by the extension agents in the study area;

METHODOLOGY

The study area is South-East Nigeria; which comprises of five states namely; Anambra, Abia, Ebonyi, Enugu and Imo states (Figure 1). The area has an approximated land area of 5824.7 sq. km and lies between longitude 6°50^1 and 8°15$^/$E and latitudes of 4°30$^/$ and 7°15$^/$N (Mamman, 2000). The South-East states have about a total of 16,381,729 people (NPC) who are mainly of Igbo tribe. The vegetation lies within the rain forest and dry savannah region of Nigeria. It has a tropical humid climate with two distinct seasons of the year namely, the rainy and dry seasons. The rainy season last from April to October and is marked by regular drop in ambient temperature. The dry season lasts from November to March. It is characterized by a hot dry spell. Within the two seasons, the temperature range varies from about 18 to 34°C. Dominant arable crops of the area include yam, rice, cassava, maize, cocoyam, groundnut, pepper and various types of vegetables. Oil palm, banana, plantain and coconut are plantation crops dominant in the area. Small ruminant and backyard poultry production dominate the livestock industry and fisheries.

Sampling technique

A multistage sampling procedure involving purposive and simple random sampling techniques were employed in selecting the respondents used for the study. In the first stage, three (3) states out of the five (5) states making the South-east Nigeria states were randomly selected. These states include Abia, Ebonyi and Enugu. Secondly, forty (40) extension agents each serving in ADP were randomly selected from each of the three (3) selected states to give a total of one hundred and twenty (120) respondents. The study was carried out using primary data source. Data were obtained primarily using questionnaires which were directed to the extension agents. The data collected were analyzed using descriptive statistics. Objectives (i) and (ii) were achieved using descriptive statistics such as frequency distribution tables, percentages and objective (iii) actualized using 4-point Likert scale.

Model specification

Model for Likert scale rating

The Likert scale mean score derived from a 4-point rating was used to elicit the extension agents utilization capabilities in using the ICT components in terms of skill and knowledge in extension delivery respectively. Likert is a tool used in making explicit decision on the feelings attached with a particular observed phenomenon from possible factors or variables. In this case of using mean score derive from 4-point Likert rating to analyse the utilization capability of the respondents, some ICT facilities were itemized and the respondents will be required to rank them based on their agreement with their extent of utilization. In using 4-point Likert scale or rating, the response ranged between very great extent (VGE) with assigned value of 4 and very low extent (VLE) assigned the value of 1 with decision rule of 2.5. The rating ranged from 4 to 1 as follows:

Very great extent (VGE) – 4
Great extent (GE) – 3
Low extent (LE) – 2
Very low extent (VLE) – 1

Likert fomular:

$$X_s = \frac{\sum fn}{Nr}$$

Where X_s = Mean score, $\sum$ = summation, F = frequency of each (4, 3, 2, 1) option, n = Likert numerical values (responses of the respondents), Nr = number of respondents to each response category (total number of respondents).

RESULTS AND DISCUSSION

Socio-economic characteristics of the respondents

The result of the socio-economic characteristics of the extension agents indicates that majority (61.67%) were males while 38.33% were females (Table 1). The age range of majority (43.33%) was between 40 to 49 years, while 1.67% fell within age bracket of 60 to 69 years. However, the mean age of the extension agents was 40 years. Also, majority (78.33%) of the extension agents were married whereas the least about 5% were divorced. Furthermore, analysis revealed that majority (85%) attended post secondary education while only 5% completed primary education. The mean level of education of the extension agents was 13 years. The monthly income of most (50%) of the extension agents ranged between ₦21,000 to 40,000 while few (1.67%) earned between ₦81,000 to 100,000. Result showed that greater proportion (68.33%) of the extension agents belonged to social organizations while few (31.67%) could not belong to any organization. Most of the respondents (70.12%) belong to multi-purpose cooperative organizations. The least (3.33%) were engaged in agro-input marketing. The mean distance covered by the extension agents to ICTs source was 3.32 km.

Existence and awareness ICTs among extension agents

The result of data analysis on the existing ICT facilities in the study area revealed that most (98%) of the 'extension agents' testified to the existence and awareness of radio, followed by phone with 86% and television with 62% response. Others were 33 and 31% who were aware of media van and prints respectively (Table 2). 23% of the respondents were aware of projector, while few (6 and 5%) had noticed the existence of internet connected computer and cassette recorder/ player respectively. This follows the findings of Isiaka et al. (2009) who reported that extension workers were moderately aware of the existing ICTs potential in extension service delivery. Although, this finding slightly differs from Adebayo and Adesope (2007) who reported high (88.5%) level of awareness among female extensionists. This suggests that the extension agents in South East Nigeria have high awareness of conventional ICTs than contemporary ones; hence, more need to be done in educating them on the existence of contemporary ICTs for improving extension service delivery. The results on the sources of ICT facilities among the respondents in the study area indicated that most (58%) of them sourced their ICTs personally from open market, specifically, the responses were radio (95.56%), internet connected computers (25%), television (84.09%), mobile phone (93.33%); whereas 50 and 64% sourced cassette recorder/players and prints/libraries respectively, from the open market as well (Table 3). This is followed by 11% of the respondents whose source is cooperative society. The least was the government by 2% response.

Table 1. Percentage distribution of the respondents (extension agents) by socio-economic characteristics.

Gender	Frequency (n = 120)	Percentage (%)
Male	74	61.67
Female	46	38.33
Age range		
20-29	12	10.00
30-39	42	35.00
40-49	52	43.33
50-59	12	10.00
60-69	2	1.67
70 and above	-	-
Mean age	40 years	
Marital status		
Single	20	16.67
Married	94	78.33
Divorced	6	5.00
Separated	-	-
Widowed	-	-
Educational attainment		
No formal education	-	-
Adult literacy	-	-
Primary education	6	5
Secondary education	12	10
Post secondary education	102	85
Mean educational level	18 years	
Secondary occupation		
Students	2	1.67
Farming	94	78.33
Agric trading	4	3.33
Teaching	-	-
Transportation	-	-
Agro-processing	20	16.67
Farming category		
Crop production	64	35.96
Livestock production	38	21.35
Fisheries	4	2.25
Forest production	-	-
Agric. trading business	12	6.74
Non-agric trading business	-	-
Transportation business	6	3.37
Agricultural processing	12	6.74
Formal private employment	2	1.12
Mining /quarrying	-	-
Public sector employment	20	11.24
Domestic duties	12	6.74
Students	6	3.37
Artisians	2	1.12
Traditional medical practices	-	-

Table 1. Contd.

Farm size		
0.0-0.5	26	39.39
0.6-1.0	15	22.73
1.1-1.5	18	27.27
1.6-2.0	2	3.03
2.1-2.5	3	4.55
2.6-3.0	2	3.03
Mean farm size	0.5	
Household size		
1-3	54	45.00
4-6	44	36.67
7-9	12	10.00
10-12	10	8.33
Mean household size		
Working experience		
<3	35	29.17
4-6	70	58.33
7-9	10	8.33
9 and above	5	4.17
Mean working experience	5 years	
Income categories		
10,000-20,000	40	33.33
21,000-40,000	62	51.67
41,000-60,000	12	10.00
62,00-80,000	4	3.33
81,000-100,00	2	1.67
Membership of social organization		
Yes	64	53.33
No	56	46.67
Primary activities of social organizations		
Multi-purpose cooperative society	20	31.25
Crop production only	18	28.13
Agro-processing	8	12.50
Agro-input market	6	9.38
Agricultural marketing	12	18.75

Source: Field survey (2012).

There is an indication that majority of the extension agents sourced their ICTs personally from the open market and this may have be attributed to the liberation of telecommunication sector in Nigeria, which have open up the sector to private telecom service providers. This has given rise to the use of mobile phone and modem for browsing from even the comfort of their homes. The result of analysis on percentage distribution of ownership/access of ICT facilities shows that most of the extension agents (98.33%) and (85.83%) owned radio and television, respectively. This was followed by 81.67% who own mobile phones and 42.50% who have libraries. Twenty-four percent (24%) owned cassette recorder/player, the least were 15% who own internet connected computer. The result in Table 4 on place of access to ICT facilities by extension agents shows that most (85.0%)

Table 2. Percentage distribution of the respondents according to awareness of existing ICT facilities.

Existing ICTs	Frequency (n = 120)	Percentage (%)
Radio	346	96
Television	242	62
Phones (handset)	306	86
Prints/libraries	112	31
Internet connected computer	35	10
Cassette recorder/player	41	11
Media van	120	33
CD-ROM	80	22
Mobile cinema	26	7
GIS (geographical information system)	12	3
Facsimile	7	1
Projector	84	23
Total	1411*	

Source: Field survey (2012). *Multiple responses recorded.

Table 3. Percentage distribution of the respondents according to sources of ICT facilities in the study area.

ICTs facilities	Frequency (n = 120)					Percentage (%)				
	Open market	ADP	Fellow farmers	Government	Cooperative society	Open market	ADP	Fellow farmers	Government	Cooperative society
Radio	172	-	6	-	4	95.56	-	14.2	-	2.22
Television	148	8	8	4	8	84.09	4.55	4.55	2.27	4.55
Inter-connect comp	8	-	4	-	8	25.00	-	2.22	-	4.44
Mobile phone	168	-	4	20	12	93.33	-	4	20	12.00
Cassette recorder/players	16	-	-	4	12	50.00	-	-	12.50	37.50
Prints/libraries	64	8	4	-	12	64.00	25.0	12.50	-	37.50
Total	576*	16	26	28	56	58	4	6	2	11

Source: Field survey (2012). *Multiple responses.

accessed ICT facilities from cybercafé/tele-centres, followed by 57.5% who accessed from home. Some (35.0%) accessed from their workplace while few (18.3%) accessed ICT facilities in the public libraries (Table 5). Adebayo and Adesope (2007) observed that 59.3% of the female extensionists have no 'personal computers' in their offices; while those who indicated that they have 'personal computers' in their offices stated that they were not connected to the internet. This is a serious situation that shows that there is still a lot to be done if the Nigerian agricultural sector must meet up the global challenges of ICT. Most of the extension agents (30%) covered a distance of 4 km in search of ICT facilities, a reasonable number (37%) covered equal or less than 2 km while some (5.83%) still move a distance of 5 km and

Table 4. Percentage distribution of EAs according to ownership/access of ICT facilities in the study area.

ICTs facilities	Frequency (N = 120)	Percentage (%)
Radio	118	98.33
Television	103	85.83
Mobile phones	98	81.67
Prints/libraries	51	42.50
Internet connected computer	18	15.00
Cassette recorder/player	29	24.17
Media van	0	0
Geographical information system (GIS)	2	1.7
Total	419*	

Source: Field survey (2012). *Multiple responses.

Table 5. Percentage distribution of extension agents according to place of ICT access.

ICTs access	Frequency (N = 120)	Percentage (%)
Home	69	57.5
Workplace	42	35.0
Cybercafé/Tele centre	102	85.0
Public libraries	22	18.3
Total	235*	

Source: Field survey (2012).

Table 6. Percentage distribution of extension agents according to distance covered to search for and purchase ICTs facilities.

Distance in kilometer	Frequency (N = 120)	Percentage (%)
Less than or equal to 1	18	15.00
<2-2	27	22.50
<3-3	32	26.67
<4-4	36	30.00
5 and above	7	5.83

Source: Field survey (2012).

and above to access ICTs (Table 6). As indicated by Salau and Saingbe (2008), extension workers ss to available ICTs is either at home, place of work or business centres.

As indicated by Tanko et al. (2013), the use of ICT facilities has continued to be hampered by the persistent problem of access, connectivity literacy and cost.

Access to ICTs

The results of the analysis on the access to ICT facilities by extension agents indicates that majority (x = 2.5) had access to television, followed by a reasonable number (x = 2.7 and 2.6) who had access to radio and phones, respectively. Few (1.4, 1.2 and 1.0) had access to libraries, and cassette recorder/players, while the least response x = 1.0 was on access to internet connected computer and GIS, respectively (Table 7). This is in agreement with the findings of Salau and Saingbe (2008) who reported that extension workers in Nasarawa state had high access to radio, television and mobile phone, while fewer (49%) had access to computers, which they regarded as key to all ICT facilities. The result in Table 6 showed that the extension agents studied, had operational literacy on utilization of ICT facilities at varying levels. Majority (48 and 35%), (33 and 41%), (34 and 31%) and (24 and 32%) of the extension agents can operate radio, phones, cassette recorder/ player and television at excellent and very good levels, respectively (Table 8). Few can operate most of the modern/cotemporary ICTs such as internet connected computer, prints/libraries,

Table 7. Mean score responses of extension agents on the extent of access to ICT facilities.

Items	Mean score (x)	Decision rule (2.5)
Radio	2.7	Accepted
Television	2.5	Accepted
Phone	2.6	Accepted
Prints/libraries	1.4	Rejected
Cassette recorder/player	1.2	Rejected
Internet connected computer	1.0	Rejected
GIS	1.0	Rejected
Average	1.8	Rejected

Weighted mean = 2.5. Source: Field survey (2012).

Table 8. Percentage distribution of extension agents on ICTs operation literacy.

ICTs facility	Frequency (N = 120)			
	Ex	V. G.	G	P
Radio	48(41)	35(30)	26(22)	9(8)
Television	24(23)	43(42)	28(27)	10(10)
Internet connected computer	0(0)	4(22)	6(33)	8(44)
Phone	32(33)	40(41)	22(22)	4(4)
Libraries	19(33)	8(14)	14(25)	6(11)
Cassette recorder/player	10(34)	9(31)	6(21)	4(14)
Projector	2(10)	4(20)	2(10)	12(60)
GIS	0(0)	0(0)	0(0)	2(100)

Source: Field survey (2012). Figures in parenthesis are in percentages.

projector and GIS. This implies that there is still a huge gap in the operational knowledge of the extension agents; hence, adequate training on efficient operation of contemporary ICTs should be organised for the agents.

Utilization of ICTs to extension services

The result of the extent of ICTs application in farm information by the 'extension agents' showed that virtually all information concerning farming were disseminated through ICTs, but at low level. However, the responses are: physical farm measurement (2.6), how to estimate farm output (2.7), best planting techniques (2.6), use of fertilizer (2.8) and method of keeping livestock (2.6) were areas of high level of application in farm information by the respondents (Table 9). While quantities of chemical to use (1.0), quantities of seeds/cuttings/stems to use (2.4) among others recorded low application. This justifies the findings of Isiaka et al. (2009) who opined that most of the extension workers were not really conscious of the fact that ICT devices such as computer, internet and GSM can be used to efficiently accomplish numerous functions in extensions service delivery, and as indicated by Wijekon and Newton (2000), without appropriate grounding in communication technology potentials, efficient utilisation among the extension workers may not be realised. The result of the mean score analysis with cut-off of 2.5 in Table 10 shows that the respondents utilized radio (x = 3.0), television (x = 2.8) and phones (x = 3.0) to a great extent in their farming activities. Others were internet (x = 2.2), cassette recorder/players (x = 2.9), prints/libraries (x = 2.8), projector (x = 1.8) and GIS (x = 1.0) (Table 10). This agrees with the findings of Salau and Saingbe (2008) who reported that 56.22% of the sampled extension workers used ICT items such as telephones, internet, radio, television, video films/camera and power point for agricultural extension activities. They concluded that more education and awareness is required to improve the level of utilization of ICTs by extension workers.

Conclusion

Based on the findings of this research work, it can be concluded that radio and television, followed by phone were the most accessed and utilized ICTs among the extension agents. However, the extent of access and utilization of contemporary ICTs such as internet is still

Table 9. Mean score responses of extension agents on the application of ICTs in farm activities.

Items		Decision
Get information on new varieties of seeds and seedling	2.6	Accept
Quantities of chemical to use	1.0	Reject
Get information on market situation	2.4	Reject
Acquire information from research institutions and farmers	2.7	Accept
Acquisition of skills	0.6	Reject
Best planting techniques	2.6	Accept
Best time to harvest farm products	2.0	Reject
Best processing methods	2.2	Reject
How to store produce	1.0	Reject
Best chemicals for storing produce	2.2	Reject
Record keeping	0.4	Reject
Get information on farm input	1.8	Reject
Communicating with extension works and farmers	2.0	Reject
Use of fertilizer	2.8	Accept
Use of agro-chemical	0.9	Reject
Information on credit needs	0.7	Reject
Searching for recent farm information	1.2	Reject
Improved breed of snail	1.4	Reject
Improved breed of poultry	0.2	Reject
Method of deworming pig	2.1	Reject
Methods of disease prevention and control	1.9	Reject
Disease/pest control	1.8	Reject
Best method of feeding livestock	2.6	Accept
Average score	1.7	Reject

Average Mean score = 1.7, decision = 2.5. , Source: Field survey (2012).

Table 10. Mean score responses of extension agents on the extent of utilization of ICTs in the study area.

ICTs facility	Mean score	Decision rule
Radio	3.0	Accepted
Television	2.8	Accepted
Internet connectivity	2.2	Rejected
Phone	3.0	Accepted
Libraries/prints	2.8	Accepted
Cassette recorder/ player	2.9	Accepted
Projector	1.8	Rejected
GIS	1.0	Rejected
Average score	2.4	Rejected

Source: Field survey (2012). Cut-off: 2.5.

very low. The study also indicated low level of operational literacy and this has provided a huge gap in the operational knowledge of the extension agents; hence, adequate training on efficient operation of contemporary ICTs should be organised for the agents. ICTs can be harnessed to have a direct impact on enhancing both rural livelihoods and the agricultural sector in which the rural poor are immersed. This can be achieved through increase human and capital investment in ICTs projects that is driven by technologies of today. Based on the findings of this study, the following recommendations have been put forward:

(1) Agricultural extension services in the south eastern Nigeria should incorporate or strengthen the use of other ICTs especially the contemporary ICTs such internet,

mobile phones etc. for information dissemination to the farmers. Extension organisations should open websites to enable them have link with others.
(2) ICT training programmes should be organised/established for training extension agents on opportunities that abound in the use of ICT facilities for extension services delivery. This can be achieved by
(3) Improvement in rural infrastructure particularly electricity supply is very essential in the study area.4) Extension agents should be encouraged to access and utilize ICTs by supporting them with financial and material resources. This can be actualised by providing them with credit grant or the equipments by the government.

REFERENCES

Adebayo E.L, Adesope OM (2007). Awareness, access and usage of information and communication technologies between female researchers and extensionists. Int. J. Edu. Develop. Inform. Commun. Technol. (IJEDICT) 3(1):85-93.

Agbamu JU (2007). Essentials of Agricultural Communication in Nigeria. Malthouse Press Limited Lagos. pp. 20-24, 91-93.

Alexander GF (2004). Information and Communication Opportunities for Technology Transfer and Linkages; Paper presented during the Expert Consultation on Agricultural Extension, Research-Extension-Farmer Interface and Technology Transfer, Food and Agriculture Organisation Regional office for Asia and Pacific, Bangkok, Thialand.

Arokoyo T (2007). ICTs Application Agricultural Extension Service Delivery. Proceedings of 12th Annual conference AESON (4th-7th July, 2007).

Asiabaka C (2009). Promoting Sustainable Extension Approaches: Farmer Field School (FFS) and Its Role in Sustainable Agricultural Development in African." http://www.codesria.org/Links/conferences/ifs/Asiabaka.pdf [accessed 2009 March 18].

CTA (2003). Information for Agricultural and Rural development in ACP Countries: New stakeholders, New Made and priority themes. Technical centres from agricultural and rural cooperation, Wagenninger, the Netherlands.

Isiaka BT, Lawal-Adebowale OA, Oyekunle O (2009). Agricultural Extension Agents' Awareness of ICT Potentials and Training Needs on Usage for Improved Extension Service Delivery in Selected Southwest States. Nig. J. Hum. Soc. Sci. Creative Arts 4(1):18-30

Kiplangot J (2003). Does Agricultural extension have a new beginning because of ICTs? Reflections on experience in sub-saharan Africa keynote paper presented at ICT observatory 2003; ICTs transforming agricultural extension wageningen, 23–25 September 2003.

Mamman AB (2000). Nigeria: A People United, A future Assured (Sokoto State). Gabumo Publishing Co. Ltd., Lagos, Nigeria. P. 298.

Meora SN, Jhamtani A, Rao DUM (2004). 'Information and communication technology in Agricultural development: a comparative analysis of three projects from India'. Agric. Res. Exten. Network. Network 135:13.

National Population Commission (NPC) (2006). Official Population Report of South East Nigeria.

Norad A (2002). Information and Communications Technology (ICTs) in Development Cooperation. Network Agency for Development Cooperation.

Obayelu A, Ogunlade I (2006). Analysis of The Uses of Information & Communications Technology for Gender Empowerment & Sustainable Poverty Alleviation in Nigeria. Int. J. Educ. Develop. 4:3-5.

Omotayo OM (2005). ICT and Agricultural Extension: Issues in transferring agricultural technology in developing countries proceedings of 3rd annual conference of AESON Ilorin. pp. 132-135.

Pigato M (2004). Information & Communications Technology Poverty & Development in Sub-Saharan Africa & South Asia, Africa Region Working Paper Series. P. 20.

Salau ES, Saingbe ND (2008). Access and Utilization of Information and Communication Technologies (ICTs) Among Agricultural Researchers and Extension Workers in Selected Institutions in Nasarawa State of Nigeria. PAT 4(2):1-11.

Tanko L, Adeniji OB, Nwachukwu H (2013). Evaluation of the access to and utilization of information communication technology (ICT) facilities among extension officers in Shiroro LGA, Niger State, Nigeria. J. Agric. Exten. Rural Develop. 5(1):8-13.

Wijekon R, Newton B (2000). Multimedia Training Support for Extension Trainers in Developing Countries. CESO Report. P. 19. The Netherlands.

Assessment of rural poultry extension services in Oshimili North Local Government Area, Delta State, Nigeria

Margaret Joke Koyenikan

Department of Agricultural Economics and Extension Services, University of Benin, Faculty of Agricultural Sciences, University of Benin, Benin-City, Nigeria. E-mail: mkoyenikan@yahoo.com.

The study assessed rural poultry extension services in Oshimili North Local Government Area of Delta State, Nigeria. Primary data were collected using interview schedule with 120 household poultry rearers that were selected randomly from four communities in the local government area. Data were analyzed using frequency counts, percentages, Pearson product moment correlation (PPMC) and analysis of variance (ANOVA). Findings showed that majority of the respondents were females 61.67 and 35% were above 50 years, and 70% were married with household size of 5 to 8 (30.83%), while 43.22% had no formal education. Also majority reared poultry on free range (53.3%), had stock sizes of ≤ 20 birds (77.50%) and 36.44% had poultry experience of above 15 years. About 92.50% had no contact with extension agent. There was poor access to improved poultry information and technologies (all < cut-off) while high needs were expressed with handling day old chicks and control of pest and diseases having mean=2.83 each and deworming (mean=2.78). Stock size (r= -0.312; p=0.001) and extension contact (r= -0.455; p=0.001) were significant negatively related to rural poultry extension services required. Scavenge-based poultry farmers indicated the greatest need for extension services (F= 5.81; p = 0.01). It was recommended that Delta State ADP and affiliate programmes should deploy extension agents/facilitators to the study area and ensure that information and technologies on poultry are addressed.

Key words: Rural poultry, extension service.

INTRODUCTION

Food security involves adequacy in quantity and in quality. Animal protein is essential in human nutrition because of its biological significance. The realization of the importance of animal protein made various governments in Nigeria to pursue programmes to boost production of livestock to ensure the attainment of Food and Agriculture Organization (FAO) recommendation of 35 g/caput of animal protein per day (Olaniyi et al., 2008). Among the different kinds of livestock that can be profitably produced in the country, agriculturalists and nutritionists have generally agreed that developing the poultry industry in Nigeria is the fastest means of bridging the protein deficiency gap presently prevailing in the country (Harban et al., 1978; Olomu, 1996).

Poultry meat and eggs are important foods for improving nutritional and health status of humans particularly the vulnerable populations. Village/rural poultry production is widely practiced in Africa. It involves any genetic stock of poultry (improved or unimproved) raised extensively, semi-intensively or intensively in relatively small numbers. The four classes of rural poultry production system on the basis of management and degree of commercialization are scavenge-based, free range, semi-intensive, intensive systems (Kekeocha, 1984) in Rushton et al. (1998). There is a minimal investment of inputs and labour but production is geared to consumption or sale (Sonaiya, 1995). Therefore, increasing its production would result in a positive impact on household food security both in increased dietary intake and income generation (Awuni, 2002). Some of the constraints facing poultry industry are shortage or high cost of good quality feed, prevalence of diseases, poor management practices, poor infrastructure and marketing, facilities, lack of credit (Imoukhuede, 1994; Omotosho, 1997). Olaniyi et al. (2008) however found that the major constraints to utilization of poultry production technologies were poor access to capital and inadequate extension contact. Fawole (2006) stated however that prospect for sustainable poultry production in Nigeria is high given locally available resources and suggested proper information dissemination, management and utilization as a way to increase poultry production in Nigeria. Williams and Williams (1984) in Age et al. (2002) reported that if the country wants to increase substantially the protein level in its peoples diet, then a more dynamic and aggressive livestock policy, especially in the areas of livestock extension would need to be vigorously pursued.

Extension is designed to help rural people satisfy their needs, interests, desires and basically to solve problems. Extension as the means by which farmers receive information and technologies is important for improving agricultural productivity and efficiency. Van den Ban and Hawkins (1998) stated that the major role of agricultural extension in many countries in the past was seen to be transfer of new technologies from researchers to farmers. Now it is seen more as a process of helping farmers to make their own decisions by providing them a range of options in a given innovation from which they can choose and by helping the farmers to develop insight into the consequences of each option. Agbamu (2005) identified some of the problems militating against extension service in most developing countries as inadequacy and instability of funding, poor logistics support for field staff, ineffective agricultural research-extension linkage, and disproportionate extension agents to farm family ratio. Others are lack of clientele participation in programme development, insufficient and inappropriate agro technologies for farmers. The problems by implication also affect the extension services to the poultry sub-sector.

Improvement in information communication technology (ICT) is likely to increase the prospect of extension services delivery through television, radio, global system mobile (GSM), the internet and the print media to complement public sector extension services. Extension services to rural poultry development could take advantage of ICTs. Dessie and Ogle (2005) reported that experience has shown that intensive persuasion was needed to convince the rural poultry producers to introduce regular watering, feeding, cleaning of birds' night shelter and to take care of the young chicks before starting any research or development programme to attain the genetic potential of the local birds. Various agencies/institutions (governmental and non government-tal) including Delta State ADP are known to deliver extension services. Efforts are being made by the state and federal governments to increase the animal protein consumption of the average Nigerians from the current average of 10 g/capu/day by 2006 (Age et al., 2006). However, there are still noticeable problems of high prices, shortage and intake of animal protein in the diet of average Nigerians. The trend and growth of the poultry industry is worrisome. Rural poultry production could not be said to have been affected by information and technologies disseminated by the extension service to improve production and thereby affect the lives of rural dwellers.

Delta State ADP has the mandate for extension delivery in the state. Oshimili LGA is expected to be covered by the ADP at the cell and block levels for the dissemination of information and technologies in all the sub-sectors of agriculture and the environment. It could not be categorically stated that the poultry producers have contact with extension agents or access improved poultry production information and technologies from any source. It is against this background that the study assessed rural poultry extension service delivery in Oshimili North Local Government Area of Delta State. The specific objectives were to:

(i) Examine the socioeconomic characteristics of poultry farmers in Oshimili North Local Government Area of Delta State
(ii) Examine the extent of contact of extension services with rural poultry farmers.
(iii) Identify the extension service-related constraints of rural poultry farmers.

H_1: There is no significant relationship between respondents' characteristics and poultry information access.
H_2: There is no significant difference between poultry extension services required by the different poultry management systems.

METHODOLOGY

The study area, Oshimili North Local Government Area (LGA) is one of the 25 local government areas in Delta State, Nigeria. Delta State is made up of three Senatorial Districts which are Delta North, Delta South and Delta Central. The State is one of the oil producing states in Nigeria located in the Niger Delta region in the south-south geo-political zone. The State covers an area of 17,698 km^2 with a coordinates of 5°30'N 6°00 with 25 local government areas. The main ethnic groups in the state are Igbo, Urhobo, Ijaw, Isoko and Itsekiri. It has a population of 4,098,291 comprising of 2,674,306 males and 2,024,085 females (NPC, 2006).

The major occupations of the people are farming, fishing and trading. The major livestock reared include poultry, piggery and goat while major crops produced are yam, oil palm, melon, cassava, maize. Rural poultry is prominent in the study area as every household seem to be involved in rural poultry production in Oshimili North LGA. The Local Government Area is made up of nine communities out of which four were selected namely Akwukwu–Igbo, Ibusa, Illah and Okpanam. Random sampling technique was used to select 30 rural poultry farmers/households from each of the communities to give a total sample size of 120. A well structured and validated questionnaire was administered to the respondents to collect the relevant data comprising of information on socioeconomic characteristics and poultry management and extension services on poultry from the respondents.

The dependent variable, access to poultry information was measured on a 3 – point Likert type scale (always = 3, sometimes =

Table 1. Socioeconomic characteristics of respondents.

Variable	Frequency	%	Mean
Age			
30 and below	23	19.17	40.06
31 – 40	20	20.00	
41 – 50	31	25.83	
>50	42	35.00	
Total	120	100.0	
Sex			
Female	74	61.67	
Male	46	38.33	
Marital status			
Single	20	16.67	
Married	84	70.00	
Divorced	-	-	
Widow	16	13.33	
Family size			
4 and below	37	30.83	
5 – 8	69	57.50	5.7
9 – 12	14	11.67	
Educational qualification			
No formal education	51	43.22	
Primary	16	13.56	
Secondary	18	15.25	
NCE	12	10.17	
OND/HND	13	11.02	
B. Sc and above	8	6.78	
Religion forbids poultry			
Yes	4	3.3	
No	116	96.7	
Total	120	100.0	
Major occupation			
Farming	54	45.00	
Trading	47	39.17	
Civil service	19	15.83	
Total	120	100.0	

Source: field survey Data, 2010.

2, never = 1). Poultry information required was also measured on a 3-point scale (not required=1, required=2, highly required=3) for the 14 items to give minimum=14 and maximum=42. Constraints encountered by the respondents were measured using a 3–point Likert type scale (very serious = 3, serious = 2, not serious = 1).

Data were analyzed using descriptive statistics mean, percentage, frequency, and standard deviation and inferential statistics, Pearson's product moment correlation (PPMC) was used to test relationships between selected socioeconomic characteristics and access to rural poultry extension services while analysis of variance (ANOVA) was used to test the difference in means of the different management systems with respect to rural poultry extension services required.

RESULTS AND DISCUSSION

Socioeconomic characteristics of respondents

Table 1 shows that 35% of the respondents were above 50 years which implies that majority of them were adults. This could be due to urban migration of the youths from the communities. The table also shows that 61.67% of the respondents were females. This indicates that majority of those who keep poultry were females. This finding is similar to the study of Ekue et al. (2002), who reported that rural poultry keeping is traditionally the role of women in many developing countries. The table also reveals that majority (70.0%) of the respondents were married, 57.50% had household size of 5 to 8 persons and 43.22% had no formal education. The result is similar to the findings of Moges et al. (2010) which found that most of the rural household poultry rearers in Ethiopia were illiterate. This implies that their capacity and ability to seek and understand information and knowledge rela-ted to their poultry business might be negatively affected. Furthermore, 96.7% do not have religious inclinations that forbid poultry rearing.

Poultry production information

Table 2 shows that 36.44% had poultry experience of greater than 15years, 77.5% had stock sizes of 20 and below. This agrees with Adene et al. (2006) who found that the household stocks ranged from 5 to 38 for chickens in Kano State, Nigeria while an average of 18.8 birds per household was reported by Khalafalla et al. (2001) in Sudan. This implies that majority of the respondents were small scale producers. Table 2 also shows that 45.0% of the respondents were engaged in farming, 80.8% were depended on family labour, 55.0% engaged in poultry production mainly for home consumption and sales. This agrees with the find-ings of Sonaiya (1990) and Dessie and Ogle (1996) that poultry farming objectives in Nigeria were mainly for home consumption and for sales hence there is prospect for poultry production to grow and ensure food security and alle-viate poverty if adequately practiced. Majority (92.5%) of the respondents had no contact at all with any extension agent. This implies that farmers' access to improved technological information is limited.

Table 2 further reveals that majority (53.3%) of the respondents practiced free range system. The result is similar to Jens et al. (2004) who asserted that nearly all rural and peri-urban families in developing countries keep

Table 2. Poultry production information of the respondents.

Variable	Frequency	%	Mean
Poultry experience			
5 and below	29	24.58	
6 – 10	24	20.34	
11 – 15	22	18.64	11.3
>15	43	36.44	
Total	118	100.0	
Stock size			
10 and below	47	39.17	
11 – 20	46	38.33	
21 – 50	10	8.33	
51 – 100	3	2.50	58.417
101 – 500	9	7.50	
>500	5	4.17	
Total	120	100.0	
Sources of labour			
Personal	15	12.5	
Family	97	80.8	
Hired	8	6.7	
Total	120	100.0	
Reason for engaging on poultry			
Home consumption	46	38.33	
Sales only	8	6.7	
Consumption and sales	66	55.0	
Total	120	100.0	
Poultry mgt system practiced			
Scavenged base	25	20.8	
Free range	64	53.3	
Semi-intensive	16	13.3	
Intensive	15	12.5	
Total	120	100.00	
Contact with extension services			
Not at all	111	92.5	
Yearly	3	2.5	
Quarterly	3	2.5	
Monthly	3	2.5	
Weekly	0	0.0	
Total	120	100.00	

Source: field survey Data, 2010.

a small flock of free range chickens. This implies that majority of the respondents were low income earners. Also the result is similar to Abubakar et al. (2007) who reported that village chicken are reared under free range system where they scavenge around the homestead and the surrounding.

Table 3. Sources of improved practices on poultry n=120.

Sources	Frequency	%
Inputs/sellers	21	17.50
Para-veterinary services	13	10.83
NGOs	2	1.66
LGC Agricultural personnel	1	0.83
ADP extension agent (EAs)	4	3.33
Television	2	1.66
Radio	3	2.50
Newspapers	2	1.66
Bulletins	2	1.66
Outdoor broadcast	-	-
Role play	-	-
Cooperative societies	-	-
Family and friends	5	4.16

Source: Field survey Data, 2010, Multiple responses.

Sources of improved technologies and practices on poultry

Table 3 shows that input sellers and Para-vets indicated by 17.50 and 10.83% of the respondents respectively were the major sources of improved poultry technologies and practices in the study area. This is an indication that poultry production will be mainly traditional practices which are not likely to meaningfully improve food security and reduce poverty.

Access to and need for improved poultry technologies and practices

As shown in Table 4, the respondents' access to identified technologies/practices was very low ranging from debeaking (1.03) to use of compounded feeds for birds (1.52) with all the mean scores less than mid point (2.00) and the standard deviation showing high dispersion from the mean. This might be due to the high proportion that had no contact with extension agents (92.50%) in Table 2 and poor coverage of the sources of improved practices in Table 3.

Table 4 also shows that majority of the respondents highly needed extension services in virtually all the areas such as on handling of the products and day old chicks, and control of pests and diseases (Mean score = 2.83). Low need was expressed for use of foot bath and marketing of the products since they have a mean score of (2.77) as compared to provision of water, washing of drinkers and clearing of poultry houses with mean scores less than 2.00.

Constraints to rural poultry production

Table 5 reveals that inadequate information and inadequate access to capital with (mean score = 2.96), inadequate extension agent contact with a mean score of (2.90), incompetence of extension agents (2.75) were the serious constraints while marketing of products (1.47), and pests and diseases (1.13) with mean scores less than 2.00 were less serious. This is similar to the findings of Olaniyi et al. (2008) who reported that the major constraint to utilization of poultry production technology was access to capital and inadequate extension contact.

Relationship between respondents' characteristics and poultry extension services required (correlations)

The correlation results between respondents' characteristics and poultry extension services required. The result shows that only stock size (r = -0.312) and contacts with extension agent (r = -0.455) had significant relationships with the poultry extension services required while age (r=0.166), family size (r=0.036), education (r=-0.075) poultry experience (r=0.175) were not significant at 0.05 level. The implication is that the respondents with less stock size tend to require more extension services than those with larger stocks. Also, those with more extension agent contact required less poultry extension services.

Poultry extension services required by different management systems (ANOVA)

Respondents who practiced scavenge-based system were those with the greatest need for poultry extension services (means=37.68) followed by those who operates free range system (mean=35.73). Those with the least requirement for the extension services were respondents practicing intensive management system (mean=30.29). The ANOVA result (F=5.81, p<0.01) is significant implying that there is a significant difference in the requirement of poultry extension service by respondents practicing the different poultry management systems.

Conclusions

Poultry extension was minimal in the study area due to poor contact with extension agents and poor coverage of improved poultry practices by the available information sources. The respondents who practiced scavenge based system requires more extension and those with small stock size had less contact with extension agent thus needed poultry extension services more. Rural poultry is constrained in the area by inadequate information, and lack of interest on the part of extension agents among others.

RECOMMENDATIONS

The following recommendations are made:

Table 4. Access to and need for improved poultry technologies and practices n=120.

Technology/practice	Access		Need	
	Mean	SD	Mean	SD
Washing of drinkers	1.08	0.876	1.86	0.853
Deworming	1.35	0.761	2.78*	0.537
Debeaking	1.03	0.851	2.74*	0.572
Control of pests and diseases	1.32	0.866	2.83*	0.461
marketing of products	1.04	1.00	2.77*	0.514
Brooding	1.28	0.823	2.55*	0.646
Use of foot bath	1.09	0.782	2.77*	0.530
Improved poultry breeds	1.38	1.15	2.71*	0.541
Feeding of the birds	1.52	0.856	2.76*	0.565
Culling	1.26	0.843	2.67*	0.599
Clearing of poultry house	1.21	0.752	1.49	0.789
Egg collection	1.09	1.03	2.68*	0.622
Handling especially day old chicks	1.24	0.811	2.83*	0.461
Provision of water for birds	1.15	0.824	1.87	0.849

Source: Field survey Data, 2010, *mean ≥2.0=access/required.

Table 5. Extension service-related constraints to rural poultry production (n = 120).

Constraints	Mean	Standard deviation
Inadequate information on improved practices	2.96*	0.239
No linkage to capital/money	2.96*	0.239
Inadequate extension agent contact to demonstrate skills and motivate farmers	2.90*	0.353
Poultry needs not identified	2.75*	0.506
No linkage to market produce	1.47	0.621
Pest/disease not prioritized	1.13	0.501

*serious (mean > 2.00) Source: Field survey Data, (2010).

1. There is need for Delta State ADP to deploy extension agents to the area and ensure poultry production practices are covered to address the required information and practices.
2. There is need to link poultry farmers up to credit facilities in order to develop rural poultry through technology adoption;
3. Extension service could also assist in organizing poultry farmers into groups and cooperatives in order to increas access to credit and other production (inputs);
4. There is need for other government and non-governmental (NGOs) bodies in the study area should provide extension support and enabling environment for rural poultry to thrive.

REFERENCES

Abubakar MB, Ambali AG, Tamjdo,T (2007). Rural chicken Production: Effects of Gender on Ownership, and Management Responsibilities in Some Parts of Nigeria and Cameroon. Intern. J.Poult. Sci. 6 (6): 413-416.

Adene DF, Oguntade AE (2006). The structure and importance of the commercial and village-based poultry industry in Nigeria. FAO (Rome) Study, their health and in their diets. The processors under October 2006.

Agbamu JU (2005). Problems and Prospects of Agricultural Extension Service in developing countries. Agricultural Extension Society of Nigeria, Ilorin, Nigeria. pp. 159 – 169.

Age AI, Obinne CPO, Idu EE (2006). A comparative assessment of livestock Extension Policy Implementation in Benue and Nassarawa State, Nigeria. Proceedings of 11th Annual conference AESON[3rd -6th April, 2006] pp. 1-2.

Awuni JA (2002). Strategies for the Improvement of rural chicken production in Ghana,In: Characteristics and Parameters of Family Poultry Production in Africa. IAEA, VIENNA 33–37.

Dessie T, Ogle B (1996a). A Survey of Village Poultry Productionin the Central Highlands of Ethiopia. Part I of M.Sc. Thesis. Swedish University of Agricultural Sciences, Department of Animal Nutrition and Management.

Ekue FN, Pone KD, Mafeni MJ, Nfi AN, Njoya J (2002). Survey FAO (1993): Livestock for Food, Income, Employment and Sustainable Agriculture. FAO Animal Production and Health Division, Rome, Italy.

Fawole OP (2006). Poultry farmers' utilization of information in Lagelu local government area of Oyo state, Nigeria. Int. J. Poult. Sci. 5: 497-501.

Harban S, Early NM (1978). Livestock and poultry production, New Delhi: Prentice Hall. http://www.husdyr.kvl.dk/htm/php/tun9/5 Saleque.htm) http://www.husdyr.kvl.dk/htm/php/tune99/24 Branckaert.htm). pp. 462– 63.

Jens Christian R, Anders P, Charlotte V, Ainsh MC, Lone F (2004). Keeping Village Poultry. A technical manual for small-scale poultry production. Copenhagen, Denmark.

Kekeocha CC (1984). Systems of poultry keeping. In Poultry production handbook,. London and Basingstoke, UK, Macmillan pp. 14-29.

Khalafalla MK, Bessei W, Adam MA (1995). A Development Strategy for Improving Backyard and Semi-intensive Poultry Production in Sudan. Sonaiya (Ed). Proceedings ANRPD Workshop, Addis Ababa, Ethiopia, June 13 -16, 1995. pp 124 – 127.

Moges F, Mellesse A, Dessie T (2010). Assessment of village chicken production system and evaluation of the productive and reproductive performance of local chicken ecotype in Bure district, North West Ethiopia. Afr. J. Agric. Res. 5(13):1739-1748.

Olaniyi OA, Adesiyan OI, Ayoade RA (2008). Constraints to Utilization of Poultry Production Technology among farmers in Oyo State, Nigeria. J. Hum. Ecol., 24(4): 305-309.

Olomu JM (1996). Nature's most nearly perfect food. Inaugural Lecture series 45 of the University of Benin, Nigeria

Omotosho OA (1997). Towards Increasing Research Production in Nigeria Poultry Industry. Proceedings of the first National conference on Productivity Ibadan, nigeria.Macmillian Publishers.

Sonaiya EB (1990).The context and prospects for development of smallholder rural poultry production in Africa. CTA-Seminar proceedings on Smallholder Rural Poultry Production 9-13 October 1990, Thessaloniki, Greece, 1: 35-52.

Sonaiya EB (1995). Feed resources for smallholder poultry in Nigeria. World Animal Rev. 82 (1): 25-33.

Van den Ban AW, Hawkins HS (1998). Agricultural Extension, Second Edition, Blackwell Science publication, Oxford, pp. 267- 268.

Peri-urban development, livelihood change and household income: A case study of peri-urban Nyahururu, Kenya

Nicodemus Mandere Mandere[1,2]*, Barry Ness[1] and Stefan Anderberg[1]

[1] Centre for Sustainability Studies (LUCSUS), Lund University P.O. Box 170, SE-221 00, Lund Sweden.
[2]Department of Physical Geography and Ecosystems Analysis, Lund University, Sweden.

Peri-urban development has attracted increased attention in recent years particularly due to conflict/competition between new (urban) and traditional (rural) land uses as a result of peri-urban expansion. Much of the research in peri-urban development is concentrated in peri-urban concepts and definitions, environmental impacts and also impact on agriculture. Little attention is put in the assessment of the peri-urban development on household livelihood and income. This study was therefore conducted with the objective of assessing the impact of the peri-urban development dynamics to household income using the case study of peri-urban Nyahururu, Kenya. The analysis shows a decline in full time farming households from 90% in the 1960s to 49%; an indication of the declining economic significance of agriculture. The decline in significance of agriculture was mainly due to rapidly shrinking household agricultural land as well as low and fluctuating agricultural output prices which reduced the profitability from agricultural production. The decrease in agricultural land was due to the sale of land for residence/business premises and also land bequests to children. In return, households have adopted diverse non-farm activities whose earnings proved to be of varying importance to the annual household income. However most of the households engage in low income productive non-farm activities – nevertheless, the number of households engaged in high income productive non-farm activities was comparatively higher (10% more) compared to the most rural parts of the district. The infrastructural developments coupled with emerging business enterprises were found to be the main factors that enhanced the opportunities for household engagement in high income productive activities. However most of these developments are limited to the financially constrained informal sector and hence can not provide sufficient high income opportunities to lift majority of the population from poverty. Therefore, the possibility for peri-urban development to accomplish a reduction in poverty for the households will not only depend on the infrastructural developments but also on the socio-economic opportunities that arise from the developments – which will be dependent on the developers involved and the government policy. In addition, despite the declining economic significance of agriculture in the study area, we emphasize the importance of government intervention to enhance agricultural productivity and control agricultural land conversion for food security reasons.

Key words: Kenya, peri-urban, household land holding, land use change, agriculture, non-farm income.

INTRODUCTION

Increased attention in recent years has been given to peri-urban development (c.f. Bryant et al., 1982; Anthrop, 2000; Wiggens and Proctor 2001; Maconachie and Binns 2006; Busck et al., 2006; Qviström, 2007). A number of alternative terms have been used to describe the same geographical area these are; the urban fringe, the periphery, inner rural, and rural commuting zone (Bryant et al., 1982). Busck et al. (2006), Anthrop (2000) and Ode and Fry (2006) describe the peri-urban areas as those

*Corresponding author: E-mail: nicodemus.mandere@lucsus.lu.se.

areas adjacent to built up areas of high population concentrations (that is, urban), they specified that they are the zones where traditional farming activities come into conflict with alternative economic, residential and recreational interests. Wiggins and Proctor (2001) describe the peri-urban as the zone engaged in intense interactions with the urban. They also add that it is the area of daily commuting from village to city Central Business Districts (CBDs) for work. Adell (1999) portrays the peri-urban as being significantly pre-urban where the issues of job creation, transport, housing and environ-mental issues are important. Finally, with African peri-urban focus, Maconachie and Binns (2006) characterize the peri-urban zone as the area with a "blurring" between rural and urban. They also emphasize this area as possessing great dynamism with a focus on competition for basic resources.

From the above literature definitions of peri-urban there are a few unifying components that have become apparent and have formed the basis of the peri-urban despite an inability to reach an all-encompassing definition. First, peri-urban areas are predominantly defined, in almost all the definitions, as places of conflict or competition which exist in between new (urban) and traditional (rural) land uses. Second, the outer limits of the peri-urban zone are demarcated by maximum daily commuting distances into CBDs of the urban areas determined by the means of transportation available for large portions of the population. Based on these two understood definitions, the peri-urban areas in this study are defined as those areas which are transitioning between the urban and traditional landscapes as determined by daily commuting distance to the CBDs of the nearby city and or town.

Urban growth and peri-urban development

During the last half of the 20th century in Western Europe, relatively steady economic and population growth and stable patterns of governance have led to outward urban growth, creating what has been referred to as dynamic and/or multiple use landscapes (Busck et al., 2006) in these areas. Urbanization in Africa, on the other hand, has generally been more rapid and chaotic than in Europe with deficiencies in regulation and infrastructural development (Maconachie and Binns, 2006). The rapid urban growth is often in conjunction with inadequate governance systems, infrastructural development and land administration and most often, lack of industrial and economic growth has led to what is often called the African urban crisis (Rakodi, 2005). According to Keiner et al. (2005) it is predicted that by the year 2020 more than half of the world's 8 billion human inhabitants will live in cities. This growth will result in an increase in the number of mega-cities (that is cities with over 10 million inhabitants), from 19 to 27, and in the number of large cities (that is, cities with populations greater than 1 million), increasing from 405 in 2000 to over 900 by 2020 (Redman and Jones 2005; Keiner et al., 2005; van Ginkel and Marcotillio, 2005). Urban population growth results to increased population densities within established urban areas as well as in the outward thrust of urban agglomerations (that is peri-urban regions). It is therefore anticipated that this enormous urban growth will result in outward population thrust of the urban agglomerations (that is peri-urban regions).

Following the outward urban population thrust, many recent rural areas around the cities will convert to peri-urban status. Like other peri-urban regions, their land use is expected to change from one currently dominated by agriculture to a multiple land use landscape (Antrop, 2000; Gallant et al., 2004; Busck et al., 2006). According to Carr (1997) peri-urban development is a transition development stage between rural and city. In line with Lanjouw et al. (2001) peri-urban areas have comparatively better infrastructure such as roads, electricity and telephone than their rural counterparts. However, Lee (1979) and also Lanjouw et al. (2001) acknowledge that level of infrastructure and land use changes vary from region to region on the basis of the initiatives and developers involved. According to Chaudhry (2007), limited economic opportunities and lack of basic infrastructure are main contributors to rural poverty. Therefore, it can be argued that the infrastructural development in the peri-urban areas would create new economic opportunities through creation of business opportunities and new jobs. Consequently, these will reduce poverty in the particular peri-urban areas (that is, former rural). Current peri-urban research is concentrated on assessing peri-urban land use change and environmental impacts (Busck et al., 2006, Maconachie and Binns, 2006), impact on agriculture (Houston, 2005; Vagneron, 2006; Thapa and Murayama, 2007) as well as peri-urban concepts and definitions (Kule, 2008). Little emphasis is placed on studies that evaluate the impact of peri-urban development on household livelihood support and income. This study was therefore conducted with the objective of assessing the impact of peri-urban development on household income using peri-urban Nyahururu, Kenya as the case study. To achieve this objective, the study strives to answer three research questions: 1) Have changes occurred in the land use in the region of peri urban between 1960s and now; 2) if the changes have occurred what are the contributing factors; 3) what is the impact of the land use changes on the residents' household income? This study examines the periods from the 1960s to now since many studies (e.g. Macharia, 2003) indicate that the most rapid urbanization in Kenya started being recorded in the post-colonial period that is after 1963. During the colonial period, the movement of Africans to the urban areas was restricted -only a few with passes were allowed to move and settle in urban areas where they provided labour force for the white settlers. However, after Kenya gained independence in 1963, the policy was changed and all people were free to move to any part of the country without any restrictions. This change in policies resulted in many people moving to various urban centres in search of employment and business opportunities, thus leading to what Macharia (2003) refers to as "boom urbanization". Nyahururu town like many other urban areas was affected by the change of the policy and has therefore expanded rapidly during this period.

Peri-urban Nyahururu

This study was conducted in the peri-urban Nyahururu for two reasons: 1) it is characterized by the dynamic urban sprawl that exists in an area with high agricultural potential, which implies competition between agriculture and urban development; 2) the region where the peri-urban is situated has been the focus for an earlier study on determinants of household diversification (Mandere et al., 2009). In the earlier study, the respondents were drawn from both the rural and the peri-urban parts of the district. The results of this earlier study indicated that larger household land ownership was one of the most important factors that lead to higher diversification and subsequently higher revenues for the households situated in more rural areas. However, for those households located in the peri-urban setting, the level of diversification and household income did not show a clear relationship with how much land they owned. Despite that, the outcome of the earlier study also revealed that households in the peri-urban area had access to comparatively better infrastructure than those in more rural areas; this study did not go deep into a further investigation of the unique factors that determine the land use change and household income in the peri-urban setting. As stated above, the aim of this study is to further investigate this issue. Therefore, conducting this study in the peri-urban Nyahururu could help to build upon and expand the findings of the previous study.

The peri-urban Nyahururu is an area around the town of Nyahururu. The town is the capital of Nyandarua District in the Central Province of Kenya. The geographical location of the town of Nyahururu is shown in Figure 1. The core urban centre of the town was founded in 1929 following the establishment of the railway transport system through the area. Since then the town has been expanding and by 1999, the population of the urban core reached 37,412 with a population density of 151 people per square kilometer (250 km^2 in total) (Central Bureau of Statistics, 2001). The Nyandarua District has fertile soils of volcanic origin with a high potential for agricultural cultivation (Ministry of Finance and Planning 2002). Except for the urban areas, the dominant livelihood activity in the District is agriculture. Most of the farmers in the District are smallholders, that is; farmers whose household land holding is less than 10 ha. The majority of the rural population in Nyandarua depends on agriculture for direct employment, income and food. Most of the agricultural production is rainfed. The nearby urban centres are important market points for most of the agricultural produce from the District (Ministry of Finance and Planning, 2002).

MATERIALS AND METHODS

The data for this study was collected through questionnaires and interviews with individual households in the peri-urban Nyahururu. This was mainly because there were no secondary data for the peri-urban. The lack of secondary data was an indication that the peri-urban development in the region has received little attention and, or that the process was as a result of a recent spurring rapid development.

As a first step, discussions were held with local government officers from Nyahururu town. The aim was to learn about where the peri-urban area was perceived to start and end in terms of road distance from the centre of Nyahururu town. Using indicators of peri-urban development such as multiple land use, that is intermix of agriculture, buildings, road infrastructure and also daily commuting distance to and from Nyahururu town, the start and end boundaries were approximated. The starting boundary of the peri-urban Nyahururu was found to be 9 km road distance from Nyahururu town centre while, the end boundary was estimated at 15 km road distance from the town centre. The town of Nyahururu has four main roads leading out of the town: Nyahururu-Ol Kalou - to the south; Nyahururu-Nyeri - to the east; Nyahururu-Rumuruti - to the north; and Nyahururu-Nakuru - to the west. Along all the four roads there is identifiable peri-urban development. This development provides grounds for interviews along the four transects. However, the study concentrated on the peri-urban areas in the direction of the first two roads. The reasoning behind this focus is that the peri-urban areas along the two roads lie within Nyandarua District where earlier studies on determinants of rural diversification have been conducted. Concentrating on the two transects could therefore give the opportunity to build upon and augment our findings from the previous studies. For this reason the other two transects whose peri-urban areas are located within Laikipia District were left out of observations.

Data collection

Semi-structured questionnaires were administered to 52 respondent households along the two transects earlier discussed. The respondents were selected at random along each road. During the interviews, the principal researcher read each question from the questionnaire in an open manner allowing for any other questions that arose during the interview process. The answers were recorded down in writing. According to Bowler (2005, 232), rurality is "a system with three elements – users of space, used space and use of space". This definition implies that in order to assess the impact of peri-urban development, one has to evaluate the trend of the three elements over time and the resulting effects within this

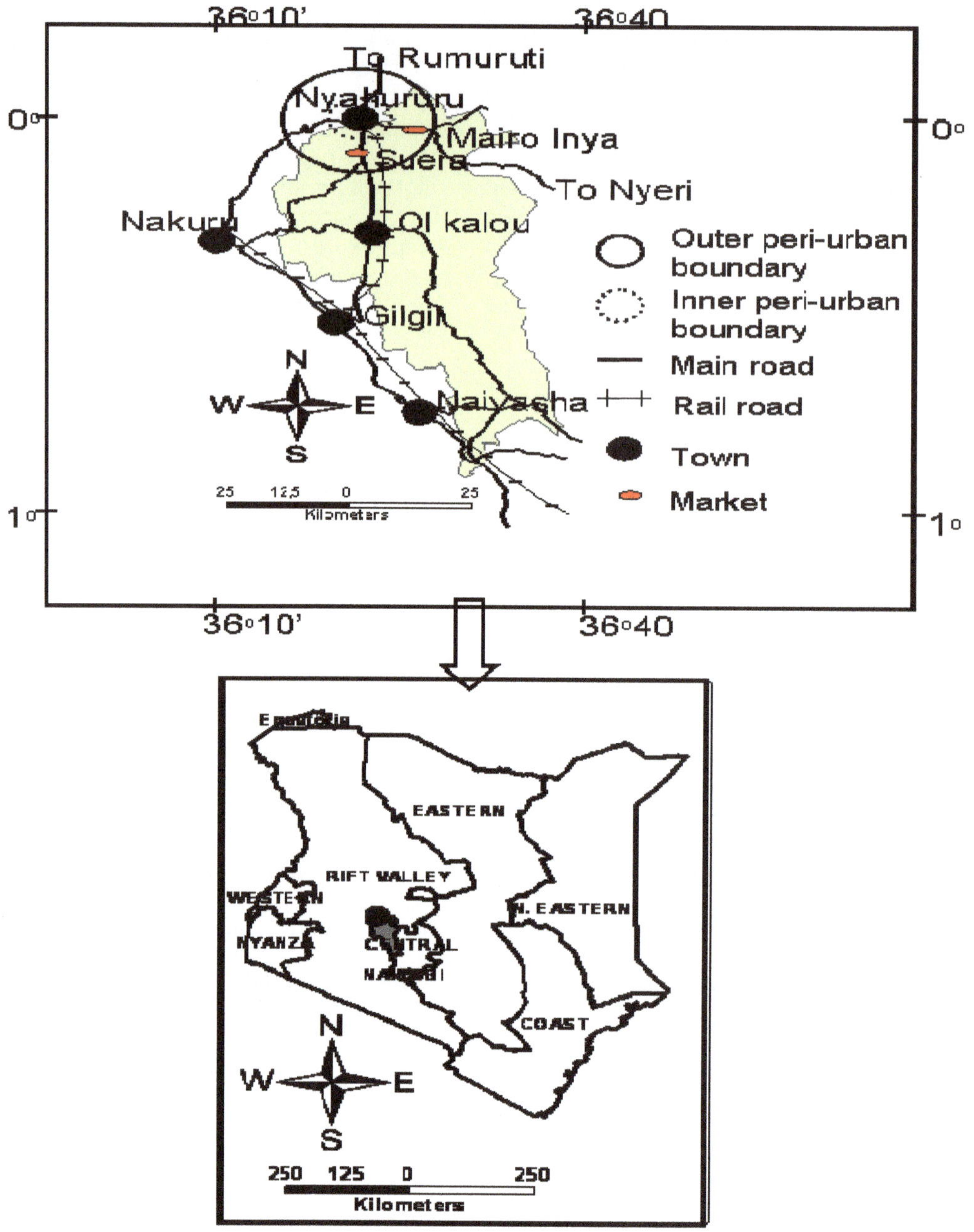

Figure 1. Geographical location of peri-urban Nyahururu.

framework. Based on these criteria, a questionnaire was designed to investigate the dynamics in regards to who is using the land (population dynamics), the used area (household land size holding) and how the area is used (household land use changes). Different parts of the questionnaire covered these different questions.

The data was qualitatively analyzed by content analysis method. The collected data were studied carefully and data coded and categorized based on the topical issues covered in the questionnaire which provided the data for the results and discussion of this study. The respondents constituted 41 men and 11 women; the age range was between 32 to 88 years. The men dominated the interview process because in the region and Kenya in general the men are the heads of the households and whenever they are present, they are given the opportunity to talk on behalf of the household (this then implies that due to this cultural inclination, gender balance - which was an aim of the researchers - was not achieve). We only managed to interview a comparatively small part of the female population, mainly those who represented the household heads that were absent during the time of the interview as well as a smaller percentage who were household leaders of households with single parents. The respondents were from seven different Kenyan tribes that is Kikuyu (71%); Luhya (8%); Luo (6%);

Kisii (6%); Kalenjin (4%); Somali (4%); and Turkana (1%). The Kikuyu tribe that dominated the interview are the natives of the study area. However 20% of the respondents from the Kikuyu tribe had immigrated to the region from parts of Central province such as Nyeri and Muranga. This implies that respondents that were native to the peri-urban region were around 55% of the total respondents. Most of the respondents were educated; 33% had attained college or higher level training while 48 and 15% had attained secondary and primary education levels respectively. Only 4% of the respondents did not attend any school at all.

RESULTS AND DISCUSSION

The results and discussion are presented in the following order: population dynamics; household land size holding; household land use changes; what factors increase the possibility for better household livelihood and income; and implications of the study's findings.

Population dynamics

In the last two decades there has been a rapid growth in population in the town of Nyahururu. According to the respondents the population growth in the town was mainly as a result of groups of migrants from several parts of Kenya and neighbouring countries such as Tanzania and Uganda. The respondents explained that the mass migration to the town was as a result of the positive economic development following the expansion of the service and manufacturing industries. The respondents mentioned tourism, banking, health care, education, roads and railway infrastructure as main service industries that have served as a magnet to the increasing population. The Kenya co-operative creameries and floriculture were mentioned as other industries that have accelerated rapid population growth.

The respondents also observed that the population of the peri-urban Nyahururu has grown rapidly, specifically beginning in the 1990s. They attributed the growth to births and also to outward thrust of the urban population (that is, migration from the urban core to the peri-urban region). According to the United Nations (2005) the population growth rate in Kenya is around 2.5% for the 5 year period from 2005 to 2010. The population growth has increased because of high birth rates and low mortality (United Nations 2005). Migrations from the urban core to the peri-urban region as explained by the respondents was driven mainly by the search for cheaper land to construct a residence or business premise and the hope to find a larger land area to accommodate both residence and gardening to supplement household food demand as living costs and food prices have escalated. Those immigrants who moved from the rural areas in search for employment and decided to directly settle in peri-urban region, cited the cheap cost of accommodation and access to the urban core - where non-farm jobs most often are found - as the main reasons for settling in the peri-urban area. The common thread is that both reasons are economically motivated, that is, based on a desire to improve livelihoods at minimum cost. These are people that Burnley and Murphy (2004:38) refer to as 'forced relocators'. They are poor or middle class people seeking employment opportunities or cheaper livelihood and they can find cheap and perhaps the only alternatives for housing in the urban fringe. Their reasons for settling in the peri-urban area are thus in sharp contrast to the dominating reasons for moving to peri-urban regions in the industrial world. For instance, Busck et al. (2006) found a longing for quietness and attractive landscapes to be the catalyst for moving to peri-urban area surrounding Copenhagen, the capital city of Denmark. Such migrants can be categorized as 'free agents' (Burnley and Murphy, 2004:38), that is those after a specific quality of life, such as serenity. Immigrants to peri-urban regions in Kenya are 'forced relocators' while those in Copenhagen are 'free agents'. This is a reflection of the different economic and social context.

The migration to the peri-urban Nyahururu has created a population composition that is cosmopolitan. According to the respondents, the incoming migrants who are from many parts of Kenya belong to different tribes than the native populations of the peri-urban region. Foreigners, particularly from neighbouring countries such as Uganda and Tanzania, are also common in the peri-urban region. These immigrants are thus fairly diverse in terms of culture, skills, economic wealth and social classes, and a new multi-ethnic community with a diverse livelihood portfolio has developed. Consequently, the land use in the area has been transformed from one dominated by agriculture to multiple land use including several non-farm economic activities as presented in subsequent sections. Besides the land use changes, this new multi-ethnic community was found to be living harmoniously together without much intertribal tensions. Actually new lifestyles where members of these new communities are getting intermarried as well as adopting each others feeding habits were common.

Household land size holding

The pressure from the peri-urban expansion due to immigrations to the region has contributed to decreasing household land holding size due to the sale of part of the land to new developers for residence, business use or both residence and agriculture. Ten percent of the respondents had actually sold part of their own farmland to new developers. The land was further divided because of land bequests to children. This is also an important contributor to the decrease of farm sizes according to majority of the respondents. The average household land holding size in the 1960s was estimated to be around 4 to 5 ha. Currently the figures the respondents gave for average household land holding size range from 0.125 to 3 ha, the mean of which is 1.2 ha. This average household land holding size is much less compared to the 2.5 ha observed by (Mandere et al., 2009) as the average household land size for Nyandarua District where the peri-urban is situated. According to Kenya National Bureau of Statistics (2006), the birth rate for the whole of Nyandarua District is uniform at 2.2% per year. Therefore, the finding that average household land size in the peri-urban area is small when compared with the whole Nyandarua District underscores the significant contribution of peri-urban expansion to diminishing household land holdings. Published literature from other developing countries similarly reveals that peri-urban sprawl resulted in declining household land holding in the peri-urban Kano, Northern Nigeria (Maconachie and Binns 2006) and also in the peri-urban regions around Dar es Salaam, Mwanza, Moshi, Arusha, Mbeya and Lindi cities of Tanzania (Lanjouw et al., 2001).

The respondents who had sold part of their land to new developers provided varying reasons as motivations behind the land sales. A majority of them sold part of their land out of urgent need to raise additional finances for basic household needs, specifically for school fees for their children. According to the responses, there are many willing buyers of land in the region due to land speculation and hence the land market is easily available. However most of them emphasized that despite the easily accessible land market and increasing land prices they only sell land as a last resort. Nonetheless, the land is privately owned and the government has currently no policies to regulate land sale and land subdivision, considering that most people in the district are poor (Ministry of Agriculture and Ministry of Livestock and Fisheries Development, 2004). Similar to the observation by Kombe (2005) in Tanzania, land is sold by willing buyer-willing seller; selling land and subdivision is likely to continue, and this may result in very small and economically non-viable household land holdings for many households.

Household land use changes

Under the household land use changes, the following were assessed: agriculture; non-farm income generating activities; as well as the factors that increase the possibility for better household livelihood and income.

Agriculture

Agriculture still remains one of the predominant economic sectors in the peri-urban Nyahururu. However, the outcome of the interviews also reveals that its economic significance is declining. This is reflected by the declining number of households that engage in agriculture as a full time activity. Over 90% of the respondent households were full time farmers in the 1960s but this has since fallen to 49% of the respondent households, with the remaining households only cultivating their land on a part time basis. The respondents offered two reasons for the declining economic importance of agriculture sector: 1) reducing household agricultural land holding size; 2) low returns from investments in agriculture mainly due to declining yield per unit area and also high cost of agricultural inputs coupled with low market price for the output.

Due to the sale of part of the household land to the new developers and also to land bequests, the household land area that is available for agriculture has declined rapidly. In response to declining agricultural land, most farmers have shifted from traditional extensive agriculture towards more intensive agricultural practices. The respondents were in agreement that historically agricultural production systems in the peri-urban region were dominated by extensive farming. A fallow system was practiced where crop land was rotated with grazing fields, a practice that optimized soil nutrient utilization and regeneration. However in the last two decades, the economic difficulties of small farms have stimulated new agricultural management practices in order to increase yield levels and incomes. The respondents were unanimous in their responses explaining that the fallow system has largely been abandoned because of land limitations, leading to a more intensive system where crops are grown in the same piece of land year after year. Similar trends were also recorded on livestock grazing fields. The finding that the farmers in the peri-urban Nyahururu have shifted from extensive to intensive agriculture. This is in contrast with other research findings in the peri-urban area around Copenhagen, Denmark (e.g. Busck et al., 2006) and in the Netherlands (e.g. Berg and Wintjes, 2000). This is because of past agricultural policies, which have resulted to the significant reduction of rents from small scale farming - leading to the fall in small scale production. The demise has in many cases caused a shift from intensive to extensive land management practices. Instead of farm properties in the European periphery being used for (intensive) agricultural production, new owners with little or no experience in agricultural production, financially stable people in search of properties in socially and environmentally attractive areas with ample space for leisure and hobby activities are been attracted (Busck et al., 2006).

The investigation also revealed a shift in the kinds of crops currently grown in the peri-urban region compared to the traditional crops of the 1960s. The traditional and

Table 1. Traditional and present crops in peri-urban Nyahururu.

Traditional crops	Currently cultivated crops
Sweet potato	Maize
Cassava	Irish potato
Maize	Beans
Arrowroot	Peas
Wheat	Tomato
Pyrethrum	Carrot

currently grown crops in the peri-urban region as obtained from the respondents are shown in Table 1. The table indicates that most of the traditionally cultivated crops such as sweet potato, cassava, arrowroot and wheat have disappeared. In their place Irish potato, beans, peas, tomatoes and carrots have been adopted. The main reasons expressed for the cultivation of new crops were that these new crops attracted the attention of the market while the demand for traditional crops has diminished dramatically. The increased market demand for the new crops is viewed as a result of migrants into the area whose diets are different from the native inhabitants of the peri-urban region. Despite that, most of the natives would like to continue cultivating some of their native crops, due to the rapidly shrinking household agricultural land most of them have been forced to shift in response to market demand as a survival strategy. Those few who are still cultivating the native crops do so in very small quantities and often in their home gardens. Among the currently cultivated crops, maize and beans are the most dominant crops and are cultivated by over 80% of the respondents. The two crops are predominant since they are staple food for most people living in the peri-urban region and the Nyahururu town including migrants from other parts of Kenya and neighbouring countries. In addition, both maize and beans are easily sold for extra household income. Other crops currently grown such as Irish potato, peas, tomatoes and carrots are cultivated in small quantities by about 30% of the respondents. These latter crops are considered specialized crops because they mature fast, have readily available market although their market price is highly fluctuating. The unstable market price discourages the majority of respondents from cultivating these crops because these crops mighnot provide steady incomes. Furthermore the volatile market value encourages the tendency of households to give priority to crops that provide steady food supplies. This led to the transition to a more intensive production system since the field soil nutrients have been exhausted. Over the years, this has yielded an overall decline in agricultural production and created the need for larger amounts of fertilizers. The need to use larger amounts of fertilizer in order to boost crop yields has driven up the production cost which together with low and often highly fluctuating market prices for the agricultural outputs has lead to a decline of the economic significance of agriculture to most households. Consequently most of the households seek additional livelihood income by engaging in non-farm income activities.

Non-farm income generating activities

As earlier stated, in response to the declining economic significance of agriculture, most of the respondent households have adopted non-farm income activities for additional income. In addition, a number of new migrants to the peri-urban region, particularly those lacking the farming skills and those without access to farming land, are mainly engaged in non-farm income generating activities. The non-farm activities that the households participated in to create an additional source of income were diverse; they include business activities, professional employment, non-farm wage, and farm wage labour on the nearby agricultural farm.

The importance of non-farm incomes to the overall household incomes varied widely among respondents on the basis of the type of non-farm activity they were engaged in. The results are shown in Table 2. The table indicates that many households took part in more than one non-farm income generating activities. It also shows that households who get part of their income from business activities and professional jobs have a dramatically higher income than the rest of the households. Furthermore it can be seen that the households that are employed in non-farm activities such as business and professional jobs draw over 80% of their annual incomes from the non-farm activities. Those engaged in professional jobs were all college graduates, while those running business were a mixture of college, secondary, and primary school graduates. This indicates that educational level is a basic requirement to gain access to professional jobs, while possibilities of venturing into business is primarily determined by the possibility of access to the necessary financial means. Interestingly, for those households involved in non-farm wage labour and farm labour as their non-farm income activities, their annual income was much lower. For these households, the share of non-farm income to the total annual household income was less than 50% - thus implying that agriculture is a very important economic sector in the peri-urban environment. The size of land owned by household was important in determining household activities and income level for these latter categories of households. In general, the larger the household land ownership, the more the diversification particularly in crops and the higher the annual household income. Employment in the non-farm wage

Table 2. Income share for each activity as a percentage of household annual income

Business (%)	Professional jobs (%)	Crops (%)	Livestock (%)	Farm labour (%)	Non-farm labour (%)	Average household income (KES/yr)	Respondents (%)
36	60	4	0	0	0	586000	7
0	95	5	0	0	0	416000	3
81	0	7	12	0	0	238950	3
90	0	10	0	0	0	216000	10
0	0	70	30	0	0	85000	33
0	0	55	22	23	0	80000	3
0	0	52	0	21	27	48000	10
0	0	65	0	0	35	46000	6
0	0	100	0	0	0	46000	16
0	0	70	0	30	0	43000	9

labour, and farm wage labour activities within the scope of this study was dominated by people with primary and secondary education levels.

According to Bryceson (1996; 2002), it is the non-agricultural activities that lead to more lucrative, poverty alleviating strategies in Africa. However, this study may at least in part contradict the conclusions made by Bryceson; it is clear that not all non-farm activities lead to a higher income. Judging from the responses, it is only the professional jobs and business that are likely to lead to poverty reduction as the results indicate that these types of jobs lead to higher income. These are the kind of non-agricultural activities that according to Lanjouw et al. (2001) would be classified as high labour productivity activities. However, the non-farm wage labour and farm labour would be categorized as low labour productivity activities. Our findings further indicate that in the peri-urban Nyahururu, only a few respondents (23%) are involved in high labour productivity activities while the rest are either engaged in low labour productivity activities or are not engaged at all in non-farm income activities. Like most other parts of Kenya (Nyang'au, 2002), most of the employment opportunities are found in the informal sector. The professional jobs are limited both in the peri-urban and the core urban area-the opportunities for professional jobs are mainly found in government or private institutions such as hospitals, schools, banks, processing and other service industries where professional skills are of vital necessity. But these latter institutions are few and hence limit professional employment opportunities in the region. As argued by Boardi et al. (2005), salaries in the informal sector are very low and are often not adequate to alleviate poverty amongst the poor. In addition, according to (Ministry of Agriculture and Ministry of Livestock and Fisheries Development, 2004) most people in Nyandarua district are poor and access to any credit is constrained due to lack of collateral and high interest rates. It could therefore be argued that it is the domination of employment in the informal sector coupled with a possible lack of sufficient finances that prevent the majority of households from engaging in high productive activities that would yield a higher income. Nonetheless, the proportion of those who worked in high labour productivity is 10% higher compared with observations (Mandere et al., 2009) for the whole of Nyandarua where the peri-urban region is situated. With this in mind, one interesting question needs to be further looked into namely; what factors increase the possibilities for a higher income and a subsequent better household livelihood?

What factors increase the possibility for better household livelihood and income in the peri-urban setting?

In the process of developing non-farm income generating activities, a majority of immigrants and wealthier natives invest in commercial properties and a wide range of business activities. The newer businesses are often located in the periphery of the urban core because of road infrastructure and in order to provide farmers with outlets to sell products or to purchase what is needed for farming. They also give peri-urban dwellers access to food supplies and other basic necessities without traveling to the urban core. The businesses are commonly owned as single person enterprises, only a few of them are owned by a group. The size of the businesses varies greatly; however most of them are relatively small enterprises employing only a few people besides the owners. In addition, a majority of the businesses are of the informal type and involve low capital investments. This kind of business development has been growing rapidly in the peri-urban area during the last two decades and dispersed market centres, e.g. Mairo Inya and Suera have emerged within the area.

The development of these market centres has lead to the establishment of infrastructures such as electrification, telephone services, new schools and health centres. This development has created new employment opportunities and thereby increased immigration to the area. Existing road infrastructure coupled with the new infrastructural development (as mentioned above) do not only lead to further expansion of the new market centres but also increase chances for enhanced household income through the subsequent creation of employment . Lanjouw et al. (2001), mention infrastructure, and particularly roads, as one of the main factors increasing probability for an engagement in business and thereby the possibilities for a higher income. Heyer (1996) also Ellis (1998) emphasize the vital role of access to infrastructure and services as a key to poverty reduction in Africa by fostering adoption of non-farm activities. The infrastructural developments in the peri-urban area may therefore partly explain why the number of households in the high income productive activities in the peri-urban was 10% more than the rural parts of the district where infrastructure is often lacking. However, as earlier stated, a majority of the household in the peri-urban region have no access to the higher income productive activities. This could indicate that pre-dominate informal sector employment and income opportunities do not foster sufficient socio-economic opportunities for most households – and that this informal sector does not allow the majority to improve their income levels. According to Kombe (2005), it is the socio-economic opportunities that are available together with peri-urban expansion that are key determinants of benefits to household income levels. These socio-economic opportunities are primarily based on the pattern and rate of peri-urban development (lanjouw et al., 2001; Kombe, 2005). According to lanjouw et al. (2001), the pattern and rate of peri-urban expansion is dependent on the developers involved and their initiatives. We would therefore like to argue that the possibility for peri-urban development to accomplish a reduction in poverty for the households, will depend on the level of infrastructure and availability of high productive job opportunity-this in turn is dependent on government policy as well as private sector developers who will be attracted to the region.

Implications of the study findings

Since our findings have shown that majority of the peri-urban residents lack the opportunity for high income generating activities; we argue that although more people in the peri-urban area are engaged in high income productive activities than the rural parts; there is need for government intervention to ensure that the benefits trickle down to majority of the poor. The government may intervene by providing the necessary strategies to attract developers with sufficient funds to invest in high income generating opportunities and also to increase benefits from the engagement in the informal sector businesses. This is of vital necessity considering that the economic significance of agriculture is declining and that conversion of agricultural land to non-farm activities is rapid and likely to accelerate. The rapid conversion of the agricultural land to non-agricultural land use in the study area poses a threat to future food production and food security in the region since the production in this peri-urban area is of central importance of food supply for the expanding urban population. Government intervention to control the land sale and conversion rate seems necessary to secure food production. The need for government intervention can not be overemphasized as the finding indicates that peri-urban expansion is in a rapid process in many parts of the world and this threatens agricultural land and local food security (Buxton et al., 2006). Peri-urban sprawl has caused a reduction of agricultural land in Nigeria (Maconachie and Binns, 2006), Tanzania (Lanjouw et al., 2001) as well as in Australia, South America, North America and Asia (Buxton et al., 2006) indicating that it is not only a local problem but an issue of global significance.

Bunce and Walker (1992) assert that declining agricultural returns are main contributors for conversion of farmland to other uses in peri-urban regions. But Barr (2005) argues that this conversion is driven by land speculation arising from proximity to the urban cores. Our studies indicate that both these factors are important drivers for land conversion. Declining agricultural returns acts as a catalyst to the sale of part of the farmland to new developers and or for households to convert part of their agricultural land for non-farm activities, while land speculation creates the increased land demand and hence provide readily available land market. Since it is almost impossible to control land speculation, to protect agricultural land the government intervention should be aimed at increasing profit from agricultural production. There are many approaches to increasing agricultural productivity and profits among which include subsidies on agricultural inputs (Boadi et al., 2005; Buxton et al., 2006) availability of credit and financial services (Ministry of Agriculture and Ministry of Livestock and Fisheries Development, 2004; Boadi et al., 2005) as well as price guarantees (Boadi et al., 2005). Besides price guarantees, all the fore-mentioned measures are part of the Kenyan government strategy to help improve agricultural production and income for farming households (Ministry of Agriculture and Ministry of Livestock and Fisheries

Development, 2004). Other strategies being implemented in efforts to revitalize agriculture in Kenya are lowering cost of agricultural inputs by encouraging competition in farm input supplies, reducing taxation on agriculture and improving agricultural extension services. Our finding on economic significance of agriculture is still on the decline in spite of these efforts. This indicates that the policy instruments have not worked sufficiently in addressing the problem in the study area. A further research to assess the pitfalls of the current strategies to revitalize agriculture is therefore of vital necessity. The outcome of such a study will assist the policy makers to improve on the agricultural production and profitability through adoption of improved strategies.

Conclusion

The analysis has shown that the peri-urban Nyahururu has undergone some vigorous developments in population composition; household land size holding; and household land use changes leading to declining significance of agriculture; adoption of diverse new non-farm activities; and improvement livelihood and income. The decline in significance of agriculture was mainly due to rapidly shrinking household agricultural land as well as low and fluctuating agricultural output prices. This in return reduces the profits from the agricultural production - consequently the number of households engaged in full time farming has reduced by 41% since the 1960s. Thus, non-farm activities have become common among most households. However most of the households engage in low income productive activities- nevertheless, the number of households engaged in high income productivity was comparatively higher (10% more) compared to the most rural parts of the district.

The infrastructural developments (e.g. roads, schools, electricity, telephone etc) together with emerging business enterprises were found to be the main factors that enhanced the opportunities for household engagement in high income productive activities. However most of these developments are limited to the financially constrained informal sector and hence can not provide sufficient high income opportunities to lift majority of the population from poverty. Therefore, the possibility for peri-urban development to accomplish a reduction in poverty for the households will not only depend on the infrastructural developments but rather the socio-economic opportunities that arise from the developments-which will be dependent on the developers involved and the government policy. In addition, despite the declining economic significance of agriculture in the study area, the importance of government intervention to enhance agricultural productivity and control agricultural land conversion for food security reasons is emphasized.

REFERENCES

Adell G (1999). Literature review: Theories and models of the peri-urban interface-a changing conceptual landscape. London: Development Planning Unit, University College, London

Anthrop M (2000). Changing patterns in the urbanized countryside of Western Europe. Landscape Ecol. 15: 257-270.

Barr N (2005). The changing Social Landscape of Rural Victoria. Department of Primary Industries, Tatura, Victoria.

Berg Ld, Wintjes A (2000). New rural lifestyle estates in The Netherlands. Landscape and Urban Planning, 48: 169-176.

Boadi K, Kuitunen M, Raheem K, Hannine K (2005). Urbanization without development: environmental and health implications in African cities. Environ. Dev. Sustain. 7: 465-500.

Bowler I (2005). Rural alternatives. In Kule L (2008). Concepts of rurality and urbanity as analytical categories in multidimensional research. Proc Latvian Acad. Sci. Section B. 62: 1/2.

Bryant CR, Russman LH, McLellan AG (1982). The city's countryside: land and its management in the rural-urban fringe. London: Longman.

Bryceson DF (1996). Deagrarianization and rural employment in sub-Saharan Africa: A sectoral perspective. World Dev. 24(1): 97-111.

Bryceson DF (2002). The scramble in Africa: reorienting rural livelihoods. World Dev. 30(5): 725-739

Bunce M, Walker G (1992). The Transformation of Rural Life, in Bowler IBC and Nellis CM (Eds.). Contemporary Rural Systems in Transision, Volume 2, Economy and Society, Redwood Press, Melksham, pp. 49-61.

Burnley I, Murphy P (2004). Sea Change. Movement from Metropolitan to Arcadian Australia, UNSW Press, Sydney.

Busck GA, Pilgaard KS, Præstholm S, Reenberg A, Primdahl J (2006). Land system changes in the context of urbanisation - examples from the peri-urban area of greater Copenhagen. Geografik Tidsskrift/Danish J. Geogr. (106): 21-34

Buxton M, Tieman G, Bekessy S, Budge T, Mercer D, Coote M, Morcombe JA (2006). Change and continuity in peri-urban Australia, state of the peri-urban regions: a review of the literature (monograph 1). Melbourne, RMIT University.

Carr M (1997). New patterns: Process and change in human geography. Walton-on-Thames: Nelson, p. 525.

Central Bureau of Statistics (2001). Population and housing census: Population distribution by administrative areas and urban centres, Vol. 1. Nairobi: Ministry of Planning and National Development.

Chaudhry MA (2007). Introduction managing agriculture and rural development for poverty reduction. Nat. Res. Forum, 31: 250-252.

Ellis F (1998). Household strategies and rural livelihood diversification. J. Dev. Stud. 35(1): 1-38.

Gallant N, Shoard M, Andersson J, Oades R, Tudor C (2004). England's urban fringes: Multi-functionality and planning. Local Environ. 9(3): 217-233.

Heyer J (1996). The complexities of rural poverty in sub-Saharan Africa. Oxford Dev. Stud. 24(3): 459-476

Houston P (2005). Re-valuing the Fringe: Some findings on the value of agricultural production in Australia's peri-urban regions. Geogr. Res. 43(2): 209-223.

Keiner M, Koll-Schretzenmayr M, Schmid WA (2005). Introduction. in Keiner M, Koll-Schretzenmayr M and Schmid WA eds. Managing urban futures: sustainability and urban growth in developing countries. Hampshire: Ashgate Publishing.

Kenya National Bureau of Statistics, 2006. Basic Report: Kenya Integrated Household Budget Survey-2005/06 (revised edition. Nairobi: Ministry of Planning and National Development.

Kombe WJ (2005). Land use dynamics in peri-urban areas and their implications on the urban growth and form: the case of Dar es Salaam, Tanzania. Habitat Intanational, 29: 113-135.

Kule L (2008). Concepts of rurality and urbanity as analytical categories in multidimensional research. Proc Latvian Acad. Sci. Section B. (62): 1/2.

Lanjouw P, Quizon J, Sparrow R (2001). Non-agricultural earnings in

peri-urban areas of Tanzania: evidence from household survey data. Food Policy, 26: 385-403.
Lee L (1979). Factors affecting land-use change at urban-rural fringe. Growth Change, 10(4): 25-31.
Macharia K (2003). Migration in Kenya and its Impact on the labour Market. American University, United States of America. Paper prepared for Conference on African Migration in Comparative Perspective, Johannesburg, South Africa, 4-7 June 2003.
Maconachie RA, Binns T (2006). Sustainability under threat? the dynamics of environmental change and food production in peri-urban Kano, Northern Nigeria. Land Degrad. Dev. 17: 159-171.
Mandere NM, Anderberg S, Armah FA, Elimqvist B (2009). Determinants, income and perception of sugar beet adoption in Nyandarua District of Kenya. Submitted to Geojournal.
Ministry of Agriculture and Ministry of Livestock and Fisheries Development. (2004). Strategy for Revitalizing Agriculture 2004-2014. Nairobi, Government Press.
Ministry of Finance and Planning (2002). Nyandarua district development plan 2002-2008: Effective management for sustainable economic growth and poverty reduction. Nairobi: Government Press.
Nyang'au IM (2002). The role of informal sector in Rural Development: A case study of carpentry and tailoring in Gucha district, Kenya. Swedish University of Agricultural Sciences, Department of Rural Development Studies Uppsala. A Master thesis.
Ode A, Fry G (2006). A model for quantifying and predicting urban pressure on woodlands. Landscape and Urban Planning, (77): 17-27.
Qviström M (2007). Landscapes out of order: Studying the inner urban fringe beyond the rural-urban divide. Geografiska Annaler: Series B, Human Geogr. (89): 269-282.
Rakodi C (2005). The urban challenge in Africa. In Keiner M, Koll-Schretzenmayr M and Schmid WA (eds) Managing urban futures: sustainability and urban growth in developing countries. Hampshire, Ashgate Publishing.
Redman CL, Jones NS (2005). The environmental, social, and health dimensions of urban expansion. Population and Environment, (26): 505-520.
Thapa RB, Murayama Y (2007). Land evaluation for peri-urban agriculture using analytical hierarchical process and geographic information system techniques: A case study of Hanoi. Land Use Policy, 25: 225-239.
United Nations (2005). world population prospects: the 2004 revision. New York United Nations Department of Economic and Social Affairs, United Nations.
Vagneron I (2006). Economic appraisal of profitability and sustainability of peri-urban agriculture in Bangkok. Ecol. Econ. 61: 516-529.
van Ginkel HJ, Marcotillio PJ (2005). Asian urbanization and local and global environmental challenges. In Keiner M, Koll-Schretzenmayr M and Schmid WA (eds). Managing urban futures: sustainability and urban growth in developing countries. Hampshire: Ashgate Publishing.
Wiggens S, Proctor S (2001). How special are rural areas? the economic implications of location for rural development. Dev. Policy Rev. 19: 427-436.

Agricultural extension reforms and development in Uganda

Nana AfranaaKwapong[1] and Ephraim Nkonya[2]

[1]Humboldt University Berlin, Germany.
[2]International Food Policy Research Institute (IFPRI), Washington D.C.

This study was done with the objective of contributing to the policy debate on the changing landscape of agricultural rural services in Uganda. The study examines the perception of Agricultural Extension providers on the new changes in provision of agricultural extension services. Data was collected from twenty two key informants. Result show that political interference is negatively affecting the image of the National Agricultural Advisory Services (NAADS) and undoing some of its achievements. This suggests the need to re-evaluate the current political support given to agriculture to ensure that it builds on past achievement of NAADS and addresses its weaknesses.

Key words: National agricultural advisory services, agricultural extension, political interference, Uganda.

INTRODUCTION

A number of countries have taken bold steps to modernize agriculture through enhancing agricultural extension, a sector that is key to increasing the farmer knowledge to new agricultural technologies. Uganda is one of the SSA countries that invested significantly in improving provision of agricultural extension and other agricultural development programs. Uganda has recently initiated three major rural development programs, which build on the progress of past efforts. The Prosperity for All (PFA) was Uganda ruling party's election manifesto in 2006, which implemented programs comparable to the Plan for Modernization of Agriculture (Joughin and Kjær, 2010). PFA set a goal of enabling households to earn an annual income of UGX 20 million (approximately US$10,000 per year).

The second major agricultural program is the Africa-wide Comprehensive Africa Agriculture Development Program (CAADP), which Uganda has committed to implement and has prepared its investment strategy. The major goal of the country level CAADP investment strategy is to enhance agricultural-led economic growth and to achieve the millennium development goals (MAAIF, 2010). The CAADP has set a goal of achieving a 6% agricultural annual growth and allocation of at least 10% of government budget to the agricultural sector. In 2007/08, Uganda's agricultural growth was only 2.6% (Ibid), underscoring the weak growth of the sector. The investment strategy, which Uganda has prepared, will be implemented through its Development Strategy Investment Plan (DSIP).

The third strategy is the medium term development framework – the National Development Plan (NDP) – was initiated in 2008 following the expiration of the Poverty Eradication Action Plan (PEAP), which was implemented from 1997 to 2008. Evaluation of the PEAP found that one of its major weaknesses was the low agricultural productivity due to low investment in the sector, which employs 73% of the population and contributes 20% of the GDP (UBOS 2010). To address this shortcoming, the NDP gives more emphasis on the agricultural sector. To implement this focus, the Agricultural Sector Development Strategy Investment Plan (DSIP) was formulated to serve as the broader framework of the agricultural sector investment and development strategies. DSIP, which is a medium term plan running from 2010/11 to 2014/15 is designed to harmonize and consolidate all past agricultural development strategies, namely, the Plan for Modernization of Agriculture (PMA) and the Prosperity for All (PFA) (MAAIF, 2010).

It is important to reflect on the past agricultural development strategies in order to learn from their strengths and weaknesses. Of particular importance for this study are the agricultural extension services, which have seen dramatic changes under the PEAP and the new development strategies. The Government of Uganda initiated agricultural extension reforms that included the demand-driven National Agricultural Advisory Services (NAADS) program. The main objective of the reforms is to enhance agricultural technology advisory services. This had the key strategy to implement the PMA strategy, which was Uganda's overarching poverty reduction strategy with an objective of transforming agriculture from subsistence to commercial farming. The NAADS program, whose implementation started in 2001, was PMA's main pillar, which attracted significant investment by the government and donors. NAADS targets the development and use of farmer groups, and in the process empowers them to procure advisory services, manage linkage with marketing partners. The NAADS program has been one of case studies of decentralization of agricultural services that uses the new demand-driven advisory services approach, in which private-sector agricultural extension providers are given a key role in providing agricultural advisory services in sub-Saharan Africa (SSA).

This study was done with an objective of contributing to the policy debate on the changing landscape of agricultural rural services in Uganda. The study focuses on the agricultural extension services, which is the major rural services under the PMA, PFA and other government programs. The present study examines the perception of Agricultural Extension (AEA) providers on the new changes in provision of agricultural extension services.

The changing landscape of provision of agricultural rural services in Uganda

As is the case in other countries, the performance of the publicly provided and funded traditional agricultural extension services in Uganda was poor (MAAIF and MFEPD, 2000). To address this problem, the NAADS program was introduced by the Act of 2001, which gave it a mandate to develop a demand driven, farmer-led agricultural service delivery system targeting the poor subsistence farmers, with emphasis to women, youth and people with disabilities. Its development goal is to enhance rural livelihood by increasing agricultural productivity and profitability in a sustainable manner. NAADS also aims to empower farmers to participate in the decision making process of type of technologies to be promoted by the providers of advisory services in their sub-county. This demand-driven approach differed significantly from the traditional supply-driven. The first phase of NAADS (2001-2009) was introduced in 2001 in six pilot districts. The program was rolled out to 545 sub-counties or 83% of all the sub-counties in 2006/07. NAADS, a twenty five year program in 2011, had now been rolled out to 79 of the 80 districts and to 710 sub-counties in the country. In the State of Nations address on 01 June 2010, the President stated that the program had been rolled out in all 80 districts, 929 sub-counties and 137 urban councils. Hence virtually, NAADS has reached all sub-counties and the biggest challenge is whether it has reached the farmers, in other words, whether it has met its expectations.

NAADS operates through farmer groups at village level. The farmer groups in a given sub-county form the farmer forums. Each farmer group prioritizes three enterprises and the advisory service needs. The priority enterprises and advisory service needs are sent to the farmer forum, which determines three priority enterprises in the sub-county. NAADS supports the selected priority enterprises and the required advisory services needed to address the identified constraints and advisory service needs. Following selection of the three enterprises, NAADS provides technologies for demonstration on a member of a farmer group's (or host farmers) field-technology development site (TDS). The host farmer is chosen by fellow members of the group, and private service providers are contracted to carry out the demonstrations and advise farmers at these TDSs.

Until early 2008, advisory services were provided by private providers, who included non-governmental organizations (NGOs), private extension agents with no affiliation to NGOs. In the sub-counties where NAADS was not operating, the public extension agents continued to provide agricultural extension services. In the sub-counties where NAADS was operating, the public extension agents regulated and facilitated private extension service providers. NAADS changes in 2008 reinstated the public extension service provision. The government directed the districts to stop contracting private extension workers. This change meant that the public extension workers now provide most of the advisory services with the private extension workers contracted to provide only specialized services. A circular from the Ministry of public services dated 14th January 2010 directed that all agricultural extension staff at the sub-county level be converted to NAADS. The conversion exercise was to cover all extension staff based at the sub-county level on permanent and pensionable terms. The public extension workers contracted under NAADS were to be well enumerated and facilitated under the NAADS program. This directive set a goal of achieving the conversion by April 2010. However, this directive was not fully implemented.

Like NAADS phase I, NAADS phase II (under DSIP) provides advisory services in conjunction with the specialized export crop authorities (coffee, cotton, and tea), dairy development authority, and the genetic information resource center and data bank. Additionally, cooperatives and NGO will continue to provide advisory services under the supervision of public extension services.

Under NAADS phase II, four major areas show the similarities and differences with the first phase. NAADS phase II plans to provide advisory services by:

1. Empowering farmers to make and implement decisions on farm management and livelihoods. This component is meant to consolidate the achievement under NAADS phase I, under which a large number of farmer groups were formed. One of the strategies for enhancing farmers' capacity to make and implement decisions is formation of the high level farmer organizations, which will help farmers groups to enhance their capacity to make and implement decisions and to have vertical and horizontal linkage along the value chain. This will also help farmers to have greater voice in marketing and bargaining power.
2. Improving access to new agricultural technologies and information: Unlike the major thrust of NAADS phase I, this component will form a stronger research-extension linkage through formation of adaptive research support teams (DARST) in each district, under which the linkage of extension service providers and farmer groups with the zonal agricultural research and development institutions will be strengthened. Under DARST, participation of extension and farmer groups in decision making of type of research at zonal research and development institutions will increase. The DARST will also organize technology demonstration in villages in order to enhance farmer access to new technologies. Enhancing access to new technologies and information will also involve linking farmers to credit institutions for farmers willing to take up credit.
3. Delivering appropriate advisory services and information. This component aims to strengthen the farmer groups formed under NAADS phase I and to form new ones. Like NAADS phase I, this program will be implemented using the public-private partnership of providing advisory services. However, the public extension service providers will take a greater role. Unlike NAADS phase I however, delivery of advisory services will use a variety of methods including Farmer Field Schools (FFS). Different advisory services will also be used including extension services through mobile phones, films and radio programs. Instead of using the farmer groups, the TDS will be located at farms of model farmers.
4. A component which shows a significant departure from NAADS phase I is the agribusiness development and value addition. This component is a significant departure from the NAADS phase one, which largely focused on agricultural production. Under this component, NAADS will enhance provision of marketing services and value addition. DSIP states that the private sector will play a leading role in implementing agribusiness and value addition. Secondly, NAADS phase II has a particular emphasis on advisory services of sustainable land management (SLM) technologies. Benin et al. (2010) and Nkonya (2008) identified weak advisory services on SLM as one of NAADS weaknesses. This appears to be a step towards addressing this weakness.

METHODOLOGICAL APPROACH

Qualitative approach is used to achieve the objective of examining the perception of Agricultural Extension (AEA) providers on the new changes in provision of agricultural extension services. Respondents were selected from five districts, namely Kabale, Ntungamo, Nakasongola, Sironko and Lira. The districts selected were based on the period of implementation of the NAADS program in the districts. The respondents included District Agricultural Officers, farmer associations, NGO agricultural extension workers, NAADS coordinators and rural service providers who were knowledgeable on the changes in the agricultural extension reforms in Uganda. A total of 22 key informants were interviewed. The farmer associations accounted for 27% of interviewees while the NAADS coordinators and districts agricultural officers each contributed 18% of the participants. The results presented in this paper represent the interpretation of the personal opinions of the respondents. This is noted as a weakness which limits the generalization of the results.

RESULTS AND DISCUSSION

Strengths of the current agricultural extension system

The current agricultural advisory services are mainly provided by public extension services and by NGOs and private providers. The NGO and private providers provide specialized advisory services while the public AEAs provide the traditional advisory services. The key informants who participated in this study identified six strengths of the current system and we discuss them below.

Wider coverage was the most frequently cited strength of the current agricultural extension system (Table 1). Public AEAs are more available where they are recruited in the sub-counties and in terms of coverage they are

Table 1. Strengths and weaknesses of the current agricultural extension system.

Variable	Public AEA	Private AEA	Total
Strengths			
Wider coverage by public AEA	5	2	7
Well trained & Higher qualification of public AEA	5		5
Private AEA capacity to offer specialized services	2	3	5
Timely provision of extension services and better payments by private AEA	2	3	5
Better monitoring and supervision by public AEA	2	2	4
Trainings	1		1
Weaknesses			
Low salary and poor facilitation of public AEAs	5	1	6
Poor monitoring and supervision	3	3	6
Private AEAs profit oriented	3	2	5
Inadequate staffing	4	1	5
Lack of skills on modern farming methods	2	1	3
Inefficiency in recruitment and procurement process	1	1	2
Costly provision of extension services	2		2
Political interference	1		1
Misappropriation of funds	1		1
Weak linkage between MAAIF and district as result of decentralization	1		1
Limited sanctions	1		1
Few enterprise selection	1		1
Model farmer approach limiting wider impact	1		1
Non cooperation from farmers	1		1

more able to cover wider operational areas even though they may not be as efficient as the NGOs and private extension providers. The public extension workers are based in the sub-counties and are more in touch with the farmers, and the farmers can approach them at any time when their services are needed. Distribution of AEA and accessibility however differs. Access to agricultural extension services across districts shows that the central region has the highest density of AEAs per 100,000 rural households. A hundred thousand rural people in the central region are served by 15 AEAs while the corresponding number for the northern region is only 8, which is 86% smaller than the case in the central region[1].

Other strengths of the agriculture extension system reported by the key respondents was that, the public AEAs are available to do follow up after trainings or provision of services if they are well facilitated. They are salaried workers who are stationed in the districts and sub-counties and available for monitoring projects. Unlike the contracted NGO/private extension workers who do not continue with follow ups when their contracts end. On-the-job trainings of public AEA was also reported as one of the strengths of the agriculture extension system.

Figure 1 also shows a similar trend - with districts closer to Kampala having fewer rural people served by one AEA. Kalangala in the central region reported the highest density of AEAs (58 AEAs per 100,000 rural people) while Kaabong - a new district in the northern region – reported the lowest density (1.8 AEAs per 100,000 rural people). This demonstrates the poor agricultural extension services in the remote areas. This is consistent with findings of Jagger and Pender (2006), who observed that access to programs and organizations is concentrated in areas with high market access.

Well trained and higher qualification of public AEAs, private AEAs capacity to offer specialized services, and timely provision of extension services by private AEAs was each reported five times by the key informants (Table 1). The Public AEAs are well trained from recognized institutions at University or Diploma level. They are skilled experienced professionals who know their duties well and follow their professional ethics.

This is consistent with Nkonya et al. (2012) and Swanson (2008) who find share of government affiliated AEA with degree or post-graduate education highest compared to other private or NGO AEAs, reflecting positive results of governments investment in government AEA when it was the major service provider of agricultural extension.

The key informants observed that using NGO/private service providers for specialized activities is better since they possess specialized skill which the government

[1] Equivalent density for east and western regions is 10 and 8.4 respectively.

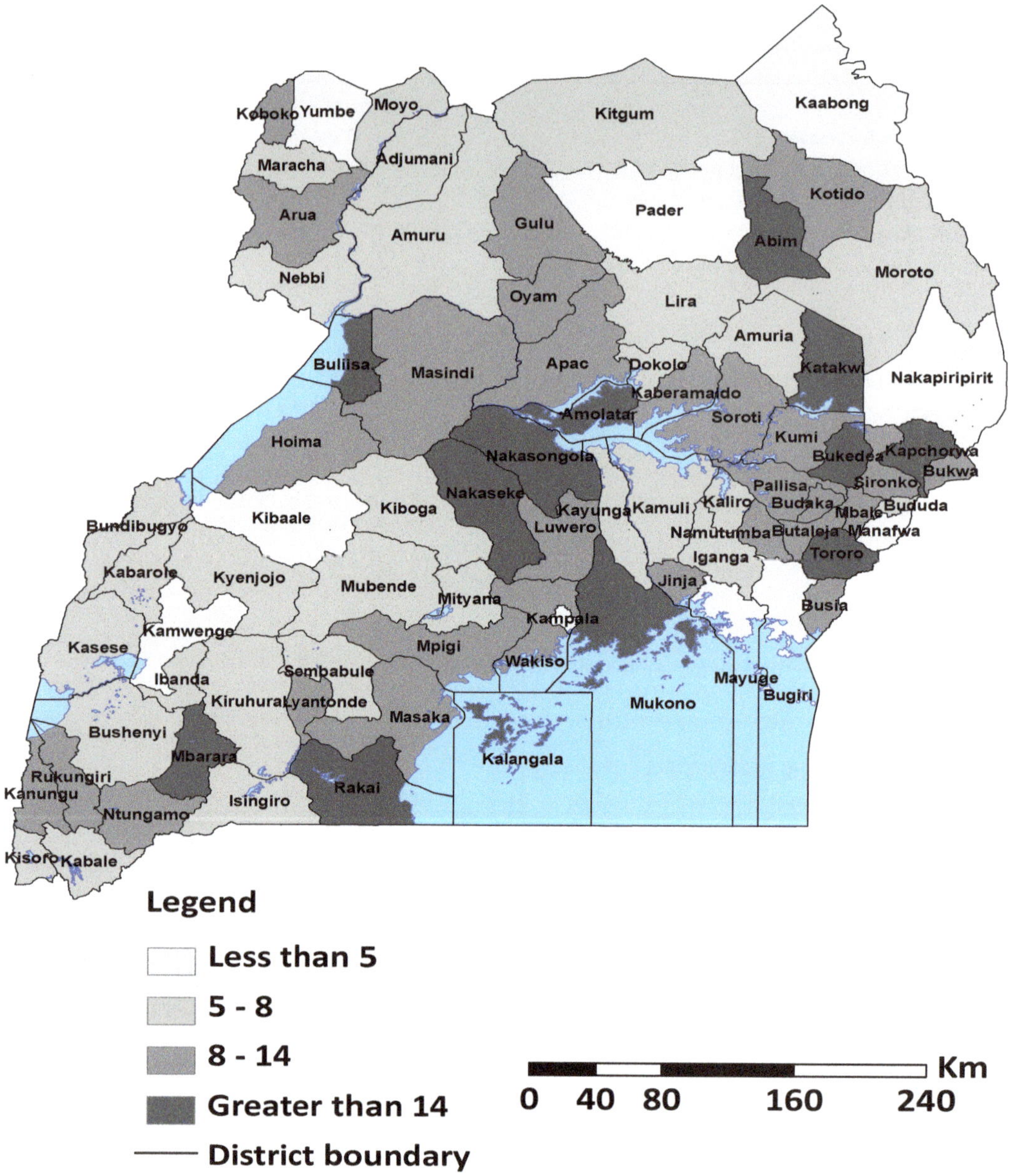

Figure 1. Number of agricultural extension agents per 100,000 rural people.

extension worker may not have. Farmers demand specialized services which can better be provided by the NGO/private extension providers and monitoring can be done effectively by the farmers themselves. The key informants also reported that the NGO/private extension providers are able to complete their work on time according to their terms of contract. If NGO/private AEAs are contracted through NAADS, there are very specific outputs required which makes them more able to provide expected deliverables according to their terms of reference.NGO/private Service providers provide better services since they receive good payments which are higher than that of the public extension workers.

The key informants also observed that private extension workers are easier to use in mobilizing farmers. Additionally, key informants observed that AEA affiliated with NGOs work more closely with farmers in their area of jurisdiction than is the case with the public AEAs. However, their coverage is always limited with budget. For example, Rutatora and Mattee (2001) observed that NGOs have become a major provider of agricultural extension services in Tanzania but their coverage is limited and always closer to urban areas. Jagger and Pender (2006) also observed lower concentration of NGOs in remote areas in Uganda. The NGO affiliated AEAs have done well in improving the fruit and vegetable sector, capacity building, building farmer groups and forming higher level organization.

Other strengths of the agriculture extension system reported by the key respondents was that, the public AEAs are available to do follow up after trainings or provision of services if they are well facilitated. They are salaried workers who are stationed in the districts and sub-counties and available for monitoring projects. Unlike the contracted NGO/private extension workers who do not continue with follow ups when their contracts end. On-the-job trainings of public AEA was also reported as one of the strengths of the agriculture extension system.

Weaknesses of the current agricultural extension system

One of the most frequently cited weaknesses by both the public and private AEAs were low salary payments and poor facilitation of public AEAs (Table 1). The key informants reported that the contracted private extension workers earn more money than the public extension workers. This creates a disincentive to the public AEAs who in some cases have to supervise the private NGO/AEAs who earn more than they do. The public AEAs also have poorer working facilities than NGO/private AEAs.

Poor monitoring and supervision was also frequently cited by key informants as a major weakness (Table 1). Monitoring of the activities of the private extension workers is poorly done and this leads to poor advisory services from private providers who may not be committed to providing quality advisory services. The problem is compounded by corruption, which the key informants observed to be common due to the competitive nature of awarding contracts. The key informants also observed that there are no consistent follow ups after the contracts of the NGO/private extension workers are signed. This has been due to the limited capacity of the districts production department and the NAADS secretariat.

The private AEAs contracted are profit oriented and are not concerned much about the successful impact of the services they provide (Table 1). The key informants observed that, unlike the public AEAs who are committed to their work, the private AEAs look out for their personally monetary benefits. Inadequate staffing at the districts and sub-counties was also reported by the key informants. This had resulted in a few AEAs serving a large number of farmers. The key informants noted that with limited staff poorly facilitated only a few farmers were reached.

The implementation of the NAADS approach of contracting NGO and private extension workers to provide specialized services has been faced with a number of challenges. The selection process of the NGOs and private AEA for provision of specialized services starts with, identification of the specialized services needed in the selected enterprises. This is done at the NAADS sub-county farmer forum. The district then places an advertisement, which states the kind of advisory services required and the professional qualification of the providers. Interested applicants pick up application forms and apply for the position to the sub-county. The sub-county farmer's forums are involved in the selection process. They are supported by a technical team composing of the staff from the production department, financing department, audit department, NAADS coordinator and sub-county chief. A suitable qualified applicant is selected and awarded the contract. Payments are made in phases with an initial payment to begin the work and other payments upon provision of reports and after approval of the Subject Matter Specialist (SMS) that the assigned task was well performed. The farmers' forum has members with low level of education, limited capacity to evaluate and select the advisory service applicants properly. Selection therefore is sometimes biased and may not necessarily be based on merit. Once the farmer's forum rejects an individual, the technical advisors cannot guarantee the selection of suitable candidate even if that applicant is highly qualified. Corruption is also a major problem in the selection process. As it will be seen in the discussion below, there is also lot of political interference and nepotism in the selection process. Aside the biases in the selection process, the selection process takes a long period of time. The selection committee after selecting the preferred candidate, sends the information to district NAADS coordinator and then to the NAADS secretariat for approval. The process takes a period of time before the extension worker is contracted. In some sub-counties that are very remote, NAADS had failed to recruit staff as there were few applicants out of whom there were no qualified applicants or in some cases, no applicant submitted application. This is consistent with Jagger and Pender (2006) and Rutatora and Mattee (2001). NAADS may not have full knowledge of the background of the NGO/private extension agents bidding for service provision. Applicants sometimes falsify documents submitted and the NAADS selection committee may not be able to verify the documents. This may result in awarding contracts to non-qualified persons. The corruption problem in NAADS recruitment process has been widely cited as a major problem (Parkinson, 2008; Feder et al., 2010).

The key informants also mentioned that it was costly for the government to pay both the private and public extension workers. However, this weakness is contrary to the strength discussed above that pluralistic extension services creates potential for providing specialized advisory services. The Private AEAs also provide advisory services on technologies and/or services prescribed in the terms of reference. So if farmers ask for other advisory services on technologies or services, the private provide would always decline providing them even if they are able to do so. This suggests a considerably large number of private providers to give specialized services.

Other weakness of the current extension system (Table 1) each mentioned included the political interference which will be discussed below, misappropriation of funds, weak linkage between the Ministry of Agriculture (MAAIF) and districts since decentralization of powers. Supervision of the district staff by the central government has been weakened since the district officials are answerable to the district and not the ministry. Also included in the weakness is the limited sanction for public AEA. One key informant observed that public extension workers are permanent and pensionable and even in case of mismanagement it takes a long time for a public extension officer to be disciplined and even during investigations he/she still may receive salary.

In addition, it was noted the NAADS program (phase 1) looked at only a few selected enterprises leaving out other important services. Advisory services on other enterprises which farmers are engaged in but not selected under the NAADS program are not given. This approach has however been revised under NAADS phase II - as discussed earlier – through prioritization by Village Farmer Forum (VFF) increasing the number of selected enterprises per sub-county.

NAADS approach of using selected host farmers (or model farmers) and technology development sites to carry out demonstrations and advisory services were noted by one key informant to have the weakness of not reaching out to a large number of farmers. Another key informant also cited non-cooperation from farmers. Overall, the key informants showed key strengths and weaknesses of the new extension services.

Enhancing advantages and address disadvantages of the current extension approach

In addressing the issue on what needs to be done to enhance the advantages of the current extension approach and who should be responsible for each of the action, the following suggestions were made (Table 2):

There is the need to build the skills of the new public AEAs to provide services on agro-business enterprise and other specialized skills which farmers may need. This was the most frequently cited measure by the key informants. Training of AEAs could be done by providing refresher courses and trainings for the extension workers to update their knowledge on modern farming methods. The key informants also suggested that the salary payments of public AEAs be increased to motivate them in providing quality services. The public AEAs should be well facilitated and should have access to transport facilities. Also, key informants noted that, monitoring and supervision of the activities of the AEAs should be properly monitored.

Currently, there are farmers in NAADS sub-county communities who do not participate in NAADS since they do not belong to farmer groups. To address this shortcoming, key informants suggested more vigorous sensitization of farmers to join farmers - an aspect, which is consistent with the RDS policy. The sensitization should also promote PMA's prime objective of commercializing agriculture such that the farmers manage their farms as business rather than simple way of life.

Key informants also suggested that the recruitment process of NGO/private AEAs should be strengthened to address the corruption and low capacity of farmer forum to vet the applicants. NAADS should also re-examine the hiring process to ensure a competitive recruitment process in order to appoint competent service providers. Members selected to sit on the farmer's forum should have a certain minimum level of education. They should be able to understand the qualification of the applicant and use the right criteria in selecting the most qualified applicant.

Additionally, key informants suggested the need to increase the agricultural extension budget as the government implements the DSIP plan, which aims to achieve the Maputo Declaration of allocating at least 10% of government budget to agriculture. With an increase in budget, the number of staff could be increased to serve more farmers.

The current political interference and the rapidly changing agricultural policy landscape have created confusion on the NAADS approach. Hence there is need for clarifying the NAADS approach. Farmers should be well informed on the NAADS program, its objectives and implementation strategies and how the farmers can benefit from the program. Extension workers together with other stakeholders like politicians and religious leaders should all be involved in educating farmers on the NAADS program. There is also the need to increase the Technology Development Sites and consequently model farmers in order to increase access to new technologies.

Other suggestion to enhance the agriculture extension program and address its problems included the need to give NGO/private extension workers long term contracts and to better coordinate and supervise their work. Current short-term contracts compromise the propensity of NGO/private AEAs to invest in improving their advisory services. There is also the need to increase the enterprises selected per subcounty and increase the number of farmers benefiting per subcounty. The Ministry of Agriculture (MAAIF) should restructure programs to ensure that enterprises not covered under NAADS program are either integrated or create a separate program to support those enterprises. Also suggested was the need to increase the number of beneficiaries from the program and encourage farmers co-funding.

Successes and failures of NAADS program in selected districts

Successes of NAADS in selected districts

The most frequently cited success of NAADS was

Table 2. Measures to enhance advantages and address disadvantage of the current agricultural extension system.

Parameter	Public AEA	Private AEA	Total
Capacity building	7	4	11
Increase salary and facilitation of public AEAs	7	1	8
Improve monitoring and supervision	3	3	6
Sensitization of farmers	4	2	6
Selection of competent private AEAs	1	4	5
Increase funding	4		4
Increase staffing	3		3
Limit political interference	1		1
Increase TDS		1	1
Long term contracts for Private AEAs		1	1
Increase enterprise selection	1		1
Farmer to co-fund	1		1
Increase number of beneficiaries	1		1
Improve quality of services	1		1

Table 3. Successes of NAADS programs in selected districts.

Successes of NAADS	Kabale	Lira	Nakasongola	Ntungamo	Sironko	Total
Group formation	x		x	x	x	4
Higher adoption rates of technologies	x	x		x		3
More commercialized	x				x	2
New enterprises	x		x			2
Empowered to demand advisory services	x					1
Linked to SACCOS			x			1
Better food security			x			1

formation of farmer groups. Four of the give districts reported that NAADS enhanced farmer group formation (Table 3). For example, in Nakasongola three farmer cooperatives for milk and poultry were established and are operating successfully. This is consistent with the design of NAADS, which provides advisory service through farmer groups. Provision of rural services through community groups is increasingly becoming popular in community-driven development (CDD) and other participatory development programs (Mansuri and Rao, 2004). Working in groups has helped farmers to exchange information and pool production and marketing resources.

The second most cited NAADS success was higher adoption of agricultural technologies – a success that was reported by three of the five districts visited for the study (Table 3). This is consistent with Benin et al. (2010) who observed higher adoption rates of NAADS beneficiaries. In Kabale for example, key informants reported that a lot of farmers were now using improved seeds and breeds.

Introduction of new enterprises were reported in two districts. In Nakasongola, key informants reported that the introduction of improved dairy breeds has recorded much success. Dairy which was the sixth priority enterprise of farmers in Nakasongola but now it is the second priority enterprise after cassava. Other priority crops in the districts are poultry, mangoes and oranges.

Consistent with the PMA objective, two districts also reported commercialization as a success story of NAADS. Introduction of new enterprises was also reported as a NAADS success in two districts. Benin et al. (2010) also found a significant increase in new enterprises for farmers in NAADS sub-counties (Figure 2).

Other success stories of NAADS – each reported by only one district – include: improved food security, farmer empowerment, and linking farmers to microfinance (SACCOS) institutions. The qualitative results show that the successes perceived by key informants were consistent with NAADS structure of providing advisory services through groups, advising farmers on new technologies, introducing new technologies and commercialization of agriculture. The small number of farmers reporting empowerment is a problem which raises concern about how NAADS achieved its prime

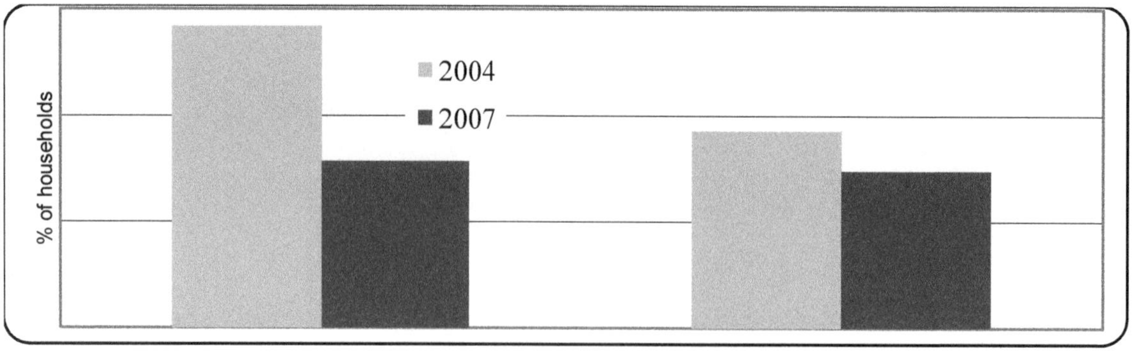

Figure 2. New enterprises adopted by NAADS and non-NAADS farmers in 2004 and 2007.
Source: Benin et al. (2010).

objective of empowering farmers. As it will be seen below, some of the factors which led to limited empowerment include corruption, political interference and procurement loopholes, all of which are likely to dampen empowerment efforts.

Failures of NAADS in selected districts

Political interference was the most cited failure of the NAADS program. All of the five districts mentioned political interference as NAADS weaknesses (Table 4). This is consistent with the Joughin and Kjær (2010), who observed an increased political interference which increased after the introduction of the PFA program. Given that the PFA was contained in an election manifesto, it took a political tone and approach. NAADS was meant to be an implementing instrument of PFA. Consequently, it attracted significant political attention from the president. The PFA was initiated with an objective of working within the existing government programs - especially PMA and NAADS, which serve the major sector. Technically, it was supposed to harmonize the rural development programs but it took a different approach. Unlike PMA which gave the private sector a leading role in its implementation, the PFA approach gave the government a leading role in provision of extension services, credit provision and agricultural input distribution.

As seen in Table 5, the frequency of President Yoweri Museveni's speeches criticizing NAADS after formation of PFA (2007 - 2011) was quite high. On average, the president mentioned NAADS in public speeches reported in the *New Vision* newspaper once in every two months. This is quite a high frequency for one agricultural program to attract such high profile attention. Worse still, most of the president's comments were negative. The tone of the president's speeches seem to ignore NAADS achievements discussed above and cited by other studies (Benin et al., 2010; Scanagri, 2005; OPM, 2005; Benin et al., 2007). Such political interference has tarnished NAADS' image among farmers. Given that NAADS is still charged with implementation of NAADS phase II under DSIP, there is need for restoring NAADS' image by addressing its past weaknesses discussed earlier. This requires an unbiased political attention which builds on NAADS strengths and addressing its weaknesses. There is also need on NAADS part to clarify its approach in order to clarify the changes precipitated by the rapidly changing agricultural policy landscape.

Corruption was reported by two of the five districts. This is consistent with other studies which have reported corruption in contracting service providers (Parkinson, 2008; Feder et al., 2010). Corruption was also one of the reasons of reinstating the old public extension services. Other weaknesses reported by at least two districts include low capacity of farmer forum to recruit competent service providers, weak or lack of advisory services on marketing and inadequate inputs.

CONCLUSIONS AND POLICY IMPLICATION

This study was done with the objective of contributing to the policy debate on the changing landscape of agricultural rural services in Uganda. The study examines the perception of Agricultural Extension (AEA) providers on the new changes in provision of agricultural extension services. Most studies focus on the perspectives of the beneficiaries, that is, farmers and not the service providers, that is, AEA who are the implementers and administrators of the program. Our findings provided timely in-depth information on the AEAs perspectives of the programme and how to improve on the program and achieve better results. Evidence show that the NAADS program has wider coverage even though distribution of agricultural extension agents differs across regions with the central region having the highest density of agricultural extension agents per 100,000 rural households. NAADS has enhanced the formation of farmer groups. This is consistent with the design of NAADS, which provides advisory service through farmer

Table 4. Failures of NAADS programs in selected districts.

Failures of NAADS	Kabale	Lira	Nakasongola	Ntungamo	Sironko	Total
Political interference	x	x	x	x	x	5
Corruption	x			x		2
No market advisory services	x			x		2
Low capacity of farmer forums to recruit providers	x		x			2
Inadequate inputs			x		x	2
Covers only priority enterprises				x		1
No benefit to large scale farmers & youth	x					1
Poor enterprise selection		x				1
Inadequate staffing			x			1
Procurement loopholes			x			1
Farmer high expectation					x	1
Low quality inputs					x	1
Late delivery of inputs	x					1
Weak monitoring of advisory service providers			x			1
Paying matching farmers for farmers with no direct benefit is problematic		x				1

Table 5. Frequency of President Yoweri Museveni's speech mentioning NAADS, 2007-2011.

Date	Key message of speech	President's speech mentioning NAADS
14.01.07	Museveni orders NAADS to be investigated following complaints that its impact on modernizing the farming community is not felt.	*"I am going to study, investigate, engage NAADS in a dialogue and come up with a stand on its future activities. I have not been aware about the way they have been implementing their programs. We should find an appropriate policy towards modernizing agriculture since it is the backbone of our economy,"*
05.09.07 10.09.07	Government spends lot of money on NAADS program but has achieved little impact due to misuse of funds by officials. Museveni suspends NAADS funds until cabinet sits and reviews the program aimed at improving service delivery and accountability of donor funds. NAADS officials to be probed	*"We want to have a change so that the money benefits the people instead of being spent on things you do not see"* *"Imagine all that money has been coming to one sub-county. Many officials have been misusing this money, organizing one seminar where one person talks and he gives an accountability of sh2m,"*

Table 5. Contd.

17.09.07	Museveni advises more funds from the NAADS program should be allocated for the purchase of farm inputs	*… the President suggested that the funds be used to buy farm implements. Museveni asked MPs to spearhead the campaign against poverty by actively training the 'wananchi' using NAADS materials and local extension workers*
23.09.07 07.10.07 08.10.07	NAADS program will not be stopped since it is a good program introduced by the NRM government. The program has been halted to undergo a review process of its performance and impact. Legal actions to be taken against officials who mismanaged funds.	*"…since it was the NRM government that introduced the program to help modernize agriculture and enhance production, it could not simply close it".* *"…the Government would audit the utilization of funds…, including taking legal action against those involved in swindling the funds".*
06.04.08 14.08.08 22.08.08	President's poverty tours to promote government's PFA program.	*"The crusade we have now is to change the living situation of our people so that they can use their small pieces of land to earn high incomes,"* *"… poverty was still a problem because "people are sleeping and even NAADS that we sent to wake them up joined them in sleep."* *"I am happy that you have woken up and are engaging in commercial agriculture. I have seen the projects and I have instructed NAADS to construct dams in places where these projects are so that you can do some irrigation during the dry seasons,"*
08.09.08	Museveni tells NAADS to support farmers move from subsistence farming to commercial farming.	*"NAADS should do more work by moving into the villages to help farmers who are currently engaging in subsistence farming...We need to stop producing for home consumption only.* *"I am going to talk to the people managing NAADS to strengthen their monitoring and advisory roles. I will ask them to increase the number of model farmers in the district," he said. "Don't get scared of telling me the truth about the performance of NAADS. We need to find out how they are working so that we advise them to improve,"*
10.02.09	The president warned the agriculture minister, Hillary Onek, against mishandling the implementation of the NAADS program.	*Museveni said he was writing to "clarify one point one more last time." He instructed Onek to stick to the six homesteads per parish, instead of scattering resources to many homesteads. "This is not acceptable to me. NAADS are again bringing confusion with 'demonstration' farmers, 'model' farmers, 'lead' farmers, etc,"... Resources should only be spent on six homesteads…Any NAADS official that fails to implement this or any other provision in the NAADS program will face severe sanctions"*

Table 5. Contd.

14.09.09 01.10.09 04.10.09 23.12.09 17.01.10	Museveni sets up special taskforce to investigate, cause arrest and prosecute people involved in theft and misuse of funds meant for NAADS. This follows accusation by the President of NAADS official misappropriating public funds and inflating agricultural inputs.	*"The special committee attached to the President's office will verify the standards of government projects to ensure they conform to stipulated goals,"* *"NAADS officials approve companies to tender cow dung and grass, just to steal money. What sort of officer can approve this? Is this person fit to be in public service? Why should such a person be on the payroll of the Government?"* *"I am going to send my spies to verify the authenticity of the list (lists of beneficiaries submitted to him during his tours of Prosperity-for-All projects). If the listed beneficiaries are non-existent or the figures attached to the names of farmers don't match, the Police will take the NAADS officials to the university of understanding (Luzira prison)". "If you see Police detectives combing the villages, don't get alarmed. They would be carrying out investigations on my instructions."*
22.12.09 30.12.09	Museveni announced that NAADS program will be reformed in 2010 to benefit more farmers. More farmers were to receive inputs	*"Next financial year we are going to transform NAADS into a poverty alleviation program. We are going to spread wealth so that at least each family gets two or three goats or pigs or sizeable number of poultry … This will cure the envy and arguments that the NAADS program had benefited only the rich farmers and a few people"*
23.06.10 07.07.10 09.08.10	Museveni again stopped the releasing of NAADS funds pending a review of the program.	*"I have been thinking of ways of sending this money to the poor. I will continue withholding sh120b NAADS funds unless I am satisfied that it will now reach the poor,"* *"Those who have not yet got NAADs money should not worry. The program is not ending tomorrow. I will make changes to ensure funds reach as many farmers as possible…I have told them, hold on, don't distribute it. Let me go round the country. When I come back, I will tell you how to use it."*
22.07.10 18.08.10	Museveni decides to stop NAADS tendering system. The award of tenders to supply inputs to farmers under NAADS is was not cost-effective. Funds to be channeled directly to farmer to make their own procurement	*"We started NAADS to make poor Ugandans rich, but they are supplied inputs at high prices which they cannot afford. I am going to stop the system such that farmers can buy inputs for themselves...These farmers only lack the money, but they can buy the farm inputs from their villages,"*
04.10.10 10.12.10 10.01.11	Museveni Lifts Ban on NAADS Money Disbursement for the scheme to facilitate the PFA program. Funds will directly be sent to villages or zones.	*"I have moved across the whole country assessing the implementation of Naadsprogramme and I am closing my tour in Kampala District. I have identified the problems in Naads and together with other stakeholders, we have resolved on how to go about them," he said adding "So now , the funds I had blocked are going to be released but this time round all those implementing the programme must religiously follow the set new guidelines ,"*

Source: The New Vision Online (Archives January 2007 - March 2011).

groups. Provision of rural services through community groups is increasingly becoming popular in community-driven development (CDD) and other participatory development programs (Mansuri and Rao, 2004). Working in groups has helped farmers to exchange information and pool production and marketing resources. The NAADS program has also promoted the adoption of agricultural technologies.

However, political interference is negatively affecting the image of the National Agricultural Advisory Services (NAADS) and undoing some of its achievements. This shows a clear case example of how a well structured agricultural program could fail to achieve the desired results due to politicization of the program. NAADS was used as one of the major implementation strategies of the prosperity for all (PFA), a political manifesto of the ruling party in the 2006 election. After formation of PFA in 2005/06, the frequency of the president speeches on NAADS increased tremendously. Between January 1, 2007 to March 2011, frequency of president's NAADS speeches reported in New Vision was once in every two months. Most of the comments on NAADS were negative. The tone of the president's speeches seem to ignore NAADS achievements discussed above and cited by other studies (Benin et al., 2010; Scanagri, 2005; OPM, 2005; Benin et al., 2007). Such political interference has tarnished NAADS' image among farmers. Funding for NAADS from the government was also suspended several times in an attempt to coarse the program to implement some of the president's agendas. The political nature of the PFA program and its emphasis on the greater role of government involvement in provision of agricultural services and subsidies was contrary to the pluralistic and demand-driven approach under NAADS. There is need of re-evaluating the current political support given to agriculture to ensure that it builds on past achievement of NAADS and addresses its weaknesses. There is also need for clarifying the NAADS approach. The political interference and the rapidly changing agricultural policy landscape have created confusion on the NAADS approach, which in itself was still not well-understood. Farmers should be well informed on the new NAADS implementation strategies and how the farmers can benefit from the program. Particularly, the role of the private sector in provision of advisory services and access to input program all need to be well-articulated to ensure that they do not compromise the potential of private AEAs to offer specialized advisory services required under DSIP and involvement of the private sector in agricultural processing and marketing.

Abbreviations

AEA, Agricultural Extension Provider; **CAADP,** Comprehensive Africa Agriculture Development Program; **CDD,** Community Driven Development; **DSIP,** Development Strategy Investment Plan; **NAADS,** National Agriculture Advisory Program; **NDP,** National Development Plan; **NGO,** Non-Governmental Organization; **PEAP,** Poverty Eradication Action Plan; **PFA,** Prosperity for All; **SACCO,** Savings and Credit Cooperative; **TDS,** Technology Development Site.

Conflict of Interest

The authors have not declared any conflict of interest.

REFERENCES

Benin S, Nkonya E, Okecho G, Pender J, Nahdy S, Mugarura S, Kato E, Kayobyo G (2007). Assessing the impact of the national agricultural advisory services in the Uganda rural livelihoods. IFPRI Discussion Paper 724. IFPRI, Washington, D.C., USA.

Benin S, Nkonya E, Okecho G, Randriamamonjy J, Kato E, Lubade G, Kyotalimye M,Byekwaso F (2010). Impacts of and returns to public investment in agricultural extension: the Case of the National Agricultural Advisory Services (NAADS) Program of Uganda. IFPRI mimeo.

Feder G, Anderson J, Birner R, Deininger K (2010).Promises and realities of community-based agricultural extension. IFPRI Discussion Paper 00959.

Jagger P, Pender J (2006). Impacts of programs and organizations on the adoption of sustainable land management technologies in Uganda. In: Pender J, Place F and Ehui S (eds.), Strategies for sustainable land management in the East African Highlands. International Food Policy Research Institute, Washington DC.

Joughin J, Kjær AM (2010). The politics of agricultural policy reforms: the case of Uganda. Forum for Development Studies. Danish Centre for International Studies and Human Rights.

MAAIF (Ministry of Agriculture, Animal Industries and Fisheries) (2010). Agriculture for food and income security. Agricultural Sector Development Strategy Investment Plan: 2010/11-2014/15.

MAAIF, MFEPD (2000). Plan for modernization of agriculture: Eradicating poverty in Uganda. Ministry of Agriculture, Animal Industry and Fisheries and Ministry of Finance, Planning and Economic Development, Kampala.

Mansuri G, Rao V (2004). Community-based and driven development: A critical review. World Bank Research Observer 19(1): 1–39.

Nkonya E, Kwapong NA, Bashaasha B, Mangheni M, Kato E (2012). Pluralist and demand-driven and traditional supply-driven agricultural extension services in Africa: Which reaches more farmers and women? The case of Uganda.

Nkonya E, Pender J, Kaizzi K, Kato E, Mugarura S, Ssali H, Muwonge J (2008). Linkages between land management, land degradation, and poverty in Sub-Saharan Africa: The Case of Uganda. IFPRI Research Report #159, Washington D.C.

OPM (Oxford Policy Management) (2005). Evaluation report: the plan for modernization of agriculture. OPM, Oxford, UK.

Parkinson S (2008). Learning participation in rural development: A study of Uganda's National Agricultural Advisory Services. Guelph, Ontario: University of Guelph.

Rutatora D, Mattee A (2001). Major agricultural extension providers in Tanzania. Afr. Study Monogr. 22(4):155-173.

Scanagri Consulting Company (2005). Midterm evaluation of Uganda's National Agricultural Advisory Services (NAADS) program. Draft Consultancy report submitted to the Government of Uganda.

Swanson B (2008). Global review of good agricultural extension and advisory service practices. Food and Agricultural Organization of the United Nations, Rome.

UBOS (Uganda Bureau of Statistics) (2010).Uganda National Household Survey, 2009/2010. Socio-economic module. Kampala, Uganda.

Contributions of mediating and independent variables to the adoption of co-financing public extension visits

D. B. AFFUL, A. OBI and F. S. LATEGAN

Department of Agricultural Economics and Extension, University of Fort Hare, P/Bag 1314, Alice, 5700, South Africa.

Most public extension organizations worldwide operate on inadequate budgets. Extension field surveys, therefore, need to be as cost-efficient as possible. The study investigated the comparative contributions of mediating and independent variables to the adoption of co-financing of the delivery of public extension visits. A semi-structured, self-administered questionnaire was used to collect information from 97 small-scale crop farmers in the Free State province of South Africa, from 1 September to 7 October 2010. Non-probability sampling techniques were employed to select respondents. The mediating variables were found to contribute more to the adoption of co-financing for the delivery of public extension visits than the independent variables. Need tension, especially the difference between aspired and present situation with regard to practice adoption and production efficiency made the most individual contributions to the variation in adoption of co-financing for the delivery of public extension visits. Adoption behaviour analysis could therefore be focussed on a limited number of mediating variables in surveys. Focussing adoption behaviour investigations on the more direct precursors of behaviour can reduce extension survey costs. This also offers opportunities for generating extra funds to deliver optimal extension visits to farmers. While quantifying the savings is the subject of future research, there are strong grounds for concluding that they are considerable.

Key words: Comparative contributions, mediating variables, independent variables, co-financing, bextension visit delivery, adoption.

...

INTRODUCTION

Public extension services have generally been criticized for many reasons including inadequate operating resources and a lack of financial sustainability (Rivera, 1991). The global budgetary constraints that have plagued public extension services since the 1980s led to calls for users of such services to contribute towards the cost of the service (Anderson, 2008). Persistent financial problems have contributed to the ineffectiveness of public sector extension (Rivera, 1991). Problems associated with ineffective public extension include poor logistic support, fewer monthly workshops, loss of personnel (Bagchee, 1994); few extension visits to farms such as less than one to one visit per month (Oladele, 2008) and irregular farm visits (Ulimwengu and Sanyal, 2011). The positive effects of extension contact on adoption and output are well-documented (Buyinza et al., 2008). Specifically, the positive influence of personal contact on the success of knowledge dissemination activities is a common and consistent finding in the literature of adoption (Hoag, 2005; Research Utilization Support and Help, 1996). More visits by extension agents are known to increase the effective price received by farmers and net return on production (Maheswari et al., 2008). Personnel costs however, are identified as one cause of the financial problems facing public extension services (National Treasury, Republic of South Africa, 2003). The cost of extension visits takes up by far the largest proportion (47.15%) of extension funds (Wilson and Gallup, 1955) and this also contributes to the operating financial woes of public extension. Dinar (1996) quoting Elkana and Epstein (1972) indicate that extension visits take up 39% of the total time of the extension advisor. This indicates that extension visits to farmers take up a lot of financial resources to accomplish. It is therefore, not unexpected when Dinar (1996:2) commented that among the public extension commercialization taking place in both developed and developing countries, activities being sold for charges are those associated with dissemination of information and direct contacts with producers.

In Australia, Marsh and Pannell (2000) observed that the general trend was to charge for the delivery of public extension information and not for the information itself. Discussions on user contributions for public extension delivery in South Africa have been documented in government policy papers (Department of Agriculture, 2005). The literature reviewed showed that no empirical study with regard to the comparative influence of the mediating and independent factors of adoption in the area charging for the delivery of extension visits has yet been conducted in South Africa. This study was, therefore, motivated by the need to reduce in one way, the cost of doing extension work in the area of extension surveys. This was achieved through an investigation of the comparative influence of the mediating and independent factors on farmers' acceptance to contribute financially towards the cost of farm visits by public extension workers. The mediating variables of adoption were hypothesized to contribute more to the adoption of the payment for the delivery of public extension visit than the independent variables.

THEORETICAL BACKGROUND

A producer's decision whether or not to adopt a new idea has occupied the minds of agricultural extension practitioners for decades. This pre-occupation has centred on trying to understand why producers do not adopt what seem to be "good" ideas that will benefit producers and what factors trigger the adoption of new ideas that enhance farm production and, eventually, profits. Many models of innovation[1] adoption and behaviour change have been developed over the years as a result. Among the earliest investigations into these adoption issues was that by Ryan and Gross (1943) and what is known as the classical adoption and diffusion model. Since then other models have been put forward; each trying to improve on its predecessors. These include the psychological field theory (Lewin, 1951), the classical 5-stage adoption process (North Central Rural Sociology Committee, 1961), the Campbell model (1966), the Tolman model (1967), the innovation decision process model (Rogers and Shoemaker, 1971) and its revised model (Rogers, 1983), the attitudinal determinants of behaviour (Fishbein and Ajzen, 1975). A framework that has a few variables and yet, comprehensive enough to account for all causes of adoption and non-adoption behaviour still remained elusive. Furthermore, for the field-level extension practitioner, there was need for a framework that did not only explain change but how to bring about change. In order to achieve this user-friendly framework, Düvel (1991) built on the earlier theoretical models (Rogers, 1983; Lewin, 1951) and drastically reduced the large number of intervening factors associated with behaviour analysis in the Tolman model (1967). He proposed the mediating variable concept to replace Tolman's intervening variable concept.

The mediating variables comprise needs, perception and knowledge. To allow for a wider spectrum of specific positive and negative forces, Düvel (1987) put forward a more refined list of Rogers's (1983) attributes of an innovation and, therefore, of perception. In this wise "relative advantage" was refined as relative advantages (Düvel, 1987) and the concept 'prominence' was introduced as an equivalent of Roger's innovation attribute, 'relative advantage' (Düvel, 1987). In this way the relative attractiveness of an innovation is redefined to include its relative advantages and prominence (Düvel, 1987). The relative advantages concept (advantages and disadvantages) concurs with Leeuwis and van den Ban's (2004) comment on the relationship between farmer evaluation of advantages and disadvantage of an innovation and adoption. The other dimension of perception in this redefinition is incompatibility of the innovation with the situation of the adopter (Düvel, 1987). The latter corresponds to the socio-economic, communication etc. circumstances of the adopter. Most of the factors that make a farmer unable or incapable of adoption that is, personal /environmental factors, fall into this category of variables and are of a more independent nature. They are subjectively perceived and make up what is commonly referred to as the independent variables of adoption and are, therefore, normally analysed as independent variables.

Field surveys cost time and money to accomplish. The advantage of reducing the numerous factors that affect adoption to a few that is comprehensive enough to account for behaviour is the significant saving on time and money that is associated with the approach. A comparative investigation of the influence of the independent and mediating variables on the adoption of new ideas such as the payment for the delivery of public extension visits, serves to determine which set of variables should receive

[1] The word ' innovation' is used in this study according to Rogers (1983) definition: "an idea, practice, or object that is perceived as new by an individual or other unit of adoption". This definition forms the basis on which farmers' financial contribution towards cost of extension visits (co-financing) is described as an innovation because it is new to the South African public extension service.

more attention in surveys to reduce the size of a survey questionnaire. This invariably reduces the cost of a survey without compromising quality of the results. The mediating and independent variables of adoption as enunciated here were used to investigate farmers' acceptability of payment for the delivery of public extension visits to assess their relative contributions in this regard.

METHODOLOGY

The paper investigated the relative contributions of the mediating and independent variables to farmers' acceptance to contribute financially towards the delivery of public extension visits. Non-probability sampling techniques, convenience and purposive, were used to select respondents because of a lack of farmer list from which to draw a sample. Information was collected from 97 small-scale[2] crop farmers in three of five districts of the Free State province of South Africa using a semi-structured, self-administered questionnaire from 1 September to 7 October 2010. The three districts chosen by simple random sampling were Motheo, Xhariep and Lewelputstwa. The hypothesized relationships between individual independent and individual mediating variables and the acceptance to pay for the delivery of public extension visits are presented in Tables 6 and 7, respectively. The data were analysed using SPSS. The data analyses comprised descriptive statistics and hypothesis testing by means of the Chi Square (X^2) tests of independence, and the multiple linear regression. More specifically, the multiple linear model afforded better insight into the relative importance of the factors that might explain the acceptability of the idea of farmers' contribution towards the delivery of public extension visits. In this model, both the independent and mediating variables that were found to significantly influence adoption of the innovation under study in the earlier models were entered simultaneously. In general, these procedures were used to test whether any observed differences were statistically significant.

Multicollinearity of the independent and mediating variables, reflecting variable redundancy which leads to high correlation among the included variables, was assessed by means of Variance Inflation Factor (VIF) and tolerance values to screen off variables to be deleted from the regression analysis (Pallant, 2007). Following Stockburger (1998), in which categorical variables with two levels may be directly entered as predictors or predicted variables in a multiple regression model, a multiple regression model was specified to study the relationship between the independent and mediating variables and payment for the delivery of public extension visits. The prediction of Y is accomplished by the following equation:

$$\dot{Y}_I = b_0 + b_1 X_I + \mathcal{E}_1 \qquad (1 = 1, 2, 3, \ldots n) \qquad (1)$$

Where $\dot{Y}$ is the predicted value on the dependent variable that is, payment for the delivery of public extension visits, the b values are the regression weights or the coefficients of the predictor variables, the X's represent the various predictor variables (independent and mediating variables), $\mathcal{E}_1$ is the error term and n is the number of observations.

[2]There seems to be different ways of categorizing/defining farmers in the literature (Department of Agriculture, 2005; Hadebe et al., 2004; Düvel, 2002). For this reason, an attempt was made to have an appropriate definition for the target group of farmers in this study as follows: Small-scale commercial crop farmers in this study are defined according to the categorization by Düvel (2002) and comprise, therefore, farmers who produce mainly for the market and Land Redistribution for Agricultural Development programme beneficiaries who may have own consumption and the market in view as the ultimate purpose of production.

RESULTS

The results of the multicollinearity analysis of the independent and mediating variables used to investigate farmers acceptance to pay for extension farm visits are presented in Tables 1 and 2, respectively. All nine independent variables passed the screening test (Table 1) and were entered into the first regression analysis separately; four of the sixteen mediating variables, however, showed multicollinearity (Table 2) and were deleted while the rest were entered into a separate regression analysis. The deleted variables were, influence of innovation on aspired yield (EAP: YHA), influence of innovation on gross farm income (EAP: GFI), influence of innovation on practice adoption (EAP: PA) and most important plan to achieve goal (PLAN: NC). The results of the first regression analysis of the independent variables and their influence on the adoption of co-financing of public extension visits are presented in Table 3. The results show that all variables except, "mentorship" positively correlated with the dependent variable, that is, payment for the delivery of public extension farm visits. However, only farming orientation, credibility of the extension service/agent, desired number of visits and group membership made a significant contribution to the variation in adoption of payment for the delivery of public extension visits. Together, these four variables explained 58.8% of the variation in the adoption of the payment delivery of public extension visits. The model was significant (p = 0.004).

The results of the fitting of the other multiple regression model developed to assess the aggregate contribution of the mediating variables on the adoption of payment for the delivery of public extension visits are presented in Table 4. The findings indicate that, with the exception of awareness of disadvantages of payment for the delivery of extension visits, all other variables positively correlated with payment for the delivery of public extension visits. The following mediating variables were found to significantly influence the payment for the delivery of public extension visits: "reason for farming", "farmers' goals for the next five years", "unawareness of optimum yield per hectare achievable"; the others were "need tension resulting from the difference between aspired and present yield per hectare", "need tension resulting from the difference between the aspired and present level of practice adoption", and "awareness of the disadvantages of payment for the delivery of public extension visits".

DETERMINANTS OF THE ADOPTION OF PAYMENT FOR THE DELIVERY OF PUBLIC EXTENSION VISITS

The multiple regression results of the relative importance

Table 1. The independent variables hypothesized to explain acceptance to pay for the delivery of public extension visits.

Variable	Unit	Expected sign	Description
FO		?	1 if full time farmer
ENSAI	%	+	Earnings from farming
GRPM		+	1 if belongs to a group
DNV	Number of visits	+	Desired number of visits per month
DEF		+	1 if farmer sees public extension reducing effect of drought spell on gross farm income
FENTS		+	1 If maize, sunflower or Lucerne
MENTOR		+	1 If supported by a mentor
FEXP	Number of years	?	Number of years of planting crop
CREINDEX		+	1 If farmer sees public extension as more credible

Table 2. The mediating variables hypothesized to explain acceptance to pay for the delivery of public extension visits.

Variable	Expected sign	Variable description
Reason for farming	+	If compatible with payment=1
Goal for next five years	+	If compatible with payment=1
Misperceived current adoption efficiency	-	If compatible with payment=1
Unaware of optimum yield per ha achievable	-	If compatible with payment=1
Unaware of optimum practice adoption achievable	-	If compatible with payment=1
Need tension (aspired-present yield/ha)	+	If compatible with payment=1
Need tension (optimum-aspired yield/ha)	+	If compatible with payment=1
Need tension (optimum - present adoption level)	+	If compatible with payment=1
Need tension (optimum – aspired adoption level)	+	If compatible with payment=1
Need tension (aspired – present adoption level)	+	If compatible with payment=1
Unawareness of advantages of payment	-	If compatible with payment=1
Awareness of disadvantages of payment	-	If compatible with payment=1

Table 3. Collinearity statistics of independent variables[1].

Variable	Collinearity statistics	
	Tolerance	VIF
FO	0.438	2.285
ENSA1	0.464	2.156
FENTS	0.389	2.571
FEXP	0.457	2.186
CREINDEX	0.860	1.163
DEF	0.347	2.885
DNV	0.754	1.327
MENTOR	0.843	1.187
GRPM	0.836	1.197

[1]Farming Orientation (FO), Percentage earning from farming (ENSA1), Farming enterprise (FENT), Farming experience (FEXP), Credibility of public extension service provider (CREINDEX), Effect of drought spells on gross farm income (DEF), Desired number of visits from public extension agent (DNV), Farming w3ith support of a mentor (MENTOR), Group membership (GRPM).

of the independent and mediating variables in explaining respondents' acceptance to contribute towards the delivery of public extension visits are presented in Table 5. The results (Table 5) indicate that all the mediating

Table 4. Collinearity statistics of mediating variables[2].

Variables	Collinearity statistics	
	Tolerance	VIF
Reason NC	0.757	1.321
Goal NC	0.454	2.202
Plan NC	0.025	39.597
PCEA1	0.358	2.790
OPTACH:YHA	0.507	1.974
OPTACH:PA	0.594	1.684
NTI:YHA	0.128	7.834
NT2:YHA	0.943	1.061
NT2:PA	0.187	5.347
NT3:PA	0.654	1.530
NT1:PA	0.152	6.562
EAP:YIELD	0.066	15.229
EAP:GFI	0.066	15.145
EAP:PA	0.041	24.329
UWADVC	0.154	6.474
ADISVC	0.166	6.034

[2]Need compatibility: Reason for farming (Reason NC), Reason for farming :Goal for next 5 years, (Goal NC), Need incompatibility: Most important plan to achieve goal (Plan NC), Problem perception: adoption of management practices (PCEA1), Unawareness of optimum achievable: Yield per hectare (OPTACH:YHA), Unawareness of optimum achievable: Practice adoption (OPTACH:PA), Need tension (aspired minus current level of output (NT1:YHA), Need tension (optimum minus present level of output (NT2: YHA), Need tension (optimum minus present level of practice adoption) NT2: PA), Need tension (optimum minus aspired level of practice adoption) NT3: PA, Need tension (aspired minus current level of practice adoption) NT1: PA, Influence of innovation on aspired yield (EAP: YIELD), Influence of innovation on gross farm income (EAP:GFI), Influence of innovation on practice adoption (EAP: PA), Unawareness of advantages of innovation (UWADVC), Awareness of disadvantages of innovation (ADISVC).

Table 5. Multiple regression estimates of the effects of the independent variables on the payment for the delivery of public extension visits.

Variables	Beta	T	P
Farming orientation	0.127	1.006	0.001*
Percentage farm earnings	0.173	1.488	0.142
Farming enterprise	0.236	1.862	0.067
Farming experience	0.203	-1.739	0.087
Credibility	0.549	6.429	0.000*
Drought effect on gross farm income under paid public extension	0.233	1.740	0.086
Desired number of visits	0.290	3.183	0.002*
Mentor	-0.020	-.231	0.818
Group membership	0. 149	1.724	0.009*
Constant	_	-5.266	0.000

R^2 = 0.588, Significant at 0.01*.

variables had a positive relationship with payment for the delivery of public extension visits. None of independent variables however, significantly influenced the adoption of payment for the delivery of public extension visits. The need-related factors were also dominant over the perception related variables in influencing adoption. All the three variables that significantly influenced the adoption of payment for the delivery of public extension visits were need-related, explaining 87.9% of the variation in the adoption of the innovation under study. Need tension, especially the difference between aspired and present situation with regard to practice adoption and production efficiency, made the most individual contributions at 0.353 and 0.473, respectively, to the

Table 6. Multiple regression estimates of the effects of mediating variables on payment for the delivery of public extension visits.

Variables	Beta	T	P
Reason for farming	0.074	1.746	0.026
Goal for next five years	0.099	2.300	0.025
Misperceived current adoption efficiency	0.083	1.407	0.163
Unaware of optimum yield per ha achievable	0.093	1.809	0.074
Unaware of optimum practice adoption achievable	0.110	2.320	0.023
Need tension (aspired-present yield/ha)	0.429	5.847	0.000*
Need tension (optimum-aspired yield/ha)	0.004	.093	0.926
Need tension (optimum - present adoption level)	0.117	.536	0.128
Need tension (optimum – aspired adoption level)	0.010	.236	0.814
Need tension (aspired – present adoption level)	0.291	3.484	0.001*
Unawareness of advantages of payment	0.122	1.574	0.119
Awareness of disadvantages of payment	-0.201	-4.042	0.035
Constant	-	-3.203	0.002

R^2 =0.885 * significant at 0.01 lev.

adoption of co-financing for the delivery of public extension visits.

DISCUSSION

Various adoption studies have shown inconsistent relationships between the independent variables and adoption behaviour (Chukwuone and Agwu, 2005; Campiche et al., 2004). The discussion here will, therefore, centre on those independent variables as well as the mediating variables that made a significant impact on the adoption of co-financing for the delivery of public extension visits.

Farming orientation

There is ambiguity in the literature concerning the expected difference between full-time and part-time farmers with respect to adoption (Table 6). Some of these studies include Sulaiman and Sadamate (2000) and Kenkel and Norris (1995). The finding in this study (Table 3), therefore, supports the hypothesis that farming orientation might positively influence payment for the delivery of public extension visits. In this case, full time farmers were more inclined than part time farmers to pay for the delivery of public extension visits. A possible reason could be the assumption that part-time farmers were very busy and so did not have the luxury of time to search for information for their farm work and so would be eager to invest in new ideas such as paying for the delivery of extension visits did not hold for farmers in this survey. This was supported by the fact that fewer part-time farmers (53.6% compared with 74.5% of full-time farmers) indicated they were willing to pay for the delivery of public extension visits which could afford more contacts with the extension officer. More contacts are assumed to result in the transfer of more improved farm management practices.

Full-time farmers on the other hand, depend on farming for their livelihoods. They are, therefore, willing in a sense, to invest in the delivery of public extension visits as shown by the results in this study, which could invariably result in more contacts with the extension agent for more farm management practices. This is expected to eventually improve their efficiency of production. The finding in this study, therefore, conforms to the prevalent views found in the agricultural economics and extension literature.

Group membership

The finding in this study (Table 3) that group membership has a significant positive effect on adoption was similar to what is reported in previous studies (Ajayi, 2006; Habtemariam, 2004; Gautam, 2000). This positive relationship might be due to respondents' realization of reduced cost of service to individual members in the group setting. Daramola (1989) however, did not find co-operative membership to significantly influence the probability of fertilizer adoption decisions; in fact, its influence was negative. A possible reason for Daramola's finding could be culturally-related where people tend to be individualistic because of lack of trust of other people in group settings. Another possible explanation why people might not join groups such as farmer co-operative as in Daramola's sample where 54% of respondents did not belong to co-operatives, and therefore, the reason why such groups might not influence the adoption of farm innovations, might be due to past disappointments with groups.

The finding in this study and others clearly suggest that, although group membership is a great enhancer of the adult learning and adoption process, adoption might

Table 7. Multiple regression estimates of the effects of mediating and independent variables on the payment for the delivery of public extension visits.

Variable	Beta	T	P
Farming orientation	.011	.285	.777
Credibility index	.041	1.013	.314
Desired number of visits	.009	.224	.823
Group membership	.069	1.679	.097
Reason for farming	.058	1.398	.166
Goal for next five years	.061	1.144	.256
Unawareness of optimum practice adoption achievable	.160	3.704	.000*
Need tension (aspired –present yield per ha)	.473	6.841	.000*
Need tension (aspired – present level of adoption)	.353	4.837	.000*
Awareness of disadvantages of innovation	.017	.248	.805
Constant	-	-4.000	.000

R^2 =.879, * Significant at .01.

not be guaranteed by group membership.

Desired number of visits

Past studies show different results between respondents' current number of visits received and payment for public extension services. Non-significant results have been reported in this regard (Ajayi, 2006; Daramola, 1989). Sulaiman and Van den Ban (2003), on the other hand, mention that one important condition for paid services was the farmers' insistence on field visit based advice. The emphasis on these previous investigations was on payment for general extension services based on the status quo number of visits. The accent of the present study was different because the focus was on payment for the delivery of more visits as a means to access more farm management practices. Findings in this study (Table 3) showed that famers' desired number of visits has a positive significant influence on their acceptance to pay for farm visits as hypothesized (Table 6). Most of the respondents (70 to 89.5%) who wished to receive between 2 and 4 visits per month, with a mean of 3.16 visits per month (SD = 1.213), were prepared to pay.

The mean number of visits reported here was close to the designated visits of one every two weeks (or 2 visits per month) in the Kenya extension project (Gautam, 2000) and similar to the 2 visits per month requested by livestock farmers in Turkey (Budak et al., 2010).

Credibility of public extension service provider

Findings in this study (Table 3) indicate that credibility had a significant positive effect on payment as hypothesized (Table 6). This finding concurs with Ajayi (2006). What should be of concern to policy makers is the fact that 45% of 44 respondents did not find the public extension credible and therefore, would not pay. About 54% however, would still like to pay for the delivery of public extension; perhaps this was because this was the only source they could afford compared with private extension. Policy makers should think seriously about improving the competency of field level extension practitioners and make the cost more affordable so that payment for the delivery of visits becomes more attractive to farmers.

Reason for farming

The mere assumption that human behaviour is purposeful implies that there are reasons why respondents engage in farming. The purposes or reasons could be expected to reflect the individual's needs, either directly or indirectly (Düvel, 1991). The understanding that since needs represent the basic motives governing purposeful human behaviour, they are critical in understanding respondents' behaviour concerning payment for public extension visits. Of particular importance is whether these reasons are compatible with payment for public extension visits, something that could be expected to be the case if respondents' respective objectives could be achieved with paying for extension visits. The multiple regression results of the influence of respondents' reasons for farming on their acceptance to pay for public extension visits were investigated and are presented in Table 4. This variable had a significant positive relationship with payment as expected (Table 7). This finding concurs with Afful (1995).

Goals for next five years

The assertion that behaviour is after all directed toward a goal object as a means for need (internal) satisfaction

(Düvel, 1982) and the difficulty of reliably measuring the intra-personally located needs, led to the proposition to pay attention to the farmers' goals/aspirations/objectives in situation surveys (Düvel, 1982). Respondents' goals for the next five years were thus assessed to determine any influence relationship on their acceptance to pay for public extension visits. The results show that respondents' goals for the next five years positively related to payment as expected and the influence was significant (Table 4). Afful (1995) made a similar finding regarding farmers' reasons for keeping cattle.

Need tension

As hypothesized (Table 7), findings (Table 4) indicated that need tension resulting from a difference between aspired and present output (tons) per hectare has a positive and significant effect on payment for the delivery of public extension visits. Similar results were obtained for need tension arising from a difference between aspired and present level of practice adoption. These results are in line with past studies (Msuya, 2007). There seemed, however, to be a pattern of consistent significant positive results with need tension associated with a difference between aspired and present situation but inconsistent results for need tension in relation to perceived optimum and present situation or perceived optimum and aspired situation. These aberrant findings seem to be caused by respondents' lack of agreement with what was optimum; or perhaps respondents equated the optimum with their perceived aspired level that is, did not see any difference between their aspired yields or practice adoption level and what is perceived as the optimum and thus yielding little perceptual difference. There was no indication of previous investigations in the literature reviewed concerning the inconsistent results reported here of the influence relationships between these two dimensions of need tension and adoption. For this reason, it might be unjustifiable to make definitive statements as to whether or not these findings conform to expectations. More research is needed to confirm or reject the validity of this claim.

Awareness of disadvantages

The hypothesis concerning the negative relationship between respondents' awareness of the disadvantages of payment for the delivery of public extension visits and payment was as expected (Table 7). This influence was significant but negative (Table 4). Hudson and Hite (2002) similarly found that producers who perceive that the costs of the innovation under study outweigh the benefits had a much lower willingness to pay than producers who believed the benefits outweigh the costs.

Comparative contributions of independent and mediating variables to adoption

The independent variables separately contributed 58.8% (Table 3) to explain the variation in payment while the mediating variables explained 88.5% in the variation in payment (Table 4). Ajayi (2006) similarly found that the independent variables contributed 41.2% to farmers' willingness to pay for general extension services; there was however, no investigation of the mediating variables in his study. Habtemariam (2004) also found that the aggregate effect of the independent variables on the variation in production efficiency of maize farmers was 48.9% which was still lower than the effect of the mediating variables, which was 63.6%. The findings in this study (Tables 5) further confirmed the overall study hypothesis that the mediating variables contribute more to adoption of payment for the delivery of public extension visits than the independent variables. This finding is consistent with the results of past studies (Msuya, 2007; Habtemariam, 2004). This was evident from the fact that the effect of the independent variables was masked in the presence of the mediating variables because none of them significantly influenced the adoption of payment for the delivery of public extension visits. In other words, the independent variables manifest through the mediating variables. The total effect of both mediating and independent variables in explaining the variance in adoption was 87.9%. This might be due to the masked effect of the independent variables. This study was based on non-probability sampling and this makes it inappropriate to generalize the results to the wider population of small-scale commercial crop farmers in South Africa.

Even though this limitation does not invalidate the findings in terms of their ability to answer the main research hypothesis and contribute to the body of knowledge in extension science, a replication of the study by means of probability sampling methods would make it possible to generalize these findings in the larger population of small-scale commercial crop farmers in the country.

Conclusions

The mediating variables were found to contribute more to the adoption of co-financing for the delivery of public extension visits than the independent variables. The following mediating variables were found to significantly influence the payment for the delivery of public extension visits: reason for farming; farmers' goals for the next five years; unawareness of optimum yield per hectare achievable; need tension resulting from the difference between aspired and present output per hectare; need tension resulting from the difference between the aspired and present level of practice adoption; and awareness of the disadvantages of payment for the delivery of public extension visits. Need tension, especially the difference between aspired and present situation with regard to practice adoption and production efficiency made the most individual contributions to the variation in adoption of co-financing for the delivery of public extension visits. The effect of the independent variables manifest through the mediating variables so adoption behaviour analysis could be focussed on a limited number of mediating variables in surveys. Narrowing the focus of adoption behaviour analysis to the more direct and immediate precursors of behaviour offers opportunities for a more rigorous assessment of the more relevant variables which can be changed as opposed to the more static independent variables. More importantly, focussing on the limited number of mediating variables identified in this study helps to cut down the cost of adoption behaviour surveys. This also offers opportunities for generating extra funds to deliver optimal extension visits to farmers. Cutting costs is essential in a period when public extension services face operational financial problems. The research implication of these findings is the need for future study to quantify how much income could be generated from such famer contributions for extension visits.

ACKNKOWLEDGEMENTS

The authors wish to acknowledge the following organizations for their financial support: The National Research Foundation for funding the research project from which this manuscript was written and the Govan Mbeki Research and Development Centre's fellowship that made the writing of the manuscript possible. The Free State Department of Agriculture is highly appreciated for providing logistic support during data collection of the original research.

REFERENCES

Afful DB (1995). Human constraints in stock reduction in a subsistence farming situation in South Africa, Master's dissertation, University of Pretoria, South Africa.

Ajayi AO (2006). An Assessment of Farmers' willingness to pay for extension services using the Contingent Valuation Method (CVM): The case of Oyo State, Nigeria. J. Agric. Edu. Ext. 12(2):97-108.

Anderson JR (2008). Agricultural Advisory Services, Background paper for World Development Report 2008, Agriculture for Development, World Bank, Washington DC.

Bagchee A (1994). Agricultural extension in Africa, World Bank Discussion World Bank, Washington DC. P. 231.

Budak DB, Budak F, Kaçira O (2010). Livestock producers' needs and willingness to pay for extension services in Aduna Province of Turkey. Afr. J. Agric. Res. 11:1181-1190.

Buyinza M, Banana AY, Nabanoga G Ntakimye A (2008). Socio-economic determinants of farmers' adoption of rotational woodlot technology in Kigorobya Siu-County, Hoima District, Uganda'. S. Afr. J. Agric. Ext. 37:1-16.

Campbell RR (1966). Suggested paradigm of the individual adoption process. Rural Sociol. 31:458-466.

Campiche J, Holcomb RB, Ward CE (2004). Impacts of Consumer Characteristics and Perceptions on Willingness to Pay for Natural Beef in the Southern Plains, Food Technology Report, Oklahoma State University. Food. Agric. Prod. Res. Technol. Centre pp. 1-15.

Chukwuone NA, Agwu AE (2005). Financing agricultural technology delivery in Nigeria: Would farmers be willing to pay. J. Ext. Syst. 21(2):69-85.

Daramola B (1989). The Study of Socio-economic Factors influencing Fertilizer Adoption Decisions in Nigeria: A Survey of Oyo State Farmers, Fertil. Res. 20:143-151.

Department of Agriculture (2005). Norms and Standards for Extension and Advisory Services in Agriculture, Scientific Research and Development Directorate, Pretoria.

Dinar A (1996). Extension Commercialization: How much to charge for extension services. Am. J. Agric. Econ. 78(1):1-12.

Düvel GH (1991). Towards a model for the promotion of complex innovations through programmed extension. S. Afr. J. Agric. Ext. 20:70-86.

Düvel GH (1982). Need creation and exploitation: The basis for change, S. Afr. J. Agric. Ext. 11:27-33.

Düvel GH (1987). Situation determination: From theory to a practical model. S. Afr. J. Agric. Ext. 16:1-10.

Fishbein M, Ajzen I (1975). Belief, attitude, intention and behaviour: an introduction to theory and research. Addison-Wesley Publishing Company Inc., Philippines.

Gautam M (2000). Agricultural Extension: The Kenya Experience: An Impact Evaluation. World Bank, Washington DC.

Habtemariam AG (2004). The Comparative Influence of Intervening Variables in the Adoption Behaviour of Maize and Dairy Farmers in Shashemene and Debrezeith, Ethiopia, PhD thesis, University of Pretoria, South Africa.

Hoag DL (2005). Economic principles for saving the Cooperative Extension Service. J. Agric. Res. Econ. 30(3):397-410.

Hudson D, Hite D (2002). Producer Willingness to pay for Precision Application Technology: Implications for Government and the technology Industry. Can. J. Agric. Econ. 51:39-53.

Kenkel PL, Norris PE (1995). 'Agricultural Producers' Willingness to pay for Real-Time Mesoscale Weather Information. J. Agric. Res. Econ. 20(2):356-372.

Leeuwis C, van Den Ban A (2004). Communication for Rural Innovations: Rethinking Agricultural Extension,Third Edition. Blackwell Publishing, Oxford.

Lewin K (1951). Field Theory in Social Science. Selected Theoretical Papers, Harper & Row, New York.

Maheswari R, Ashok K, Prahadeeswaran M (2008). Precision farming Technology, Adoption Decisions and Productivity of Vegetables in Resource-Poor Environments. Agric. Econ. Res. Rev. 21:415-424.

Marsh SP, Pannell DJ (2000). Agricultural Extension Policy in Australia: the good, the bad and the misguided. Aust. J. Agric Res. Econ. 44(4):605-627.

Msuya CP (2007). The Comparative Role of intervening and Independent Variables in the Adoption Behaviour of Maize Growers in Njombe District, Tanzania. PhD thesis, University of Pretoria, South Africa.

National Treasury, Republic of South Africa (2003). Inter-Governmental Fiscal Review: Agriculture. Communications Directorate, Pretoria.

North Central Rural Sociology Committee (1961). Adopters of New Ideas: Characteristics and Communication behaviour. North Central Regional Extension Publication P. 13.

Oladele OI (2008). Factors determing farmers' willingness to pay for extension services in Oyo State, Nigeria. Agric. Trop Subtrop. 41(4):165-169.

Pallant J (2007). SPSS Survival Manual: A Step-by-Step Guide to Data Analysis Using SPSS Version 17, Third Edition. McGraw-Hill, Berkshire.

Research Utilization Support and Help (1996). A review of the literature on dissemination and knowledge. www.researchutilization.org/matrix/resources/review (Accessed: 10/11/2011).

Rivera WM (1991). Worldwide Policy Trends in Agricultural Extension, Part 1. J. Tech. Trans. 16(1):13-18.

Rogers EM, Shoemaker FF (1971). Communication of innovations: A

cross-cultural Approach, Second Edition. Rinehart & Winston, New York.
Rogers EM (1983). Diffusion of Innovations, third ed. The Free Press, New York.
Ryan B, Gross N (1943). The diffusion of hybrid seed corn in Iowa communities. Rur. Sociol. 8(1):15-24.
Stockburger DW (1998). Multivariate Statistics: Concepts, Models and Applications, http://www.psyhstat.missouristate.edu/multibook/mlt00.htm (Accessed 10 May 2011).
Sulaiman R, Sadamate VV (2000). Privatizing agricultural extension in India. Policy National Centre for Agricultural Economics and Policy Research (NCAP): 10:95, New Delhi.
Sulaiman R, Van Den Ban AW (2003). 'Funding and delivering of agricultural extension in India', J. Int. Agric. Ext. Edu. 10(1):21-30.
Tolman EC (1967). A psychological model', in T. Parsons & E.A. Shils (Eds.), Towards a general theory of action. Harvard University Press, Cambridge.
Ulimwengu J, Sanyal P (2011). Joint Estimation of Farmers' Stated Willingness to pay for Agricultural Services. International Food Policy research Institute Discussion P. 01070.
Wilson M, Gallup G (1955). Extension Teaching Methods and factors that influence the adoption of agricultural and home economics practices. Circular United States Department of Agriculture, Washington DC. P. 459.

Permissions

All chapters in this book were first published in JAERD, by Academic Journals; hereby published with permission under the Creative Commons Attribution License or equivalent. Every chapter published in this book has been scrutinized by our experts. Their significance has been extensively debated. The topics covered herein carry significant findings which will fuel the growth of the discipline. They may even be implemented as practical applications or may be referred to as a beginning point for another development.

The contributors of this book come from diverse backgrounds, making this book a truly international effort. This book will bring forth new frontiers with its revolutionizing research information and detailed analysis of the nascent developments around the world.

We would like to thank all the contributing authors for lending their expertise to make the book truly unique. They have played a crucial role in the development of this book. Without their invaluable contributions this book wouldn't have been possible. They have made vital efforts to compile up to date information on the varied aspects of this subject to make this book a valuable addition to the collection of many professionals and students.

This book was conceptualized with the vision of imparting up-to-date information and advanced data in this field. To ensure the same, a matchless editorial board was set up. Every individual on the board went through rigorous rounds of assessment to prove their worth. After which they invested a large part of their time researching and compiling the most relevant data for our readers.

The editorial board has been involved in producing this book since its inception. They have spent rigorous hours researching and exploring the diverse topics which have resulted in the successful publishing of this book. They have passed on their knowledge of decades through this book. To expedite this challenging task, the publisher supported the team at every step. A small team of assistant editors was also appointed to further simplify the editing procedure and attain best results for the readers.

Apart from the editorial board, the designing team has also invested a significant amount of their time in understanding the subject and creating the most relevant covers. They scrutinized every image to scout for the most suitable representation of the subject and create an appropriate cover for the book.

The publishing team has been an ardent support to the editorial, designing and production team. Their endless efforts to recruit the best for this project, has resulted in the accomplishment of this book. They are a veteran in the field of academics and their pool of knowledge is as vast as their experience in printing. Their expertise and guidance has proved useful at every step. Their uncompromising quality standards have made this book an exceptional effort. Their encouragement from time to time has been an inspiration for everyone.

The publisher and the editorial board hope that this book will prove to be a valuable piece of knowledge for researchers, students, practitioners and scholars across the globe.

List of Contributors

Rukhsana
Department of Geography, Jamia Aliah University, Kolkata 700091, India

Jeffrey Lawrence D'Silva
Laboratory of Rural Advancement and Agricultural Extension, Institute for Social Science Studies, Universiti Putra Malaysia, Malaysia

Bahaman Abu Samah
Laboratory of Rural Advancement and Agricultural Extension, Institute for Social Science Studies, Universiti Putra Malaysia, Malaysia

Hayrol Azril Mohamad Shaffril
Laboratory of Rural Advancement and Agricultural Extension, Institute for Social Science Studies, Universiti Putra Malaysia, Malaysia

B. O. Faturoti
SARO Agro Allied Nigeria Limited, Plot C66A, Oba Kayode Akinyemi way, Amuwo P. M. B. 1837, Apapa, Lagos, Nigeria

M. C. Madukwe
Department of Agricultural Extension, University of Nigeria, Nsukka, Enugu State, Nigeria

O. Ogunedojutimi
SARO Agro Allied Nigeria Limited, Plot C66A, Oba Kayode Akinyemi way, Amuwo P. M. B. 1837, Apapa, Lagos, Nigeria

L. Anyanwu
SARO Agro Allied Nigeria Limited, Plot C66A, Oba Kayode Akinyemi way, Amuwo P. M. B. 1837, Apapa, Lagos, Nigeria

Gurmesa Umeta
Adami Tulu Agricultural Research Center, P. O. Box 35, Zeway, Ethiopia

Felekech Lemecha
Adami Tulu Agricultural Research Center, P. O. Box 35, Zeway, Ethiopia

Taha Mume
Adami Tulu Agricultural Research Center, P. O. Box 35, Zeway, Ethiopia

Pudji Muljono
Department of Science of Communication and Community Development, Faculty of Human Ecology, Bogor Agricultural University, Jl Raya Pajajaran, Kampus IPB Baranangsiang Bogor, Indonesia

Buddhi Raj Gyawali
Department of Agribusiness, P. O. Box 323, Alabama A&M University, Normal, AL 35762, Alabama, USA

Rory Fraser
Center for Forestry and Ecology, P.O. Box 323, Alabama A&M University, Normal, AL 35762, Alabama, USA

James Bukenya
Department of Agribusiness, P. O. Box 323, Alabama A&M University, Normal, AL 35762, Alabama, USA

Swagata "Ban" Banerjee
Department of Agribusiness, P. O. Box 323, Alabama A&M University, Normal, AL 35762, Alabama, USA

Mogos Y. Teweldemedhin
Department of Agriculture, Polytechnic of Namibia, Namibia

Lucia Kafidii
Department of Agriculture, Polytechnic of Namibia, Namibia

B. O. Ovwigho
Department of Agricultural Economics and Extension, Delta State University, Asaba Campus, Asaba, Nigeria

Utpal Kumar De
Department of Economics, North-Eastern Hill University, Shillong, 793022, Meghalaya, India

Manjit Das
Department of Economics, M. C. College, Barpeta, 781301, Assam, India

Emelia Guo
Institute for Continuing Education and Interdisciplinary Research (ICEIR) UDS, Tamale, Ghana

Mamudu A. Akudugu
Institute for Continuing Education and Interdisciplinary Research (ICEIR) UDS, Tamale, Ghana

Seidu Al-Hassan
Institute for Continuing Education and Interdisciplinary Research (ICEIR) UDS, Tamale, Ghana

Oto Jacob Okwu
Department of Agricultural Extension and Communication, University of Agriculture, Makurdi, Benue State, Nigeria

Shimayohol Daudu
Department of Agricultural Extension and Communication, University of Agriculture, Makurdi, Benue State, Nigeria

O. M Akinnagbe
Department of Agricultural Extension, University of Nigeria, Nsukka, Nigeria

A. R Ajayi
Department of Agricultural Extension, University of Nigeria, Nsukka, Nigeria

Kurt D. Nolte
Yuma County Cooperative Extension University of Arizona 2200 W. 28th St., Suite 102 Yuma, Arizona 85364 USA

Jorge M. Fonseca
Department of Plant Sciences, Yuma Ag Center, Yuma, Arizona, USA

Adijah M. Ali-Olubandwa
Department of Applied Community Studies and Extension, Egerton University, Kenya

N. J. Kathuri
Department of Agricultural Education and Extension, Egerton University, Kenya

Timothy E. O. Wesonga
East Africa Community, Arusha, Tanzania

B. Fungo
Faculty of Forestry and Nature Conservation, Makerere University P. Box 7062 Kampala, Uganda

L. Clark
Institute of Development Studies, London NW6 4BT, UK

M. M. Tenywa
Makerere University Agricultural Research Institute, Kabanyolo (MUARIK), Uganda

J. Tukahirwa
African Highlands Initiative (AHI), Kabale Uganda

R. Kamugisha
African Highlands Initiative (AHI), Kabale Uganda

E. Birachi
International Center for Tropical Agriculture (CIAT), Kigali, Rwanda

C. Wanjiku
International Center for Tropical Agriculture (CIAT), Kigali, Rwanda

A. R. Bizoza
National University of Rwanda (NUR), Butare, Rwanda

B. Wimba
INERA-Bukavu, Democratic Republic of Congo

P. Pali
Forum for Agricultural Research in Africa (FARA), Accra-Ghana

A. Adewale
Forum for Agricultural Research in Africa (FARA), Accra-Ghana

F. Olowole
Forum for Agricultural Research in Africa (FARA), Accra-Ghana

Uttam Kumar Baruah
Krishi Vigyan Kendra Goalpara, National research Centre on Pig, Indian Council of Agricultural Research, Thekasu part - I, Damra Road, Dudhnoi - 783124, Goalpara, Assam, India

Jyotish Barman
Krishi Vigyan Kendra Goalpara, National research Centre on Pig, Indian Council of Agricultural Research, Thekasu part - I, Damra Road, Dudhnoi - 783124, Goalpara, Assam, India

Benzamin Kaman
Krishi Vigyan Kendra Goalpara, National research Centre on Pig, Indian Council of Agricultural Research, Thekasu part - I, Damra Road, Dudhnoi - 783124, Goalpara, Assam, India

Mekbib Gebretsadik Haile
Center for Development Research (ZEF), Bonn University, 53113 Bonn, Germany

Degnet Abebaw
Ethiopian Economics Association, P.O. Box 34282, Addis Ababa, Ethiopia

J. O. Saliu
Department of Agricultural economics and Extension, Kogi State University, Anyigba, Nigeria

P. C. Obinne
Department of Agricultural Extension and Communication, University of Agriculture, Makurdi, Nigeria

S. I. Audu
Department of Agricultural economics and Extension, Kogi State University, Anyigba, Nigeria

Nicholas Ozor
African Technology Policy Studies Network (ATPS), 3rd Floor, the Chancery, Valley Road, P. O. Box 10081-00100 Nairobi, Kenya

Nnaji Cynthia
Department of Agricultural Extension, Faculty of Agriculture, University of Nigeria, Nsukka, Enugu State, Nigeria

Fonteh Athanasius Amungwa
Department of Sociology and Anthropology, Faculty of Social and Management Sciences, P. O. Box 63, University of Buea, Republic of Cameroon

Tsion Tesfaye
University of Swaziland, PO Luyengo, Hill Street, LA-8, Swaziland

Ranjan S. Karippai
Haramaya University, College of Agriculture, Department of Rural Development and Agricultural Extension, P. O. Box 138, Dire Dawa, Ethiopia

Teklu Tesfaye
Ethiopian Institute of Agricultural Research, Research and Extension Division, P. O. Box 1234, Addis Ababa, Ethiopia

Karin Eksvärd
Centre for Sustainable Agriculture, Swedish University of Agricultural Sciences, Box 7047, SE 750 07 Uppsala, Sweden

Johanna Björklund
Centre for Sustainable Agriculture, Swedish University of Agricultural Sciences, Box 7047, SE 750 07 Uppsala, Sweden

N. Ezeh Ann
Department of Agricultural Economics Management and Extension, Ebonyi State University, Abakaliki, Ebonyi State, Nigeria

Margaret Joke Koyenikan
Department of Agricultural Economics and Extension Services, University of Benin, Faculty of Agricultural Sciences, University of Benin, Benin-City, Nigeria

Nicodemus Mandere Mandere
Centre for Sustainability Studies (LUCSUS), Lund University P.O. Box 170, SE-221 00, Lund Sweden
Department of Physical Geography and Ecosystems Analysis, Lund University, Sweden

Barry Ness
Centre for Sustainability Studies (LUCSUS), Lund University P.O. Box 170, SE-221 00, Lund Sweden

Stefan Anderberg
Centre for Sustainability Studies (LUCSUS), Lund University P.O. Box 170, SE-221 00, Lund Sweden
Department of Physical Geography and Ecosystems Analysis, Lund University, Sweden

Nana Afranaa Kwapong
Humboldt University Berlin, Germany

Ephraim Nkonya
International Food Policy Research Institute (IFPRI), Washington D.C

D. B. AFFUL
Department of Agricultural Economics and Extension, University of Fort Hare, P/Bag 1314, Alice, 5700, South Africa

A. OBI
Department of Agricultural Economics and Extension, University of Fort Hare, P/Bag 1314, Alice, 5700, South Africa

F. S. LATEGAN
Department of Agricultural Economics and Extension, University of Fort Hare, P/Bag 1314, Alice, 5700, South Africa